RIVER FLOW MODELLING AND FORECASTING

WATER SCIENCE AND TECHNOLOGY LIBRARY

RIVER FLOW MODELLING
AND
FORECASTING

Edited by

D. A. KRAIJENHOFF

Department of Hydraulics and Catchment Hydrology
Agricultural University of Wageningen
The Netherlands

and

J. R. MOLL

Delft Hydraulics Laboratory
Emmeloord
The Netherlands

D. REIDEL PUBLISHING COMPANY

A MEMBER OF THE KLUWER ACADEMIC PUBLISHERS GROUP

DORDRECHT / BOSTON / LANCASTER / TOKYO

Library of Congress Cataloging in Publication Data

Main entry under title:

River flow modelling and forecasting.

(Water Science and technology library)
Includes bibliographies and index.
1. Streamflow – Mathematical models. 2. Hydrological forecasting – Mathematical models. 3. Flood
forecasting – Mathematical models. 4. Watershed management – Mathematical models. I. Kraijenhoff, D. A.
II. Mol, J. R. III. Series.
GB1207.R58 1986   551.48′3′0724   85-28216

ISBN-13: 978-94-010-8518-2   e-ISBN-13: 978-94-009-4536-4
DOI: 1007/978-94-009-4536-4

Published by D. Reidel Publishing Company,
P. O. Box 17, 3300 AA Dordrecht, Holland.

Sold and distributed in the U.S.A. and Canada
by Kluwer Academic Publishers,
190 Old Derby Street, Hingham, MA 02043, U.S.A.

In all other countries, sold and distributed
by Kluwer Academic Publishers Group,
P. O. Box 322, 3300 AH Dordrecht, Holland.

TABLE OF CONTENTS

PREFACE   ix

1. INTRODUCTION - G.A. Schultz   1

    1.1 Objectives   1
    1.2 Objectives of forecasting   2
    1.3 Criteria for successful forecasts   3
    1.4 Systems approach   4
    1.5 Principles and elements of river flow forecasting   5
    1.6 Concluding remarks   9
    Symbols   10
    References   10

2. DETERMINISTIC CATCHMENT MODELLING - T. O'Donnell   11

    2.1 Introduction   11
    2.2 Linearity/Non-linearity   12
    2.3 Analysis/Synthesis   14
    2.4 Illustrative example   14
    2.5 Linear treatment of catchment behaviour   16
    2.6 Non-linear treatment of catchment behaviour   28
    Symbols   35
    References   36

3. THEORY OF FLOOD ROUTING - J.C.I. Dooge   39

    3.1 Continuity equation for unsteady flow   39
    3.2 Momentum equation for unsteady flow   39
    3.3 Equations of characteristics for unsteady flow   41
    3.4 Boundary conditions in flood routing   42
    3.5 The finite difference approach   44
    3.6 Characteristic finite difference schemes   45
    3.7 Explicit finite difference schemes   46
    3.8 Implicit finite difference schemes   47
    3.9 Linearisation of the St. Venant equations   48
    3.10 Simplification of the St. Venant equations   50
    3.11 Comparison of hydraulic solutions   52
    3.12 Nature of hydrologic methods   53
    3.13 Linear conceptual models   55
    3.14 Comparison of linear hydrologic models   57
    3.15 Calibration of linear models   60
    3.16 Non-linear hydrologic models   62
    Symbols   63
    References   64

4. LOW FLOW SUSTAINED BY GROUND WATER - R. Mull   67

    4.1 Introduction   67
    4.2 Discussion of rainfall-discharge relations   79
    4.3 Examples   87
    Symbols   96
    References   97

5. FORECASTING MELTWATER FROM SNOW-COVERED AREAS AND FROM
    GLACIER BASINS - H. Lang   99

    5.1 Introductory remarks   99
    5.2 The snow cover and its determination   100
    5.3 The determination of the meltrates   106
    5.4 Practical methods to determine the meltrates   113
    5.5 Operational forecasting equations for glacier basins
         where past records are available   117
    5.6 Thermal and capillary retention capacity   118
    5.7 Long range, seasonal forecasting   121
    Symbols   123
    References   125

6. TIME-SERIES METHODS AND RECURSIVE ESTIMATION IN HYDROLOGICAL
    SYSTEMS ANALYSIS - P.C. Young   129

    6.1 Introduction   129
    6.2 The simplest first order, linear hydrological model   129
    6.3 More complicated linear hydrological models   135
    6.4 Recursive estimation of a simple time-series model   138
    6.5 Recursive estimation of general linear time-series models 144
    6.6 Model structure (order) identification   148
    6.7 Flow modelling for the river Wyre   149
    6.8 Time-variable parameter estimation   155
    6.9 Salinity variations in the Peel Inlet-Harvey Estuary
         Western Australia   158
    6.10 Time-series analysis and flow forecasting   164
    6.11 Flow forecasting and the Kalman Filter   168
    6.12 The Extended Kalman Filter   175
    6.13 Conclusions   176
    Acknowledgements   177
    Symbols   177
    Appendix 1. The Microcaptain Computer Program Package   178
    References   178

7. RELATIONSHIP BETWEEN THEORY AND PRACTICE OF REAL-TIME RIVER
    FLOW FORECASTING - G.A. Schultz   181

    7.1 Link between theoretical chapters and case studies   181
    7.2 Model input fields   182
    7.3 Theory versus practice in real-time river flow
         forecasting   186

7.4 Conclusions   191
References   193

8. CASE STUDIES IN REAL-TIME HYDROLOGICAL FORECASTING FROM THE UK
    - P.E. O'Connell, G.P. Brunsdon, D.W. Reed, P.G. Whitehead   195

8.1 Introduction   195
8.2 Real-time flow forecasting system for the river Dee   196
8.3 The Haddington flood warning system   210
8.4 An on-line monitoring, data management and water quality
     forecasting system for the Bedford Ouse river basin   221
8.5 Discussion   237
Symbols   238
References   238

9. RIVER FLOW SIMULATION - J.G. Grijsen   241

9.1 Introduction   241
9.2 Finite difference methods   242
9.3 Numerical properties   244
9.4 The Delft Hydraulics Laboratory method   249
9.5 Practical aspects   257
9.6 Case study: Flood control of the rivers Parana and
     Paraguay   266
9.7 Strategy for implementation of forecasting models   269
Symbols   271
References   272

10. THE FORECASTING AND WARNING SYSTEM OF 'RIJKSWATERSTAAT' FOR THE
    RIVER RHINE - J.G. de Ronde   273

10.1 Introduction   273
10.2 General description of Rijkswaterstaat and its
     warning services   273
10.3 Organization of the riverflood warning system   274
10.4 The empirical forecasting model   274
10.5 The multiple linear regression model   277
10.6 Low flow forecasting   282
References   286

11. SHORT RANGE FLOOD FORECASTING ON THE RIVER RHINE - J.R. Moll   287

11.1 Introduction   287
11.2 Flow forecasting   287
11.3 A deterministic hydrological model for the river Rhine 290
11.4 A stochastic real-time forecasting model   294
11.5 Conclusions   295
Symbols   295
References   296

12. DESIGN AND OPERATION OF FORECASTING OPERATIONAL REAL-TIME
    HYDROLOGICAL SYSTEMS (FORTH) - J. Němec   299

    12.1 Introduction   299
    12.2 Components of a FORTH system   302
    12.3 Selection of forecasting procedures   309
    12.4 Forecast updating and evaluation (WMO, 1983)   315
    12.5 Benefit and cost analysis of hydrological forecasts   318
    12.6 Examples of established FORTH systems   319
    Symbols   319
    References   319
    Annex I   320
    Annex II   322

13. CASE STUDIES ON REAL-TIME RIVER FLOW FORECASTING - G. Fleming   329

    13.1 Introduction   329
    13.2 The Santa Ynez River, California, USA   330
    13.3 Derwent River system, England   341
    13.4 Orchy River system, Scotland   354
    Symbols   365
    References   365

INDEX   367

PREFACE

Advances in computer technology, in the technology of communication and in mathematical modelling of processes in the hydrological cycle have recently improved our potential to protect ourselves against damage through floods and droughts and to control quantities and qualities in our water systems.

This development was demonstrated in a 1983 post-experience course at Wageningen University where an international group of experts reviewed successful modelling techniques and described the design and operation of a number of forecasting and control systems in drainage basins and river reaches of various sizes and under various geographical and climatological conditions. A special effort was made to bridge the gap between theory and practice; case studies showed that each forecasting system was designed to meet a set of specific requirements and they illustrated that the forecasting system can only be expected to operate reliably if, on the one hand, it is based on sound theoretical concepts and methods and if, on the other hand, it is robust so that, also under adverse conditions, it will continue to collect and process the necessary input-data and produce correct and timely signals.

We were pleased to meet with encouragement for preserving the course material and making it available to a wider public. This was effected by the team of authors who elaborated, updated and harmonized the materia in two stages; first into an issue of our university department and finally into the manuscript of this book.

A course element that could not be incorporated was a 36-minute Umatic NTSC (60 Hz) video tape, entitled: 'Salt River Project Phoenix Arizona Flow Forecasting'. It was produced in 1983 by the Salt River Project, Box 1980, Phoenix, Arizona 85001, U.S.A. Readers who are interested in this clear presentation of an enormous water management system, organized and instrumented by the latest standards, are advised to contact the Salt River Project.

We thank the Agricultural University Wageningen, UNESCO and WMO for their organizational and financial support of the 1983 post-experience course which was the origin of this book. Special thanks are due to UNESCO for its continued financial support of the conversion process of a set of lecture notes into the manuscript of this book.

the editors

D.A. Kraijenhoff van de Leur
J.R. Moll

1. INTRODUCTION

G.A. Schultz

Ruhr University
P.O. Box 102148
4630 Bochum
F.R.G.

1.1 Objectives

River flow forecasts are needed for various purposes within the framework of river basin management activities; they are needed and implemented in practice. Methods for producing forecasts, however, are usually developed in the theoretical environment. There is an old and arrogant saying among theoreticians: "today's practice is yesterday's theory". Many examples in this text-book will show that this is not necessarily true although examples where it does apply can also be easily found.

The objective of this book is to narrow the gap between theory and practice and to minimize the time delay between the development and implementation of useful new methods. There are also of course, useless new methods. This text-book tries therefore to present to the practising engineer only those new techniques which may be useful for the solution of his particular problems. The discrepancies between the needs of the practitioner and the products of theory will be discussed in more detail in Chapter 7.

This book has one more unusual feature: it avoids the use of jargon and it presents no more theory than is necessary for practical applications. The authors are aware of the fact that practitioners working in the "real world" often mistrust the products of the "other world" academia.

The prize winner of an ASCE essay contest, a practitioner, has put this in the following form (Huston 1981): 'Although the "real world people" (RWP) form 95% of the "engineering world population" and they do the work, they can read - if they find time - only publications of the 5% representing the "other world people" (OWP) and they can't understand much of it'. He gives an example: 'Should RWP want to add 1 and 1, they would do it simply by saying $1 + 1 = 2$. OWP, however, wouldn't think of doing it that way. For them to add 1 and 1 and get 2 would be unprofessional. They would go at it this way: in as much as OWP all know that $\ln e = 1$ and also that $\sin^2\alpha + \cos^2\alpha = 1$, they would say

$$\ln e + \sin^2\alpha + \cos^2\alpha = \int_0^2 x \, dx$$

D.A. Kraijenhoff and J.R. Moll (eds.), River Flow Modelling and Forecasting, 1-10
© 1986 by D. Reidel Publishing Company.

Now this, they reason, is more like it. It gives the same answer as the RWP got in their primitive 1 + 1 = 2, but in a way that is not professionally embarrassing.'

Although this is intended humorously, it is also embarrassing for those of us working in "the other world", because it contains a kernel of truth. Huston continues: 'now all of this is not to say that OWP are not useful. Some of them are actually RWP hiding out in the OW.'

1.2 Objectives of forecasting

The question arises: what do we want to forecast? The most frequent type of flow forecast is the flood forecast. Besides this, however, we are interested in low flow forecasts, in water quality forecasts and in forecasts of the hydrological effects of man-made changes in river catchments. For all of these types of forecasts examples will be given during the book.

A survey of forecasting systems in Europe carried out by WMO (see Chapter 12) has revealed that the various purposes of flow forecasts were:

1.	Flood protection	43%
2.	Energy	19%
3.	Navigation	12%
4.	Water supply and sanitation	12%
5.	Irrigation	6%
6.	Water pollution control	4%
7.	Ice problems	4%

The elements to be forecast were:

1.	Surface water level	42%
2.	Discharge	36%
3.	Volume of runoff	21%
4.	Ice, groundwater, water quality	seldom

Forecasting period:

1.	up to 24 hours	33,5%
2.	1 day to 1 week	37,5%
3.	Medium-term	15 %
4.	Long-term	14 %

Thus the most frequent desired forecast would be a flood forecast of water level for a period between 24 hours and 1 week.

A few decades ago flow forecasting was restricted to flood forecasting. This was not very efficient, due to the primitive hydrological observation and data transmission systems, and to the simple forecasting methods and computational opportunities. Developments, during the last two decades, in the fields of electronic engineering, hydrological modelling and systems theory, are now generating considerable operational benefits in the areas of flood forecasting, flood warning, reservoir management and pollution control.

"On-line monitoring of rainfall, flow and water quality variables can now be achieved with reliable instrumentation and telemetry at modest cost: ever increasing advances in the performance of on-line monitoring schemes are being achieved through the exploitation of microprocessor technology. Dedicated low-cost microcomputers can be programmed to control telemetry schemes automatically while also providing the necessary computing power to run real-time flow and water quality forecasting models. Thus the capacity of the water engineer to respond efficiently to emergency situations created by flood or pollution events has been greatly enhanced" (O'Connell, Chapter 8).

The objective of forecasting is in any case the prevention or minimization of damage (e.g. during floods, low periods or pollution problems) or the maximizing of any possible benefits. This is easy to say but difficult to quantify. This leads to the question: how do we want our forecasts and which criteria do apply?

1.3 Criteria for successful forecasts

The best possible forecast is that one which completely and identically describes a process which will occur in future. This is, however, not possible since all forecasting techniques contain an element of uncertainty. Since precise forecasts are impossible, the forecast used should be that with minimum variance of forecast errors. Although this criterion is mathematically satisfactory it does not often suit practical requirements. Rijkswaterstaat et al. officially stated during the spring flood of 1983 "if we forecast 10 cm too low, we are in error; if we forecast 50 cm too high, we did a good job". This shows the necessity of specifying the parameter to be forecast, and the form and accuracy of this forecast. This, of course, depends on the forecast purpose as well as the times and scales of river system and process involved. If human lives are endangered, it may be advisable to produce a forecast minimizing maximum errors. In cases where a certain lead time is necessary in order to carry out disaster prevention measures, the optimality criterion may be the minimum error in the estimation of time to peak.

Since the lead-time of the forecast often determines the value of the forecast, forecasting techniques extending the lead-time are of growing importance. It is advisable, therefore, to forecast a downstream flow on the basis of observed rainfall, or even of forecast rainfall, rather than on the basis of upstream river gauge measurements alone. This point will be discussed later in more detail. Here again, questions of the times and scales of the system and process are relevant.

In many cases, especially when reservoir operation is involved, the criterion for evaluating the forecast quality is not so much the forecast accuracy but rather the effects of forecast errors on the river basin management. In Germany it would be possible in many cases to reduce flood damage by pre-emptying storage reservoirs in order to catch the flood peak when it occurs. This can be done only on the basis of a flood forecast. If, however, the pre-releases are themselves damaging this may, if the flood peak is overestimated, lead to a court case where the damage costs would be claimed from the river authority responsible.

This has resulted in the fact that, in Germany, nobody is prepared to take the risk of prereleases of reservoirs on the basis of a flood forecast.

There is one more criterion to be considered. In this book, and in the literature, many different techniques are described, some of which serve the same purpose. Furthermore, there are simple and complex techniques. Although it is often assumed that the more complex model will yield better results, this is often not true - as shown by O'Connell in his case studies of Chapter 8. It is thus advisable to use the simpler and more easily understood model as long as it is not proven that a complex model is preferable.

1.4 Systems approach

The real-time flow forecast we want to produce is, in terms of systems theory (Dooge 1973), always the output of some hydrological system. If this is true, there must also be a system and a system input. If the input time series is x(t) and the output time series is y(t), the systems operation may be described by a systems function h(t).

This is true only if it is a single input, single output system and if the system's behaviour follows the theory of linear systems. As example the well-known unit hydrograph model may be mentioned in which x(t) represents the average rainfall on a river catchment, y(t) is the runoff hydrograph and h(t) is nothing but the unit hydrograph itself. If output is to be determined from input, we talk about the computation of system performance. If the system operation h(t) is to be determined from input and output, the procedure is called system identification (Fig. 1.4.1).

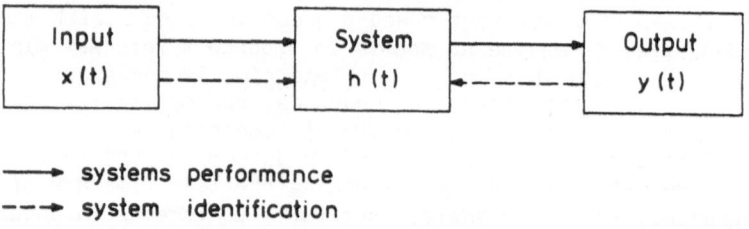

——→ systems performance
--- → system identification

Fig. 1.4.1. Systems Approach

The output y(t) is related to its causative input x(t) analytically by the convolution integral:

$$y(t) = \int_0^t x(\tau) \cdot h(t-\tau) \, d\tau \qquad (1.4.1)$$

This relationship is of general validity for all natural systems which can be approximated by linear systems. The unit hydrograph method is thus only a very special case of the general technique represented by Figure 1.4.1 and Equation 1.4.1. In Chapter 2, section 2.5, O'Donnell

presents Equation 1.4.1 in exactly the same form for this special case.
More details about linear and non-linear models as well as time-invariant
and time-variant models will also be given in Chapter 2.

The following section, 1.5, will describe, in systems terms, the
various techniques of flow forecasting to be presented in the following
chapters on modelling.

1.5 Principles and elements of river flow forecasting

The underlying theory of modelling and flow forecasting is presented in
the following chapters:

- Deterministic catchment models   (Chapter 2)
- Flood routing models   ("   3)
- Low flow models   ("   4)
- Meltwater runoff models   ("   5)
- Time series models (general)   ("   6)

These methods can be rearranged according to their relevance in a river
catchment. Starting at the catchment outlet and going towards the
divide the various techniques become relevant in the following order
(Fig. 1.5.1):

a. Low flow sustained by groundwater (Chapter 4)
b. Flood routing in a river reach   ("   3)
c. Rainfall-runoff models   ("   2)
d. Meltwater runoff models   ("   5)
e. Time series models (general)   ("   6)

1.5.1 Low flow sustained by groundwater

In Chapter 4, R. Mull deals with low flow conditions and their forecasts
(Fig. 1.5.1). In systems terms the system would be the aquifer in a
river valley, the input is groundwater recharge and the system output is
river flow fed by water from the aquifer (Fig. 1.5.2).

The system function h(t) depends on the model type. If black-box
models are applied (subsection 4.2.1), the system function is similar to
a unit hydrograph having a rather long time base. If more complex
groundwater models are used, the systems function can be derived from
the equations of subsection 4.2.2.

1.5.2 Flood routing in a river reach

In Chapter 3, J. Dooge deals with the theory of flood routing which
determines the unsteady flow conditions within a river reach, e.g.
between two river gauges (Fig. 1.5.1). In this case the system would be
the river reach between two cross-sections. Input is the flow hydrograph
into the upstream cross-sections, output the outflow from the downstream
cross-section of the river reach (Fig. 1.5.3). The systems function
h(t) again depends on the model type. Dooge in his Chapter 3 presents
various types of systems functions. An example here is the systems

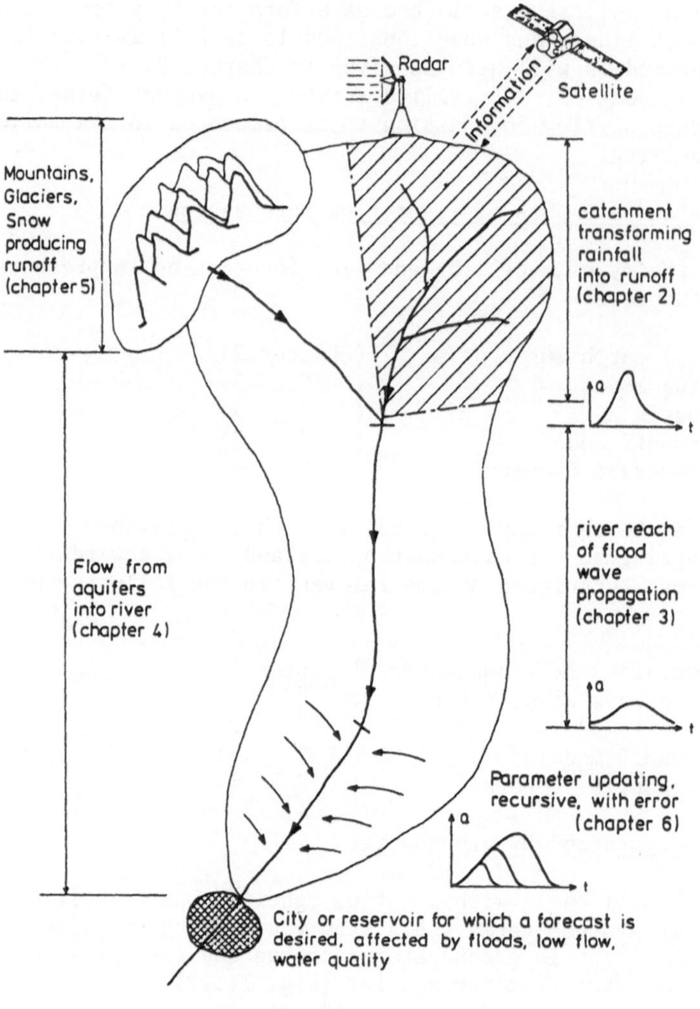

Radar
Satellite
Information
Mountains, Glaciers, Snow producing runoff (chapter 5)
catchment transforming rainfall into runoff (chapter 2)
Flow from aquifers into river (chapter 4)
river reach of flood wave propagation (chapter 3)
Parameter updating, recursive, with error (chapter 6)
City or reservoir for which a forecast is desired, affected by floods, low flow, water quality

Fig. 1.5.1. River catchment and flow forecasting techniques.

Input
x(t)
Groundwater recharge
System
h(t)
Aquifer
Output
y(t)
Low flow in river

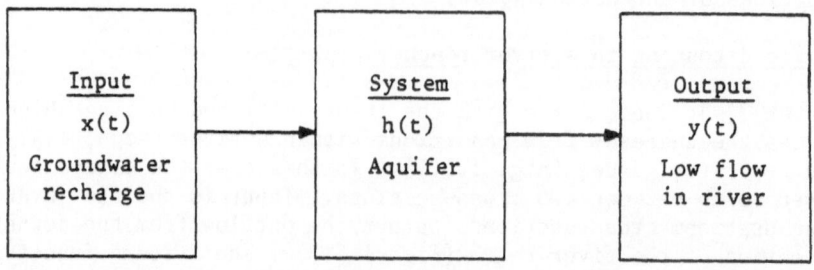

Fig. 1.5.2. Low flow sustained by groundwater.

function of the well-known Kalinin-Miljukov model (Eq. 3.13.9 in Chapter 3):

$$h(t) = (t/K)^{n-1} \exp(-t/K)/[K (n-1)!] \qquad (1.5.1)$$

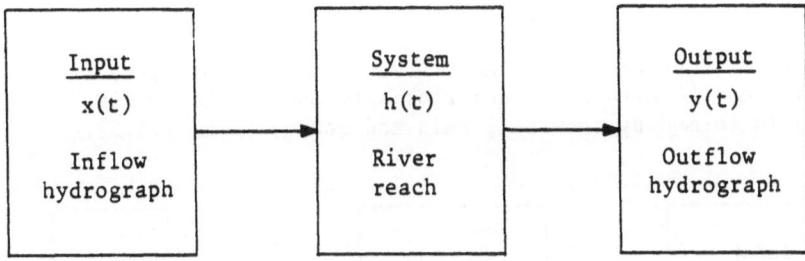

Fig. 1.5.3. Flood routing

1.5.3 Rainfall-runoff models

Chapter 2 by T. O'Donnell is concerned with deterministic catchment models and their relevance for flow forecasting. In this case the system represents the catchment, input is rainfall and/or snowmelt, output is the runoff hydrograph (Fig. 1.5.1 and 1.5.4).

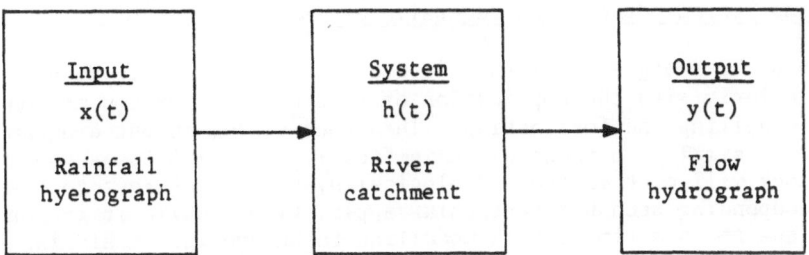

Fig. 1.5.4. Rainfall-runoff models

Chapter 2 presents various deterministic catchment models, each having its own systems function h(t). As example here only one should be mentioned, i.e. the systems function of the well-known two parameter Nash model given in Equation 2.5.5:

$$h(t) = \frac{1}{K\Gamma n} \left(\frac{t}{K}\right)^{n-1} e^{-t/K} \qquad (1.5.2)$$

For integer values of n, $\Gamma n = (n-1)!$ Computing Equation 1.5.2 for this case with Equation 1.5.1 of flood routing reveals that these equations are identical, i.e. the Kalinin-Miljukov flood routing model uses exactly the same systems function as does the rainfall-runoff model by

Nash (for integer n). Although these models were developed in different
environments and for different purposes, this similarity is not surprising since both use the analogy of a cascade of linear reservoirs.

It should be mentioned again that the output in all models is
computed by convolving the known input x(t) with the systems function
h(t) following Equation 1.4.1 (i.e. Eq. 2.5.1 in Chapter 2).

1.5.4 Meltwater runoff models

Chapter 5 by H. Lang deals with meltwater as the output, y(t) from a
system which may be either a glacier or snow pack or both (Fig. 1.5.1).
Input x(t) is formed by snowfall, rain and energy (Fig. 1.5.5).

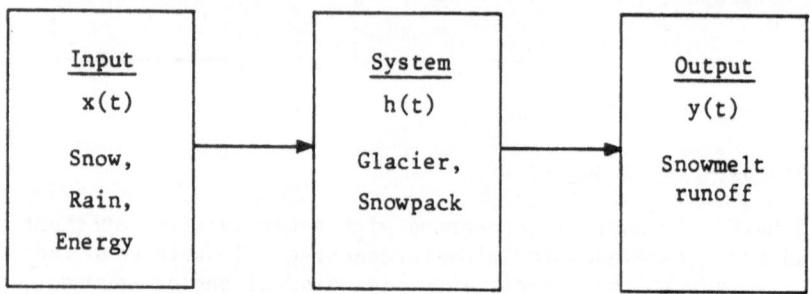

Fig. 1.5.5. Snowmelt runoff

1.5.5 Time Series methods and recursive estimation

Chapter 6 by P. Young deals with the potential of stochastic techniques
and particularly with the application of recursive estimation methods
for flow modelling and forecasting. The input-system-output diagram in
this case is similar to those given in Fig. 1.5.3 to 1.5.5 and the
system block will contain the hydrological system of interest to the
analyst, depending upon his particular application. While it is not so
easy to squeeze this time-series modelling technique into a similar
systems diagram, Fig. 1.5.6 represents an attempt to do so. The "output"
is now the model output or forecast which is, of course, compared with
the measured output in order to assess the value of the model in forecasting terms.

In this case the system is not a real world river system or part of
it but rather a calibrated model-representation of such a system. The
model calibration is done with the time series information available.

Whenever new data become available the model is recalibrated, and a
new forecast is made on the basis of new, recursively estimated model
parameters. These new forecasts are such that they also give information
on errors and reliability of the forecast. This information is of
importance e.g. for reservoir operation as stated in section 1.3.

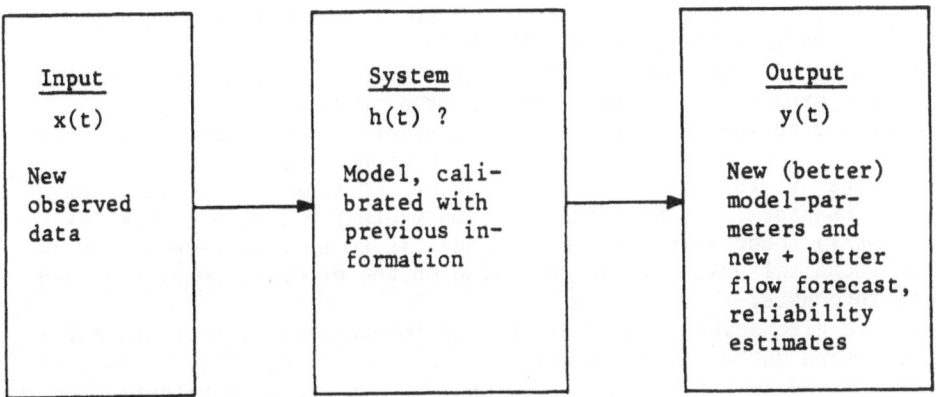

Fig. 1.5.6. Adaptive model output improvement

1.6 Concluding remarks

The five chapters 2 to 6 present the theoretical basis for various
elements of forecasts. They all deal with various catchment subsystems
and the hydrological processes occurring within each subsystem. The
system output is the desired result i.e. the forecast of river flow in
real-time. There will be not much information in these chapters about
the system inputs such as snow, rain, inflow-hydrographs to a river
reach, groundwater recharge etc. This topic will be dealt with in
chapter 7 and in most of the case studies. The special problem of the
system input, in forecasting terms, is the need for accurate hydrological
data from a wide area, and with high resolution in time and space.
Furthermore, this information must be transmitted to the forecasting
centre as early as possible in real-time, since the value of a forecast
depends on its lead-time. This means that special data collection and
data transmission systems are required. If sudden dangers may occur,
such systems have to be provided with alert systems.

In real-time flood forecasting, it is usually not the model but the
quality of the input data which forms the weakest link. It is essential
that the quality of the data be controlled. In one forecasting system,
for example, the radar signals indicated heavy rainfall. This same
phenomenon was not indicated by the terrestrial raingauges. In this
case, it was found that hot sunshine had caused significant damage to
the hardware between the radar and the computer.

Other problems are the organization, handling and processing of data
in the forecasting centre, and the dissemination of the computed flow
forecast, which is often hindered not only by technical but also by
bureaucratic obstacles (e.g. during the Mosel flood in spring 1983 in
Germany).

The reader may be confused by the great number of techniques
available for each special forecasting problem. There is the problem
of adequate model choice. Some advice for the solution of this problem
will be given in the more theoretical chapters of this volume, but much
more help will be offered along with the presentation of the case studies

particularly in Chapter 12 by Němec, where the extended experience of
WMO will come into the picture (WMO 1975).

Since most techniques for the generation of a forecast in real-time
are rather expensive - costs of equipment as well as of man power
involved - a decision has to be made 'a priori' by the user of the forecast whether the value of the forecast is higher than its cost. This
can be done by, for example, evaluation of the benefit-cost ratio as
shown in Chapter 12. Thus, the decision whether a simple or complex
(and costly) forecasting system and model is to be used depends not only
on the technical possibilities but also on the expected benefit gained
by the forecasts.

Before discussing details of special forecasting systems and models,
however, some mathematical theory must be considered. This is essential
for understanding the case studies, and for successful application of
forecasting techniques to solving particular problems. In Chapter 7 the
author will describe the relation between the theoretical models,
presented in the earlier chapters, and practical problems in the case
studies to be presented subsequently.

The objective of this book is to clarify the relationship between
theory and practice; and thus, in Huston's words - help to narrow the
existing gap between the "real world people" and the "other world people".

In Germany and Austria, post-experience courses in which both groups
participate have proven to be an effective method of bridging the gap
between 'Other World People' and 'Real World People'. An increased
awareness among 'Other World People' that the way in which new methods
have been discovered is usually not the best way to present them to
'Real World People' is of great importance.

SYMBOLS

$x(t)$	input	$L^3 T^{-1}$
$h(t)$	systems function	T^{-1}
$y(t)$	output	$L^3 T^{-1}$
K	reservoir constant	T
n	number of reservoirs	1
t	time	T

REFERENCES

Dooge, J.C.I. 1973 'Linear Theory of Hydrologic Systems', Techn. Bulle-
    tin No. 1468, Agricultural Research Service, US Dept. of Agriculture,
    Washington D.C.
Huston, J.H. 1981 'Stop the world - I want to get off!', Civil Enginee-
    ring ASCE, Dec. 1981.
WMO, 1975 'Intercomparison of Conceptual Models Used in Operational
    Hydrological Forecasting', Operational Hydrology Report No. 7, Geneva.

2. DETERMINISTIC CATCHMENT MODELLING

T. O'Donnell

University of Lancaster
Dept. of Environmental Sciences
Lancaster LA14 YQ
U.K.

2.1 Introduction

The modelling of catchment behaviour will be taken in the quantitative
sense either of reconstructing past rainfall-to-runoff behaviour or of
forecasting future runoff behaviour from design rainfalls (or from
currently occurring rainfalls). The topic has been studied in a number
of ways over many decades. The early developments often used physical
models (sand tanks, scale hydraulic models etc.), or analogue models
(electrical analogues, viscous flow analogues etc.). Such models were
based on mathematical descriptions of the relevant catchment processes,
but used physical realisations. Alongside those approaches was a variety
of empirical mathematical models, for example the Rational (Lloyd-Davies)
formula for flood peak: $Q_p = C i A$ involving rainfall rate, i, and
catchment area, A, with a coefficient, C. More recent developments have
made use of the vast power of computers to construct and use far more
realistic mathematical models of the physics of catchment processes, an
effort broadly encompassed by the description "deterministic modelling".
Later still, a powerful array of statistical modelling techniques has
become available. There is thus a wide range of tools in the river flow
forecaster's tool kit. This section of the book deals with deterministic
catchment modelling.

2.1.1 The catchment as a system

The hydrograph of total streamflow from a catchment is the overall
continuing response of that catchment to the previous history of precipitation and evaporation over the catchment. It is helpful to recognise
that a simplified representation of catchment behaviour such as the
'flow diagram' of Fig. 2.1.1 embodies all the characteristics employed
by systems engineers.
   Whether one considers the whole or its component storage parts, the
essence of Fig. 2.1.1 is of a system or sub-systems from which outputs
occur as responses to inputs. The basic representation of this essential
concept features a system 'box' that is characterised by a unique
response function, h(t). The systems approach concentrates on the

D.A. Kraijenhoff and J.R. Moll (eds.), River Flow Modelling and Forecasting, 11-37
© 1986 by D. Reidel Publishing Company.

operation performed by h(t) on an input, x(t), to produce an output, y(t).

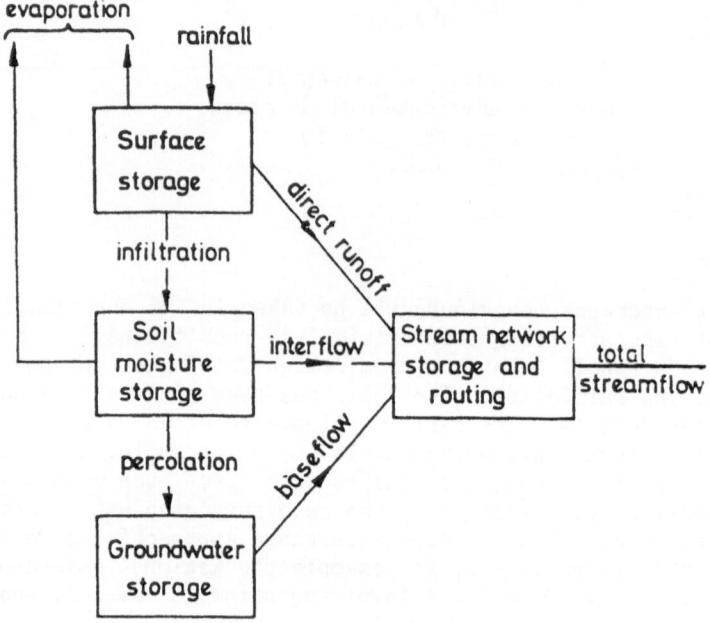

Fig. 2.1.1.   The catchment system

2.1.2 Outline of approach

There are two classes into which catchment (and other) systems may be
divided, and for which a variety of deterministic catchment modelling
techniques has been developed (Fig. 2.1.2).

Systems are either linear or non-linear; modelling techniques fall
under the headings of analysis methods and synthesis (simulation)
methods. As a prelude to details of specific examples of deterministic
catchment modelling, a brief discussion of the various divisions in
Fig. 2.1.2 follows.

2.2 Linearity/Non-linearity

A system is said to be linear if its behaviour can be described by a
linear differential equation. If x(t) represents (as a function of time,
t) an input to a system, and y(t) the resulting output from that system,
then a simple single input/single output linear system would be described
by an equation of the form:

$$A_n \frac{d^n y}{dt^n} + A_{n-1} \frac{d^{n-1} y}{dt^{n-1}} + \cdots + A_1 \frac{dy}{dt} + A_0 y = x \qquad (2.2.1)$$

If equation 2.2.1 contained terms with products of y and its differentials, or powers other than the first power, the system described by such an equation would be <u>non-linear</u>.

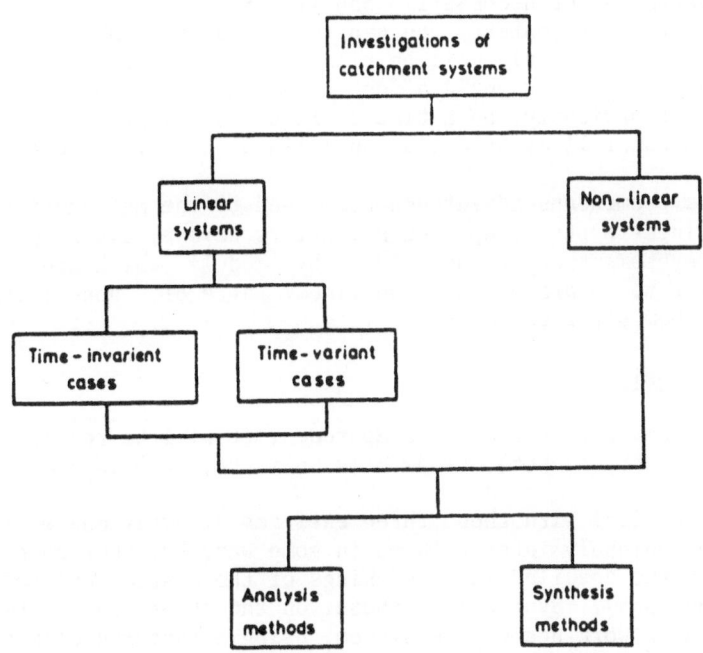

Fig. 2.1.2.   Investigations of catchment systems

The coefficients A_0 to A_n in Equation (2.2.1) may be constants, in which case the linear system is said to be <u>time-invariant</u>; or the coefficients may include one or more which vary with time, in which case the linear system is said to be <u>time-variant</u>.

The outstandingly useful property of linear systems is that they exhibit <u>superposition</u> (and hence can be handled by relatively straightforward mathematical methods). The separate outputs y_1 and y_2, due to two separate inputs, x_1 and x_2, may be directly summed (superimposed) to give the output, y_3, that would result from an input, x_3, formed by the sum $(x_1 + x_2)$.

Such direct arithmetic superposition would not apply to a non-linear system. In a linear system, y_2 would be determined entirely and solely by x_2 and would be independent of x_1. In a non-linear system, y_2 would be affected by x_1 as well as x_2.

For a time-invariant linear system, the output y_1 would always be the same whenever the input x_1 was applied. If the linear system were

time-variant, however, then y_1 would depend upon the absolute time at
which x_1 started as well as on x_1 itself. Generally, non-linear systems
are also time-variant, but one must be careful to avoid the converse:
a time-variant system is not necessarily non-linear.

Put into the catchment context, one can readily accept that any
linear catchment behaviour could well be time-variant on a seasonal time
scale (winter vegetative cover would affect behaviour differently from
summer cover). Less obviously, on a time scale of hours, processes
occurring within a rainfall event could cause linear behaviour to be
time-variant.

Because of the tremendous advantages possessed by the mathematical
tractability of linear systems, applied scientists have always sought to
approximate real systems that are non-linear by 'models' which are
linear, or at least to separate out those linear parts of a non-linear
whole, and treat them separately.

2.3 Analysis/Synthesis

Returning to the essence of the systems approach, it will be recalled
that the response function, h(t), transforms an input, x(t) into an output, y(t).

Analysis methods deal with those three entities directly and explicitly, as entities uniquely inter-related in some way, but they do so
without specifying the detailed inner workings of the system involved
(a 'black box' approach). Synthesis methods, on the other hand, attempt
to simulate the inner workings of the system, using structured elements
and mathematical relationships describing detailed physical processes
and inter-connections at work in catchments (the Pandora's box approach).
The synthesis methods make use of both recorded input and output information to modify and adjust the simulated structure and inter-connections
until an acceptable transformation of the recorded input to output is
achieved.

Fig. 2.3.1 summarizes the types of problem arising with the two
types of methods.

Under the analysis entry, three situations are shown arising from
which of the three entities, x(t), h(t) and y(t), is the 'unknown'. Of
those three, the prediction of y(t) and the identification of h(t) form
the main bulk of practical problems, with the identification problem
being the more difficult of the two. The synthesis entry stresses the
twofold problem: first the construction of a catchment simulation mechanism and then its adjustment to establish a response function that fits
catchment input and output records.

2.4 Illustrative example

All the above general introductory ideas and statements might be usefully
followed by reference to a particular example, one which in any case
enters into several of the techniques discussed later.

Consider the reservoir represented in Fig. 2.4.1 for which the
storage, S, is related to the output, y, by the general relationship:

$$S = K.y^n \tag{2.4.1}$$

TYPE OF PROBLEM		Input x(t)	Response h(t)	Output y(t)
Analysis	Prediction	√	√	?
	Identification	√	?	√
	Detection	?	√	√
Synthesis (Simulation)		√	? ?	√

Fig. 2.3.1. Problems arising with systems

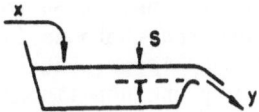

Fig. 2.4.1. A simple reservoir system

Then for an input, x, we can write from continuity:

$$\frac{dS}{dt} + y = x \qquad\qquad (2.4.2)$$

If we now take the particular case of n = 1 and K constant in Equation (2.4.1) then Equation (2.4.2) yields:

$$K\frac{dy}{dt} + y = x \qquad\qquad (2.4.3)$$

This is seen to be a linear differential equation with constant coefficients, so for n = 1, K a constant, the reservoir is a time-invariant linear system whose behaviour is described by Equation (2.4.3).

If now n = 1 but K = K(t), a function of time, then we have:

$$\frac{dS}{dt} = K(t)\frac{dy}{dt} + y\frac{dK(t)}{dt}$$

and insertion into Equation (2.4.2) gives:

$$K(t) \frac{dy}{dt} + \left\{1 + \frac{dK(t)}{dy}\right\} y = x \qquad (2.4.4)$$

This is still a linear differential equation but has time-varying coefficients. So for n = 1 and K dependent on time, the reservoir is a time-
variant linear system whose behaviour is described by Equation (2.4.4).
     Finally, if n ≠ 1 and, for simplicity, K is constant, then:

$$\frac{dS}{dt} = Kny^{n-1} \frac{dy}{dt}$$

and Equation (2.4.2) becomes:

$$Kny^{n-1} \frac{dy}{dt} + y = x$$

which is a non-linear differential equation. Thus for n ≠ 1, whatever
K, the reservoir is a non-linear system.
     Reservoir storage components such as these have been used widely in
catchment modelling. In particular linear reservoirs have been used for
unit hydrograph developments. They have also appeared (both as linear
and non-linear reservoirs) as components within non-linear synthesis
methods. (It should be noted that it requires only one non-linear component in an otherwise linear components system to make that system non-
linear.) A non-linear reservoir with $S = Ky^{\frac{1}{2}}$ is used in the Isolated
Event Model described later (in Section 2.6.2 and shown in Fig. 2.6.4).
An application of this model is described in Section 8.3.4. Lambert
(1972) introduced a logarithmic storage/outflow relationship in modelling
catchment behaviour.

2.5 Linear treatment of catchment behaviour

A necessarily brief survey of the basis of some techniques of linear
treatments of catchment behaviour will be given. A much fuller account
will be found in the authoritative monograph by Dooge (1973).
     The unit_hydrograph method is the best-known and most widely used
technique both of analysing rainfall-runoff catchment behaviour and of
designing for future floods due to possible design rainfalls. In essence,
the method is concerned with the direct_runoff path of flow shown in
Fig. 2.1.1, and involves only part of the rainfall falling on a catchment, viz. the rainfall excess. Before a unit hydrograph analysis can
be performed on a set of total streamflow and total rainfall records,
the rapid direct runoff has to be separated from the slow baseflow, and
a volume of rainfall excess equal to that of the direct runoff has to be
separated from the total rainfall. There is no unique 'correct' way of
performing those separation procedures with the result that many different methods of determining baseflow separation and rainfall loss curves
have been developed. Details of typical procedures may be found in
Linsley et al. (1958), NERC (1975), Shaw (1983) and Wilson (1983). It
is generally argued that consistent use of particular separation proce-

dures in both analysis and design is acceptable. The remainder of this survey will concentrate on the assumptions of, and the analysis/synthesis approaches to, the unit hydrograph method.

Two of the basic assumptions of the unit hydrograph method of relating a rainfall excess on a catchment to the resulting hydrograph of direct runoff are:

(1) invariance of response - the same rainfall excess, whenever it is applied, will always produce the same direct runoff hydrograph;
(2) superposition of responses - runoff due to two or more different rainfalls applied together is the arithmetic sum of the separate runoffs caused by each of the rainfalls applied separately.

Unit hydrograph theory is therefore based on the assumption that the catchment is a time-invariant linear system, at least so far as rainfall excess and direct runoff are concerned.

The T-hour unit hydrograph (TUH) of a catchment, written u(T, t), may be defined as the direct runoff hydrograph due to unit volume of rainfall excess falling uniformly over the catchment in a period of T hours. If we make T smaller, keeping the volume of rainfall excess constant at unity, in the limit (as T approaches zero) the TUH approaches the instantaneous unit hydrograph (IUH). In systems engineering terms, the IUH is the impulse response of the catchment system, i.e. the output from the system due to an instantaneous impulse input of unit size.

The impulse response of a time-invariant linear system is the key characteristic of that system, and forms the response function, h(t), referred to in sections 2.1 and 2.3. From it the response due to any other input can be derived, as illustrated in Fig. 2.5.1. The direct

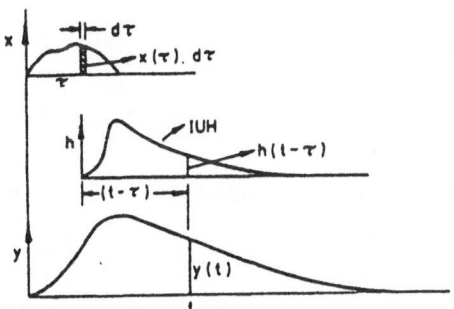

Fig. 2.5.1. The convolution operation

runoff output, y(t), is related to its causative rainfall excess input, x(t), analytically by Equation (2.5.1) (known as the convolution operation).

$$y(t) = \int_0^t x(\tau) \cdot h(t - \tau) \, d\tau \qquad (2.5.1)$$

The convolution integral in Equation (2.5.1) is the limiting case of the usual finite-period unit hydrograph procedure (illustrated in Fig. 2.5.2).

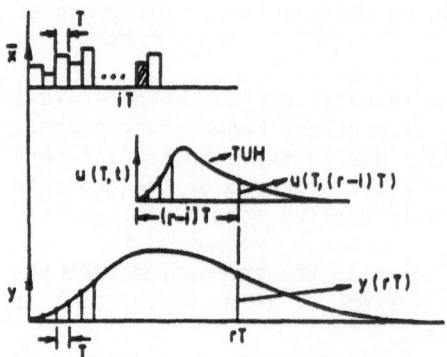

Fig. 2.5.2.   The unit hydrograph procedure

i.e. given a block rainfall histogram of mean intensities $\bar{x}_0$, $\bar{x}_1$, $\bar{x}_2$.... $\bar{x}_r$, over intervals T then:

$$y_{rT} = T \sum_{i=0}^{r-1} \bar{x}_i \cdot u(T,(r-i)T) \tag{2.5.2}$$

In the limit, as $T \to 0$, the summation of Equation (2.5.2) is replaced by the integral in Equation (2.5.1).

The finite period unit hydrograph, $u(T, t)$, is the output due to an input given by:

$$x(\tau) = \frac{1}{T} \quad \text{for } 0 \leqslant \tau \leqslant T$$

$$= 0 \quad \text{otherwise}$$

i.e. unit amount of input over a period T. From Equation (2.5.1) this can be shown to yield:

$$u(T, t) = \frac{1}{T} \int_{t-T}^{t} h(z) \, dz$$

(with the lower limit replaced by zero if $t < T$). Thus ordinates of the TUH are related to partial areas of the IUH given by 'windows' of width T ending at time t (or of width from the origin to t).

So, given the IUH, any design storm having specified volumes of rainfall excess occurring in intervals of length T can be found by first finding the TUH and then performing the usual finite-period unit hydrograph summation of Equation (2.5.2). This can be done for any desired value of T.

Clearly, then, the IUH is of great interest and the question of how to find an IUH is an important one, (the identification entry in Fig. 2.3.1). Equation (2.5.1) can readily be used to find the output, y(t), from a given input, x(t), and a known impulse response, h(t), (the prediction entry in Fig. 2.3.1). However, the problem of finding h(t) from given y(t) and x(t) is not straightforward. Earlier work on this problem invoked the use of linear catchment models which, when fitted to catchment data, were then taken to have an impulse response equal to the IUH of the catchment. This indirect synthesis approach has been followed by more direct analysis methods which endeavour to yield the IUH directly from catchment data without postulating any specific model mechanism. That order has been followed below.

2.5.1 Time-invariant linear catchment synthesis

Nash (1960) made a most significant contribution to unit hydrograph theory by showing that although it might be difficult to find the precise shape of the IUH, it is a very simple matter to derive certain overall properties of that shape, viz. its moments of area (Fig. 2.5.3).

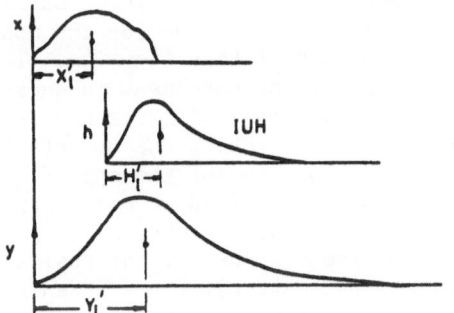

Fig. 2.5.3. Moments of area for a linear system

For a time-invariant linear system described by an equation of the form (2.2.1), Nash showed that the first moments (about the origin) of the impulse response of that system, of any input to the system and of the corresponding output from the system (H_1', X_1', and Y_1' respectively) are related by:

$$H_1' = Y_1' - X_1' \tag{2.5.3}$$

and that the second moments (about the centres of area), H_2, X_2 and Y_2 are related by:

$$H_2 = Y_2 - X_2 \tag{2.5.4}$$

Similar relations, getting somewhat more complex, exist for higher moments.

These moment relationships, stemming purely and simply from the assumption that the catchment behaves as a time-invariant linear system, involve no approximation whatsoever – they are fundamental properties of any such system, and quite definitely fall under the heading analysis.

It is a simple matter to calculate the moments of area of rainfall and runoff curves, so the moments of an IUH for a given catchment can be found simply and directly from storm data recorded on that catchment via moment equations such as (2.5.3) and (2.5.4). However, knowing the moments of the IUH is not sufficient to find the shape of the IUH. A synthesis modelling approach was needed.

In order to complete the search for the IUH, Nash (1960) set out to develop a sufficiently general linear model of catchment behaviour. Any such model has an impulse response for which some general analytical equation, expressed in terms of the parameters of the model, can be derived. From this equation, expressions for the moments of area of the model impulse response can also be derived, again in terms of the parameters of the postulated model.

For any given actual catchment data, numerical values of the IUH moments got from equations such as (2.5.3) and (2.5.4) when equated to the general expressions for the moments of the model impulse response will yield numerical estimates of the model parameters. These estimates, if substituted back into the general analytical equation for the model impulse response, will give a specific version of that equation. This is taken to be an approximation to the IUH of the catchment in question.

In other words, a general model of a class of linear systems is fitted to any specific member of the class by matching the moments of area of the general model impulse response to those of the specific system impulse response (got directly from a set of input and output data for that system). Clearly, though there is no approximation (in principle) in finding the moments of the impulse response of the real system (inevitably, data errors will enter into the matter), there is an approximation implicit in the matching process: the model response can only be as good a representation of a real system response as the model itself is a good representation of reality.

After considerable investigation, Nash (1960) chose as a sufficiently adequate general model of linear catchments a cascade of n identical time-invariant linear reservoirs in series, all having the storage characteristics:

$$S = K.y$$

This two-parameter model (n and K) calls for only two moment equations in order to solve for the values of the parameters when matching any given catchment.

The impulse response of the Nash model is:

$$h(t) = \frac{1}{K\Gamma n} \left(\frac{t}{K}\right)^{n-1} e^{-t/K} \tag{2.5.5}$$

in which Γn is the gamma function and is tabulated ($\Gamma n = (n - 1)\Gamma(n - 1)$ and, if n is an integer, $\Gamma n = (n - 1)!$). For Equation (2.5.5), the first moment (about the origin) is nK while the second moment (about the centre of area) is nK^2, so the process of finding n and K values from catchment data via Equations (2.5.3) and (2.5.4) is extremely simple:

$$n = \frac{(Y'_1 - X'_1)^2}{Y_2 - X_2}$$

$$K = \frac{Y_2 - X_2}{Y'_1 - X'_1}$$

A second and highly practical topic treated extensively by Nash (1960) is the question of estimating the IUH (and TUH's) of an _ungauged_ catchment.

Taking various physical features (area, slope, channel length etc.) of 90 _gauged_ catchments in the United Kingdom, Nash established regression equations with good correlations between these factors and dimensionless groupings of the various IUH moments (got directly from the measured rainfall and runoff data via the moment relationships). For any _ungauged_ catchment, one would enter the regression equations with values of the relevant physical factors for the catchment and so derive numerical estimates of its IUH moments. These would yield the n and K values necessary to carry out the required evaluation of the IUH (and any TUH).

Other linear time-invariant catchment models have been proposed. Diskin (1964) studied the use of two Nash-type linear reservoir cascades in parallel, one cascade representing the more rapid direct runoff component of streamflow, the other the slower interflow-plus-baseflow component. Diskin found that the part of the input to be fed into the quick branch could be correlated with the proportional size of the built-up area in a partly urbanised catchment. The price paid for attempting to model more than one route from rainfall to streamflow is an increase in the number of model parameters (the Diskin model has five parameters).

A criticism to be made of the cascade-type models is that they allow for storage effects but not for channel translation effects present in any catchment. _Time-area_ concepts have been introduced to attempt a remedy for this deficiency. The translation effects are thought of as the time-area diagram, i.e. the hydrograph, w(t), due to channel translation of an instantaneous unit rainfall input over the catchment and based on time-of-travel to the catchment outfall. This hydrograph, w(t), is then routed through linear reservoir storage effects, with the routed outflow being taken as the catchment response function, h(t).

A culmination of this viewpoint was provided by Dooge (1959) in a fundamental unifying study of the general theoretical basis of the unit hydrograph method. Dooge's general analysis of linear catchment behaviour postulates that the output from an elemental area of a catchment

incurs a translation delay time, τ, and also passes through n linear
reservoirs in its passage to the catchment outfall, n being dependent on
τ. He assumed that all elements having the same τ value have the same
chain of linear reservoirs to the outfall, and that all those reservoirs
have the same storage characteristic, K (Fig. 2.5.4). This model allowed
for a non-uniformly distributed rainfall input varying with τ. Use of
the model is not easy in computational terms; the value of Dooge's study
lies in the brilliant unification of unit hydrograph modelling approaches.

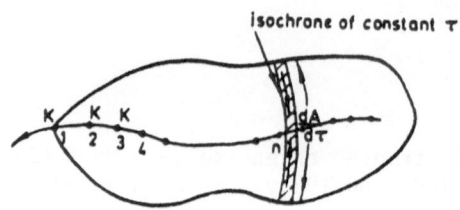

Fig. 2.5.4. The Dooge general catchment model

2.5.2 Time-variant linear catchment synthesis

In general, rather little attention has been paid to time-variant effects
in linear catchment modelling. The recognition that catchment response
may vary seasonally has resulted in the production of, for example, unit
hydrographs for 'winter' events as opposed to 'summer' events. Little
seems to have been done to systematise such studies. Further, time
variance during events has had negligible practical study. Some possible
approaches have been investigated (O'Donnell 1966, Mandeville and
O'Donnell 1973) for simple cases of time-variant behaviour. There may
well be potential for further developments in linear modelling before
abandoning the powerful property of superposition and moving into the
more intractable regions of non-linear studies.

2.5.3 Time-invariant linear catchment analysis

It was commented earlier that deriving a catchment IUH by invoking the
aid of a catchment model is bound to lead to an approximation to the
real IUH, the goodness of the approximation being dictated by the goodness of the simulation by the model of reality. It would clearly be
preferable to use direct methods of deriving an IUH (or TUH), by-passing
the need for any sort of model. Such methods would deal directly and
solely with the rainfall excess, x(t), and direct runoff, y(t), to yield
the IUH, h(t). Such analysis methods have been developed, following and
building on the historically earlier synthesis methods. The methods to
be described require only that the catchment system be linear and time-
invariant.
     The basic strategy of several of the analysis techniques has an
affinity with the 'moments of area' component of the Nash reservoir
cascade method. In fact the equations Nash derived (2.5.3 and 2.5.4)

relating the moments of x(t), y(t) and h(t) are pure <u>analysis</u> results;
the <u>synthesis</u> reservoir model followed only because analysis methods
could not proceed further.

What the moments part of the Nash model does is to <u>transform</u> the
initial x(t) and y(t) data into a more useful form, viz. their moments.
Nash's moments equations then yielded <u>directly</u> the moments of h(t), the
catchment IUH. However, it is not possible to 'untransform' the first
two moments of a function to find the function itself. Nash therefore
needed to adopt a <u>synthesis</u> approach to complete his technique, viz. to
postulate a 2-parameter model and then match the catchment IUH moments
to those of the model. The response function of the fitted model was
then taken to be the IUH of the catchment.

Schematically, a full <u>transform</u> method of <u>analysis</u> would be as shown
in Fig. 2.5.5. Firstly, the original identification problem of finding
h(t) from known x(t) and y(t) is switched into a transform domain by
finding suitable transforms of x(t) and y(t). The latter are then used
to find the transform of h(t). Finally, this transform of h(t) is
'untransformed' or inverted to yield h(t) itself.

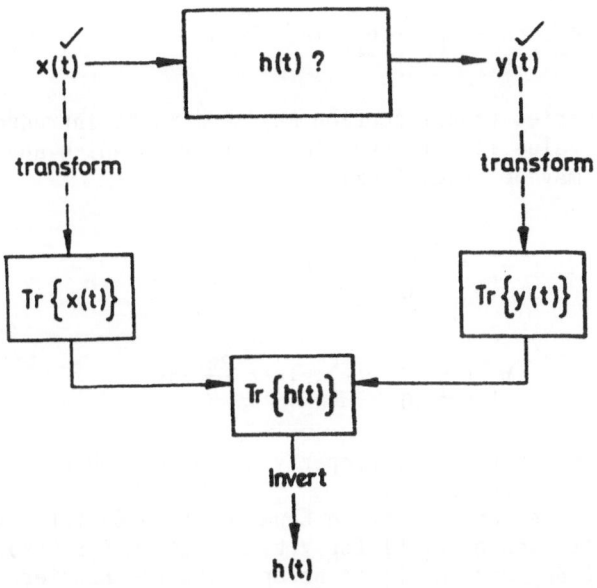

Fig. 2.5.5. The transform method for identifying h(t).

Clearly, of the many types of transform available, one needs to
choose those in which, firstly, it is possible to find directly the
transform of h(t) from the transforms of x(t) and y(t), and secondly, it
is possible to invert directly the transform of h(t).

Such transform analysis techniques have been well established and
explored in other branches of applied science. Systems engineers and
electrical circuit engineers have for many years used Laplace transforms

and Fourier transforms to establish the impulse responses of their systems. However, the first uses of transform methods for the full analysis
solution of the IUH identification problem made use of transforms other
than Laplace and Fourier (though the latter were used in later studies
(Blank et al. 1971)). One such method will be outlined. Much fuller
accounts are given by O'Donnell (1960, 1966) and Dooge (1973), the latter
providing a very comprehensive and systematic treatment of all the
transform methods used in the study of linear hydrological systems.

The first complete transform solution of the IUH identification
problem was that developed by O'Donnell (1960), later extended to direct
IUH identification (O'Donnell 1966). For the IUH case, the transforms
used were infinite Fourier series; for the TUH case, the closely related
finite harmonic series were used. Referring back to Fig. 2.5.1, consider
a time-base L, greater than or equal to the length of the direct runoff
hydrograph, $y(t)$. Then for each of the three functions, $x(t)$, $h(t)$ and
$y(t)$, in Fig. 2.5.1, one can write infinite Fourier series which match
those functions exactly at every point within the range $0 \leqslant t \leqslant L$ (including points where the functions have the value zero). For example, we
can represent $x(t)$ for $0 \leqslant t \leqslant L$ by:

$$x(t) = a_0 + \sum_{r=1}^{\infty} \{a_r . \cos r \frac{2\pi t}{L} + b_r . \sin r \frac{2\pi t}{L}\} \qquad (2.5.6)$$

While such an infinite series transformation may seem to be introducing
complication, its great value lies in the simplicity with which any of
the $\{a, b\}$ coefficients may be found from:

$$a_0 = \frac{1}{L} \int_0^L x(t)\ dt \qquad a_r = \frac{2}{L} \int_0^L x(t).\cos r \frac{2\pi t}{L}\ dt$$

$$ (2.5.7)$$

$$b_r = \frac{2}{L} \int_0^L x(t).\sin r \frac{2\pi t}{L}\ dt$$

(This simplicity stems from a powerful property of cosine and sine
functions, viz. that of orthogonality).

We may write Fourier series similar to Equation (2.5.6) for $y(t)$
and $h(t)$, but using coefficients $\{A, B\}$ for $y(t)$ and $\{\alpha, \beta\}$ for $h(t)$.
Thus each of the three functions may be transformed to its Fourier
coefficients. Of the three sets of coefficients, initially only the
$\{a, b\}$ and $\{A, B\}$ coefficients are known (from the known $x(t)$ and $y(t)$
functions). The next stage, as with any transform method, is to find
linkage equations between the transforms of $x(t)$ and $y(t)$ and those of
(the unknown) $h(t)$. This stage involves the use of the convolution
equation (2.5.1).

Substitution of the three Fourier series for $x(t)$, $y(t)$ and $h(t)$
into the convolution equation, and use once more of the orthogonality
property, yields the very straightforward relationships (O'Donnell 1966):

$$\alpha_0 = \frac{1}{L}\frac{A_0}{a_0} \qquad\qquad \alpha_r = \frac{2}{L}\frac{a_r A_r + b_r B_r}{a_r^2 + b_r^2}$$

$$\hspace{6cm} (2.5.8)$$

$$\beta_r = \frac{2}{L}\frac{a_r B_r - b_r A_r}{a_r^2 + b_r^2}$$

Thus the transforms of h(t), viz. its {α, β} Fourier coefficients, may be straightforwardly found from the known transforms of x(t) and y(t), viz. the {a, b} and {A, B} Fourier coefficients.

The final stage, inversion, is also straightforward in that it only requires the substitution of the now known {α, β} coefficients into the Fourier series for h(t):

$$h(t) = \alpha_0 + \sum_{r=1}^{\infty}\{\alpha_r.\cos r\,\frac{2\pi t}{L} + \beta_r.\sin r\,\frac{2\pi t}{L}\} \qquad (2.5.9)$$

In principle, then, the problem of IUH identification from known x(t) and y(t) by direct analysis, not requiring any synthesis model, is solved by the Fourier series transform method. In practice, some restrictions and difficulties arise. Being experimentally observed, x(t) and y(t) are known only as sampled data, not as continuous functions. Further, x(t) may only be available in histogram form (block volumes over finite periods). Thus the integrations for the {a, b} coefficients of x(t) in Equation (2.5.7) cannot be performed if x(t) is not known as a continuous function (similarly for the {A, B} coefficients of the sampled output). Also with a finite number of the two sets of sampled data points, one cannot estimate more than the same finite number of the infinite sets of Fourier coefficients for x(t) and y(t).

A detailed discussion of such restrictions and difficulties (O'Donnell 1966) shows them to be less important for practical hydrology than the direct extension of the Fourier approach mentioned earlier, viz. to the direct determination of a finite-period TUH from sampled data. Given excess rainfall input, x(t), as a histogram of block volumes over intervals T, and given only ordinates of y(t) at concurrent intervals T, then finite harmonic series fitting these data points exactly may be written, the series containing the same number of terms as there are data points. Harmonic series are also orthogonal, their coefficients being found (exactly) by analogous summations rather than the integrals of equation (2.5.7). Instead of the convolution integral (equation 2.5.1) for continuous data and the IUH, the analogous convolution summation of Equation (2.5.2) for discrete sampled data and the TUH may be used to find the linkage equations for the three finite sets of harmonic coefficients. The useful (and pleasing) result found by O'Donnell (1966) was that the harmonic TUH linkage equations had precisely the same form as the Fourier IUH linkage equations (2.5.8). Thus the transform technique yields exact TUH ordinates from (exact) x(t) histograms and y(t) sampled data, the only form in which 'real-world' data are available.

Contemplating a geometrical view of the summation of the terms in a
harmonic series reveals that the low order terms at the start of the
series dominate the fitting of the 'signal' within the data being matched,
while the high order terms towards the end of the series are likely to
be intimately concerned with fitting any 'noise' bound to be present in
'real-world' data. Hall (1977) developed a more efficient version of
the full harmonic series method, one which truncates the full series to
a relatively small number (e.g. 10) of the lower order terms. The
truncation criterion is based on a power spectrum derived from the harmonic TUH coefficients themselves as they are found from the start
onwards. Such truncation also reduces the instabilities and negative
ordinates towards the tail of a harmonically-derived TUH.

It will be noted that the initial theoretical developments outlined
above involved continuous functions, the IUH, coefficient integrations,
and the convolution operation as an integral. The more practical and
useful extension applied itself to 'real-world' sampled data, the TUH,
coefficient summations, and the convolution operation as a summation.

As a final example of time-invariant linear catchment analysis, it
is pertinent to continue with the 'real-world', sampled data, TUH problem.

Snyder (1961) seized the opportunity presented by the advent of
digital computers to develop a TUH matrix_inversion technique previously
impracticable for hand or mechanical implementation. Fig. 2.5.6 shows
the TUH convolution operation originally summarised in Equation (2.5.2)
now set out as a set of simultaneous equations.

$$y_1 \quad = X_1 \cdot u_1 + \quad 0 \quad + \quad 0 + \ldots\ldots\ldots\ldots\ldots\ldots\ldots\ldots + 0 + 0$$

$$y_2 \quad = X_2 \cdot u_1 + X_1 \cdot u_2 + \quad 0 + \ldots\ldots\ldots\ldots\ldots\ldots\ldots + 0 + 0$$

$$y_3 \quad = X_3 \cdot u_1 + X_2 \cdot u_2 + X_1 \cdot u_3 + 0 + \ldots\ldots\ldots\ldots\ldots + 0 + 0$$

$$\vdots$$

$$y_m \quad = X_m \cdot u_1 + X_{m-1} \cdot u_2 + \ldots\ldots + X_1 \cdot u_m + 0 + \ldots\ldots\ldots + 0 + 0$$

$$y_{m+1} \quad = \quad 0 \quad + X_m \cdot u_2 + \ldots\ldots\ldots + X_2 \cdot u_m + X_1 \cdot u_{m+1} + 0 + \ldots + 0$$

$$\vdots$$

$$y_{m+n-2} = \quad 0 \quad + \quad 0 \quad + \quad 0 \quad + \ldots\ldots 0 + X_m \cdot u_{n-1} + X_{m-1} \cdot u_n$$

$$y_{m+n-1} = \quad 0 \quad + \quad 0 \quad + \quad 0 \quad + \ldots\ldots 0 + \quad 0 \quad + \quad X_m \cdot u_n$$

Fig. 2.5.6. The equations relating rainfall and unit hydrograph
             ordinates to runoff.

Given m T-hour blocks of rainfall excess, X_i, in the input histogram
and n TUH ordinates at intervals T, then there are (m + n - 1) ordinates

of the direct runoff, y_i, as shown. In fact for $m \geqslant 2$, there are more equations than unknowns $((m + n - 1$ c.f. $n)$ and we have an 'overdetermined' system.

In Fig. 2.5.7, the equations are shown in matrix form. The rectangular matrix of X values has $(m + n - 1)$ rows and n columns, with a diagonal band of non-zero X_i elements enclosed by zero elements. In conventional matrix notation, the set of equations may be written:

$$|X| \cdot |U| = |Y| \qquad\qquad (2.5.11)$$

As it stands, Equation (2.5.11) cannot be solved for $|U|$ since $|X|$ is a rectangular matrix that cannot be inverted. However, by the ingenious manipulation of premultiplying both sides of Equation (2.5.11) by $|X|^T$, the transpose of $|X|$, one arrives at:

$$|X|^T \cdot |X| \cdot |U| = |X|^T \cdot |Y| \qquad\qquad (2.5.12)$$

The product $|X|^T \cdot |X| = |Z|$ is now a square matrix with n rows and n columns while the product $|X|^T \cdot |Y| = |W|$ is a vector with n rows (and one column). In effect, Equation (2.5.12) represents a reduced set of just n simultaneous equations i.e. an 'even-determined' system for the n unknowns in $|U|$. This manipulation has not 'lost' any of the information in the original over-determined system of Fig. 2.5.6. Now, however, it is possible to invert the square matrix $|Z|$ to yield:

$$U = |Z|^{-1} \cdot |W| \qquad\qquad (2.5.13)$$

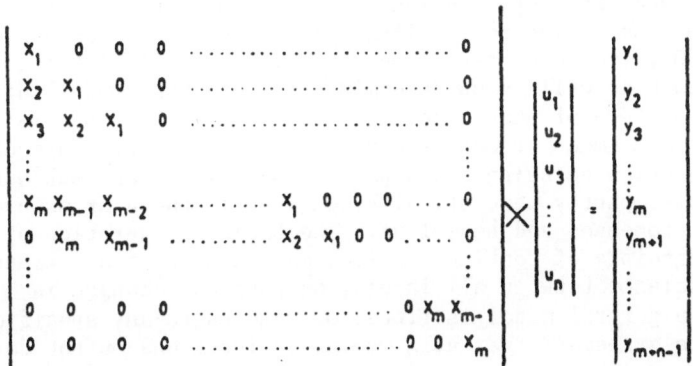

Fig. 2.5.7. The matrix form of the unit hydrograph equations.

There is a further benefit in this series of matrix manipulations. Bearing in mind that any real-world input and output data (the X_i and

Y_i) will contain errors, the solution given by Equation (2.5.13) yields
a 'best' solution in the least squares sense. Thus if $|\hat{U}|$ represents
the solution of Equation (2.5.13), then reinserting in Equation (2.5.11):

$$|X|.|\hat{U}| = |\hat{Y}|$$

gives estimates, $\hat{Y}_i$, of the original Y_i such that $\Sigma(\hat{Y} - Y)_i^2$ is a minimum.
Any other solution for $|U|$ would give a larger sum-of-squares.

Clearly, this matrix inversion technique is a very powerful one, and
is ideally suited for digital computer usage. It is the method adopted
in the U.K. Flood Studies Report (NERC 1975).

Another and considerable advantage of both the matrix inversion
method and the harmonic analysis method is that they can be used on
continuous multi-peaked hydrographs. The earlier traditional unit
hydrograph techniques can use only isolated single-peak events.

2.6 Non-linear treatment of catchment behaviour

2.6.1 Review

Attention will be concentrated on the non-linear catchment systems
synthesis route of Fig. 2.1.2. There is as yet rather little of practical application and value in the 'black box' analysis route, basically
because of the mathematical difficulties faced in non-linear analysis,
e.g. the identification of non-linear system response functions from
input/output records only.

The following survey must necessarily be brief. Fortunately there
are several good text references. Two in particular are essential comp-
lements to this account: Overton and Meadows (1976), and Fleming (1975).
Many symposia proceedings give papers dealing with not only non-linear
synthesis models and their applications but also developments in fitting
techniques. One such is the recent IAHS Oxford Symposium on Hydrological
Forecasting (1980), including papers on many (if not all) of the
other topics of this book as well as on deterministic catchment modelling.

The general pattern of such modelling techniques is to postulate a
general model of catchment processes and behaviour, the structure of the
model and its functioning being based partly intuitively and subjectively,
partly empirically, partly theoretically, on what is known or assumed
about catchment processes and behaviour. The basis of operation of such
models is the principle of continuity, i.e. maintaining at all times a
complete water balance between all inputs, outputs and changes in
storages. Such a general model is fitted or matched to any specific
catchment in some systematic way using recorded input and output data
for that catchment. The fitting is done by making adjustments to the
parameters used in algebraic equations purporting to describe the
processes and interconnections of the catchment model. Such parameter
adjustment is continued until the output computed by the model when
supplied with the recorded catchment input agrees with the recorded
catchment output (to within some specified tolerance). This general
strategy is schematised in Fig. 2.6.1.

SYSTEM   SYNTHESIS

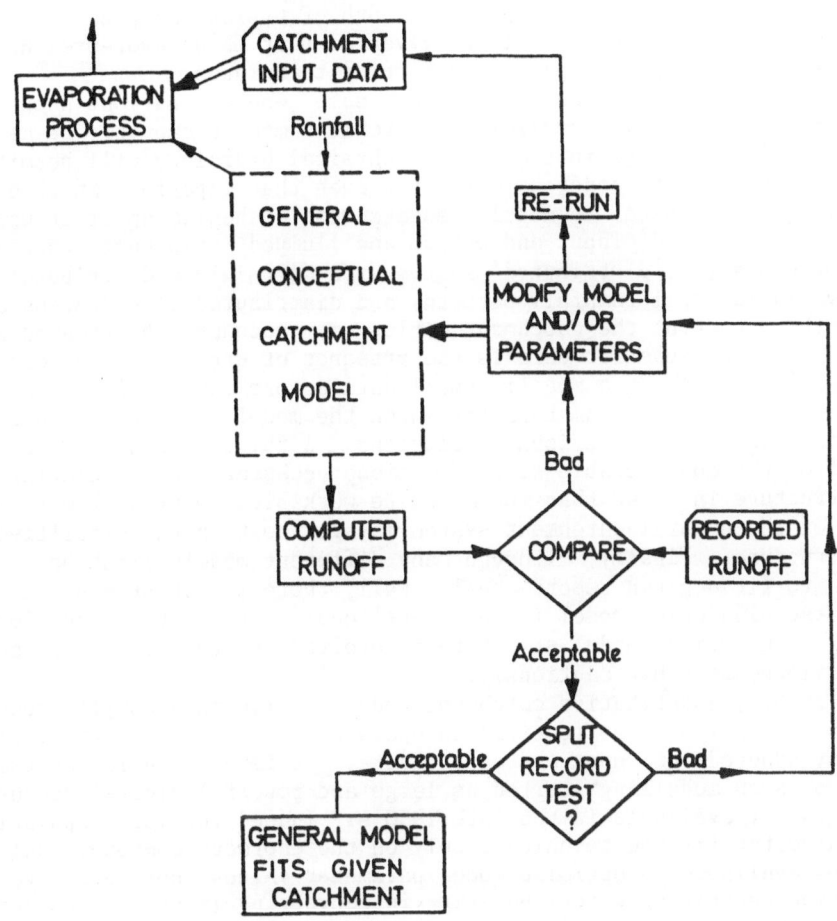

Fig. 2.6.1.   General catchment modelling strategy.

Two other important features are shown in Fig. 2.6.1:   firstly, the
model structure itself may have to be adjusted as well as its parameter
values, and secondly, good practice involves split-record testing, i.e.
using only part of the catchment data available for fitting the model
parameters, then using the remainder to test that the fitting achieved
is acceptable.

In later chapters, several detailed case studies of the use of
deterministic catchment models will be given. The aim here is to present
only some general introductory comments. Examples of models selected
from the many available will be given, mainly to display typical features
of all such models. Much fuller discussion and details may be found in
the text references. For example, Fleming (1975) gives structure dia-
grams and parameter listings for 19 catchment models along with details
of their operational functioning. Overton and Meadows (1976) not only

provide accounts of the quantitative aspects of models but also discuss
water quality modelling.

Following the general surveys below both of typical component and
overall models, some account will be given of that issue ever-present in
catchment modelling, viz., parameter fitting techniques.

To conclude this introductory review, some general cautions should
be summarised. Clearly deterministic catchment models can only be as
good as knowledge gained in the area of physical hydrology will permit;
in many features they cannot as yet match even that imperfect knowledge.
For example, it is usual to model the catchment mechanism as if it were
a system with 'lumped' input and output and 'lumped' components whereas
it is, of course, a 'distributed' system with certainly a distributed
input (variable areal rainfall pattern) and distributed flow processes
even if the output at the catchment outlet may reasonably be treated as
lumped. Another cause for care is the presence of errors in the recorded
data used for adjusting a model: small data errors may result in large
errors in those model parameters for which the model response is insensitive, thus leading to spurious conclusions. A third reason for caution
arises from the considerable simplifications necessary in postulating a
model structure in order that the model be workable: many of the fine
details of the natural catchment system may get lost in the simplified
model structure. Finally, although many different models might be
adjusted to fit a given record equally well, there is, at present, no
way of assessing which model is the 'best' one: until a technique for
evaluating an optimum model structure is evolved, synthesis techniques
must be viewed somewhat cautiously.

Inevitably, quantitative catchment models evolve into complex const-
ructions yet they must be feasible to operate and to do so repetitively
with many hundreds of input and output data. It is no coincidence that
the era of such modelling started as large and powerful digital computers
became readily available in the late 1950's. Indeed the more sophisticated parameter fitting techniques rely on the enormous computational
power now available to optimise model parameter values, not merely to do
the routine repetitive water budgeting in converting input data to output
data.

2.6.2 Examples of deterministic catchment models

The first step in the modelling strategy is to construct a generalised
flow chart showing the structure of the proposed model, along with the
inter-connections and alternative water routes between the various components of that structure. This first step must start with the clearest
specification of what the model is required to do, i.e. a decision on
the objectives of the model study. To illustrate this vital point,
Figs. 2.6.2 - 2.6.4 show three catchment model structures, each of which
was designed for a different main objective yet all of which have some
components or processes in common.

The first model, shown in Fig. 2.6.2, is one in the series of
Stanford models, developed by Linsley and his co-workers. This is a
leading example of those catchment models that attempt to be as comprehensive and all-embracing of catchment processes as possible, and at the

same time to achieve as much realism as possible. The operation of such a complex model is controlled by some 30 parameters. The detailed parameter listing and operational specification of this model will be found in Fleming (1975); only a brief summary is warranted here.

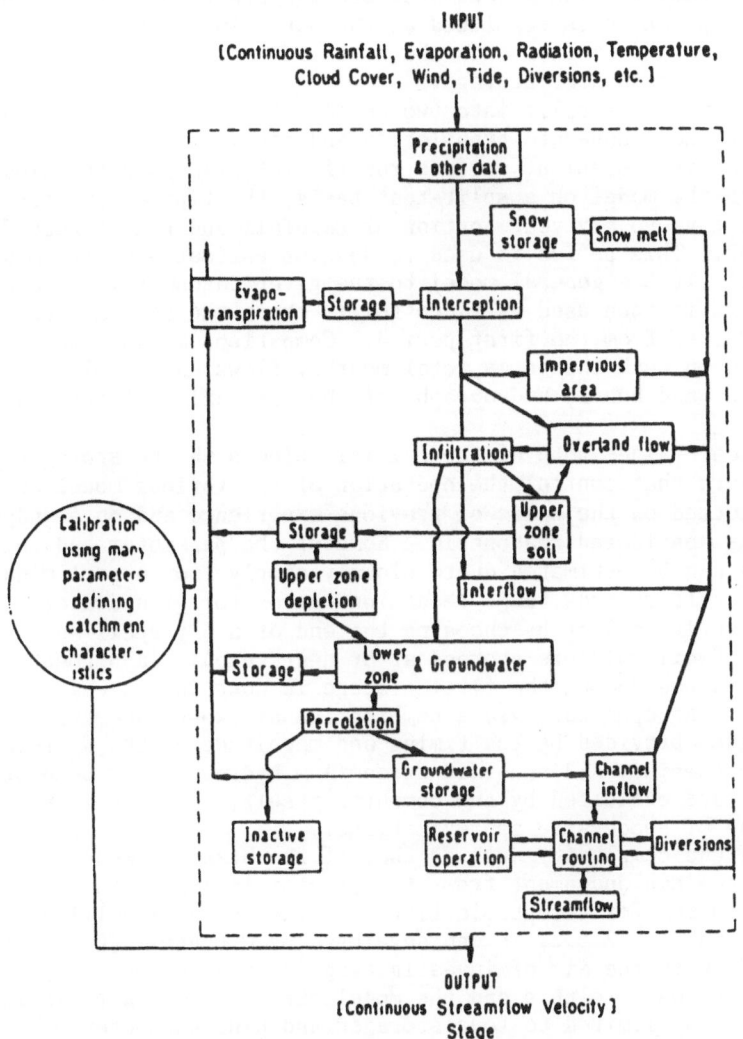

Fig. 2.6.2. The Stanford Watershed Model IV

The characteristics of the component storage and routing elements in Fig. 2.6.2 are determined by relationships, expressed in terms of certain parameters, that represent as rationally as possible the behaviour of the various segments of the hydrological cycle. After setting estimates of initial storage volumes hourly (or shorter) increments of rainfall are entered into the model. The incoming rainfall either becomes direct

runoff or is detained in upper and lower soil moisture storages, the
latter feeding a groundwater storage. The three storage zones combine
to represent the effects of highly variable soil moisture profiles and
groundwater conditions. The upper zone storage absorbs a large part of
the first few hours of rain in a storm. The lower zone storage controls
long-term infiltration. The groundwater storage controls baseflow in
the stream. Evaporation is permitted at the potential rate from the
upper zone storage and at less than the potential rate from the lower
zone storage and groundwater storage.

The direct runoff is split into two components, surface runoff and
interflow, which have separate translation and routing procedures.
Total streamflow is the sum of surface runoff, interflow and baseflow.

In applying the model on a split-test basis, the typical procedure
is to select a five to six year portion of rainfall and runoff records
for a catchment. This period is used to develop estimates of the model
parameters that fit the general model to the given catchment. A second
period of record is then used as a control to check the accuracy of the
parameters obtained from the first period. Comparison in the control
period is based on such things as total monthly flows, daily flow
duration curves, and hourly hydrographs of the two maximum floods each
water year.

The numerical values both of the initial volumes in the storages and
of the parameters that control the operation of the various model components are selected on the basis of previous experience and on a judgement of what is considered reasonable. Some of the parameter and initial
storage values can be estimated quite closely simply from a preliminary
study of the runoff records (e.g. recession curves for groundwater
storage characteristics) or by choosing the end of a dry spell as a
starting point (soil moisture storage at or near zero). Adjustment of
the parameter values during the fitting stage is done in two ways. Most
are adjusted by the operator, via a combination of experience and intui-
tion, using clues provided by the timing and magnitude of the difference
between the synthesised and recorded streamflow hydrographs. Some of
the parameters are evaluated by the computer itself, using an internal
looping routine of successive approximations.

A model of the complexity such as that of the Stanford model requires
skill, experience and judgement from its operator in making the parameter
adjustments required for acceptable fitting. The second modelling
example, shown in Fig. 2.6.3, is one developed by O'Donnell (Dawdy and
O'Donnell 1965) with the aim of investigating 'automatic' parameter
optimisation methods. To this end the model structure was kept deliber-
ately simple, being limited to four storages and nine parameters.

The emphasis was not on constructing the 'best' model of catchment
hydrology but on examining the feasibility of using the computer itself
to search for optimum parameter values (and not only to do the water
budgetting calculations). Others have followed this line of enquiry
(Clarke 1973). The many difficulties met will be discussed in the final
section below.

The final example of a catchment model, again designed for a different main objective, is the Isolated Event Model 4 shown in Fig. 2.6.4
(NERC 1975).

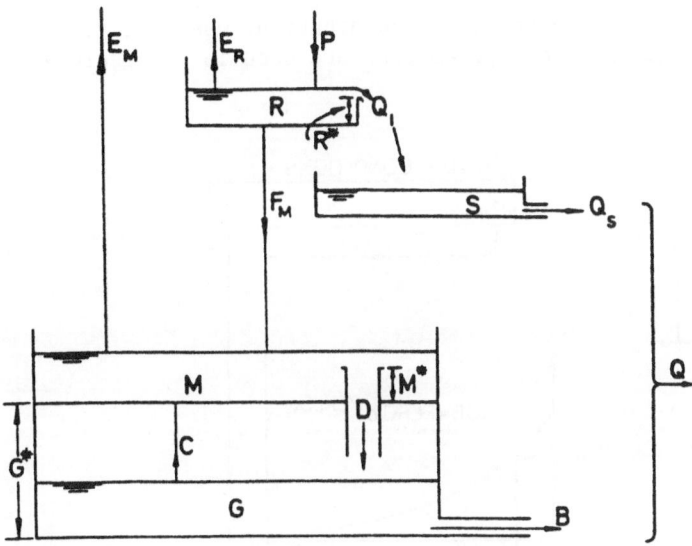

R = interception & surface storage ; S = channel storage ;
M = soil moisture storage ; G = groundwater storage.

P = precipitation ; E_R = evaporation ; R^* =
threshold value of R ; Q_1 = discharge from
interception and surface storage ; F_M = infiltration.

Q_s = discharge from channel storage.

E_M = transpiration ; M^* = threshold value of M ;
D = deep percolation ; C = capillary rise.

G^* = threshold value of G ; B = discharge from
groundwater storage.

Q $(= Q_s + B)$ = discharge at catchment outfall.

Fig. 2.6.3. The Dawdy-O'Donnell model.

As its name implies, this model was developed (alongside a unit
hydrograph approach) to deal with storm events only, thus having an
emphasis different from the previous two models which were designed to
deal with continuous uninterrupted data. Being freed from the necessity

for components to cope with the low flow behaviour essential to continuous data modelling, only four parameters are used in the isolated event
model.

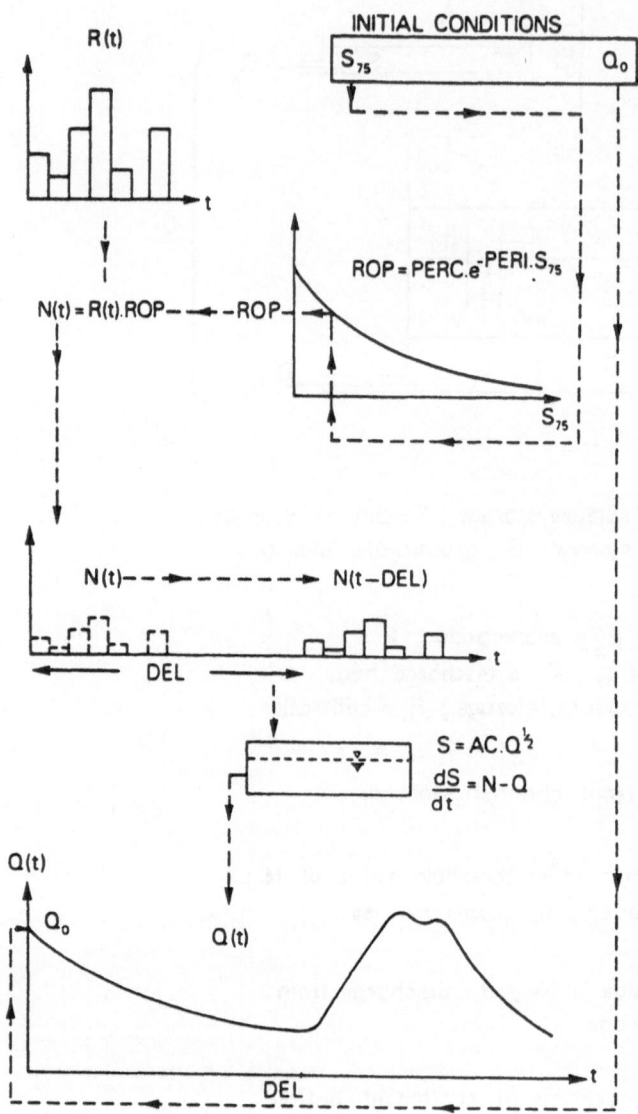

Fig. 2.6.4. Structure of the isolated event model.

Again, an automatic parameter fitting technique was used with the
isolated event model. The study included parameter fitting both one
event at a time and over groups of seasonal events.

2.6.3 Parameter fitting and sensitivity

Two aspects of catchment modelling may be readily recognised. Firstly,
there is an inevitable link between model complexity ('goodness') and
number of model parameters. Secondly, the best operators of catchment
models are their creators.

Optimisation techniques have been developed that determine values of
system parameters which maximise or minimise some function dependent
on those parameters. Such techniques are objective, and although they
may test situations that would be dismissed out of hand by a human operator, the tremendous speed with which computers can make such tests
compensates for such inefficiency.

Such techniques allow the user of a model, not as intimately familiar
with the model as its creator, to concentrate on the many other aspects
of any catchment study. Initial studies by Dawdy and O'Donnell (1965)
were followed by Mandeville et al. (1970), Ibbitt and O'Donnell (1974).
The latter paper discusses some of the many problems that may arise in
using automatic parameter optimisation methods. The optimisation procedure developed by Rosenbrock (1960) was found to perform better with
catchment models than other optimisation techniques in a comparative
study by Ibbitt (1966). It has been widely used in other catchment
modelling studies.

There is a very substantial literature on optimisation techniques
with many texts and symposia proceedings on optimisation as a discipline
in itself. It is not appropriate to attempt a survey of optimisation
methods in this limited book. One aspect, however, should be stressed
in concluding this section, viz. parameter sensitivity.

Certain model parameters will have much greater 'control' of model
performance than other parameters. That performance will be very sensitive to quite small changes in the values of such parameters. In
contrast, model performance may be almost unaffected by other parameters
that for a given record exert very little influence. An insight into
this sensitivity aspect may be gained by carrying out tests at the end
of a model parameter optimisation. If each 'best fit' parameter value
is altered, say, by 1% and the model re-run (the other parameters being
held fixed), the change in the model response can be monitored. The
operator will then be able to judge the validity of the optimised parameter values.

With recent developments in inter-active computer usage plus visual
display facilities, there is potential for a mixture of automatic optimisation alongside operator control (via model performance sensitivity
monitoring). A recent study (Klatt and Schultz 1981) provides a valuable
pointer in this direction.

SYMBOLS

i	rainfall rate	LT^{-1}
A	catchment area	L^2
Q	flow	L^3T^{-1}
x(t)	input	L^3T^{-1}

y(t)	output	L^3T^{-1}
h(t)	impulse response	T^{-1}
S	storage	L^3
K	reservoir constant	T
u(T, t)	T hour unit hydrograph	T^{-1}
T	interval length	T
n	number of reservoirs	1
τ	translation delay time	T
L	time base	T
R(t)	rainfall	L
ROP	runoff percentage	1
N(T)	effective precipitation	L

REFERENCES

Blank, D., Delleur, J.W. and Giorgini, A. 1971 Water Resour. Res., 7, 1102, Oscillatory kernel functions in linear hydrological models.

Clarke, R.T. 1973 'Mathematical models in hydrology', Irrigation and Drainage Paper 19, FAO, Rome.

Dawdy, D.R. and O'Donnell, T. 1965, J. Hyd. Div., ASCE, 91(HY4), 123, Mathematical models of catchment behaviour.

Diskin, M.H. 1964 'A basic study of the rainfall excess - surface runoff relationship in a basin system', PhD Thesis, Univ. of Illinois, Urbana.

Dooge, J.C.I. 1959, J. Geophys. Res., 64(2), 241, A general theory of the unit hydrograph.

Dooge, J.C.I. 1973, 'Linear theory of hydrologic systems', Techn. Bulletin No. 1468, Agricultural Research Service, U.S. Dept. of Agriculture, Washington.

Fleming, G. 1975 'Computer simulation techniques in hydrology', Elsevier's Series in Environmental Sciences, Den Haag.

Hall, M.J. 1977 IAHS Bull., 22(2), 313, On the smoothing of oscillations in finite period unit hydrographs derived by the harmonic method.

Ibbitt, R.P. 1970 'Systematic Parameter Fitting for Conceptual Models of Catchment Hydrology', PhD Thesis, University of London.

Ibbitt, R.P. and O'Donnell, T. 1974 IASH Pub. No. 101, Proc. Warsaw Symposium on Mathematical Models in Hydrology, 461, Designing conceptual catchment models for automatic fitting methods.

IAHS 1980 IASH Pub. No. 129, Proc. Oxford Symposium on Hydrological Forecasting.

Klatt, P. and Schultz, G.A. 1981 Int. Symp. on Rainfall-Runoff Modelling, Mississippi, Improvement of flood forecasts by adaptive parameter optimisation.

Lambert, A.O. 1972 J. Instn. Wat. Engrs. 26, 413, Catchment models based on ISO-functions.

Linsley, R.K., Kohler, M.A. and Paulhus, J.L.H. 1958 'Hydrology for Engineers', McGraw-Hill, New York.

Mandeville, A.N., O'Connell, P.E., Sutcliffe, J.V. and Nash, J.E. 1970 J. Hydrology, 11, 109, River flow forecasting through conceptual models: Part III.

Mandeville, A.N. and O'Donnell, T. 1973 Water Resour. Res., 9(2), 298, Introduction of time variance to linear conceptual catchment models.

Nash, J.E. 1960 Proc. ICE, 17, 249, A unit hydrograph study, with particular reference to British catchments.

NERC 1975 'Flood Studies Report', Vol. I, Natural Environ. Res. Council, London.

O'Donnell, T. 1960 IASH Pub. No. 51, 546, Instantaneous unit hydrograph derivation by harmonic analysis.

O'Donnell, T. 1966 Proc. Tech. Meeting, No. 21 (TNO, The Hague), 65, Methods of computation in hydrograph analysis and synthesis.

Overton, D.E. and Meadows, M.E. 1976 'Stormwater Modelling', Academic Press, London.

Rosenbrock, H.H. 1960 Comp. Jour., 3, 175, An automatic method of finding the greatest or least value of a function.

Shaw, E.M. 1983 'Hydrology in Practice', Van Nostrand Reinhold, London.

Snyder, W.M. 1961 TVA Res. Paper, No. 1, Knoxville, Tennessee, Matrix operation in hydrograph computations.

Wilson, E.M. 1983 'Engineering Hydrology', Macmillan, London.

3. THEORY OF FLOOD ROUTING

J.C.I. Dooge

University College
Civil Engineering Dept.
Earlsfort Terrace
Dublin 2
Ireland

3.1 Continuity equation for unsteady flow

The first equation required for the analysis of unsteady flow in an open channel is the continuity equation. This is usually written in differential form as

$$\frac{\partial Q}{\partial x} + \frac{\partial A}{\partial t} = r(x,t) \qquad (3.1.1)$$

where $Q(x,t)$ is flow, $A(x,t)$ is area of flow and $r(x,t)$ is the rate of the lateral inflow per unit length of channel. It is convenient to emphasize the element of forecasting by writing the continuity equation in what is known as the prognostic form i.e. with all time derivatives as unknowns on the left hand side. The prognostic form of the continuity equation is

$$\frac{\partial A}{\partial t} = -\frac{\partial Q}{\partial x} + r(x,t) \qquad (3.1.2)$$

It is important to note that the area in Equations (3.1.1) and (3.1.2) is the total storage area and includes any overbank storage as well as the conveyance area of the main channel.

3.2 Momentum equation for unsteady flow

A second equation related to unsteady flow in an open channel can be obtained by applying the principle of the conservation of linear momentum i.e. Newton's second law to the conveyance section of the channel. If the flow is assumed to be one-dimensional (i.e. no acceleration in the vertical and lateral directions) and the velocity is assumed uniformly distributed over the section, then the momentum equation for channel of uniform shape and a small uniform bed slope takes the form (St. Venant 1871):

D.A. Kraijenhoff and J.R. Moll (eds.), River Flow Modelling and Forecasting, 39-65
© 1986 by D. Reidel Publishing Company.

$$\frac{\partial z}{\partial x} + \frac{u}{g}\frac{\partial u}{\partial x} + \frac{1}{g}\frac{\partial u}{\partial t} + \frac{\tau_0}{\gamma R} = 0 \qquad\qquad (3.2.1)$$

where $z(x,t)$ is the elevation of the water surface, $u(x,t)$ is the average velocity of flow in the conveyance section, $\tau_0(x,t)$ is the average shear stress along the wetted perimeter of the cross-section, $R(x,t)$ is the hydraulic radius of the cross-section (i.e. the ratio of the flow area to the wetted perimeter), and γ is the weight density of water.

Where there is lateral inflow to the channel, the St. Venant equation must be modified to allow for the fact that momentum must be given to this increment of flow at the expense of the existing momentum of the main flow. Consequently an additional term representing this "momentum drag" must be added to the momentum equation which can be written in prognostic form as

$$\frac{\partial u}{\partial t} = -g\frac{\partial y}{\partial x} - u\frac{\partial u}{\partial x} + g\left(S_0 - S_f - \frac{ur}{gA}\right) \qquad\qquad (3.2.2)$$

where $y(x,t)$ is the depth of flow, $S_0 = -\partial z_0/\partial x$ is bottom slope and $S_f = \tau_0/\gamma R$ is friction slope. Detailed derivations of Equation (3.2.2) are available in the literature (e.g. Cunge, Holly and Verwey, 1980).

It is often convenient, for understanding or for computational purposes, to express the continuity and momentum equations in terms of some other flow variables than those of Equations (3.1.2) and (3.2.2). For example, since the continuity equation is linear in $A(x,t)$ and $Q(x,t)$ it is convenient to express the momentum equation in terms of the same variables. This can be done by using the definition relationship:

$$Q(x,t) = u(x,t).A(x,t) \qquad\qquad (3.2.3)$$

to eliminate $u(x,t)$ from Equation (3.2.2) and using Equation (3.1.2) to group and cancel terms. We obtain for the momentum equation for a prismatic channel

$$\frac{\partial Q}{\partial t} = -g\bar{y}(1 - F^2)\frac{\partial A}{\partial x} - \frac{2Q}{A}\frac{\partial Q}{\partial x} + gA(S_0 - S_f) \qquad\qquad (3.2.4)$$

where $\bar{y}(x,t)$ is the hydraulic mean depth i.e. the ratio of the area of the channel $A(y, shape)$ to the surface width $T(y, shape)$ and $F(x,t)$ is the Froude number defined by

$$F^2 = (Q^2 T)/gA^3) \qquad\qquad (3.2.5)$$

The form of momentum equation given by Equation (3.2.2) or Equation (3.2.4) is often referred to as the non-divergent form. For a prismatic channel with vertical sides ($\partial T/\partial x = 0$) the momentum equation can also be expressed in the form

$$\frac{\partial Q}{\partial t} = -\frac{\partial}{\partial x}\left[gA(\bar{y}/2 + u^2/g)\right] + gA(S_0 - S_f) \tag{3.2.6}$$

which is known as the conservative form.

3.3 Equations of characteristics for unsteady flow

The equations of unsteady flow in open channels can also be expressed in characteristics form and this formulation has advantages both for physical insight and for the computation of flow in prismatic channels. The development for the general case gives rise to complicated expressions and the following development is restricted to the case of a channel with vertical sides over the range of depth of interest, i.e. $T(x,t) = B$. The general case does not involve any new principles (Abbott 1966).

The characteristic equations are developed in the terms of the mean velocity of flow $u(x,t)$ and the wave celerity $c(x,t)$ defined by

$$c(x,t) = \sqrt{(gA/T)} = \sqrt{(g\bar{y}(x,t))} \tag{3.3.1}$$

The equation of continuity given by Equation (3.1.1) can be expressed for a rectangular channel in these variables by using Equation (3.3.1) to convert the dependent variables from $A(x,t)$ and $Q(x,t)$ to $u(x,t)$ and $c(x,t)$. The resulting continuity equation is

$$2c\frac{\partial c}{\partial t} + c^2\frac{\partial u}{\partial x} + 2uc\frac{\partial c}{\partial x} = \frac{gr(x,t)}{B} \tag{3.3.2}$$

where B is the width of the channel.

The momentum equation for a rectangular channel can be converted from $y(x,t)$ to $c(x,t)$ to obtain

$$\frac{\partial u}{\partial t} + u\frac{\partial u}{\partial x} + 2c\frac{\partial c}{\partial x} = g(S_0 - S_f - ur/Bc^2) \tag{3.3.3}$$

If Equation (3.3.2) is divided by $c(x,t)$ and added to Equation (3.3.3) we obtain

$$\frac{\partial}{\partial t}\left[u + 2c\right] + (u + c)\frac{\partial}{\partial x}\left[u + 2c\right] = g\left[S_0 - S_f + (c - u)r/Bc^2\right] \tag{3.3.4}$$

Equation (3.3.4) indicates that along a characteristic path defined by:

$$dx/dt = u(x,t) + c(x,t) \tag{3.3.5}$$

the characteristic variable $(u + 2c)$ changes in accordance with the ordinary differential equation

$$\frac{d}{dt}\left[u + 2c\right] = g\left[S_0 - S_f + (c - u)r/Bc^2\right] \tag{3.3.6}$$

Similarly it can be shown that along a characteristic path defined by

$$dx/dt = u(x,t) - c(x,t) \tag{3.3.7}$$

the characteristic variable $(u - 2c)$ changes in accordance with the ordinary differential equation

$$\frac{d}{dt} [u - 2c] = g[S_0 - S_f - (c + u)r/Bc^2] \tag{3.3.8}$$

The paths defined by Equations (3.3.5) and (3.3.7) are known respectively as the positive and negative characteristic directions and the variables $(u + 2c)$ and $(u - 2c)$ as the positive and negative characteristic variables. It can be shown (Abbott 1966) that the characteristic directions are the directions along which discontinuities in the derivatives of the solution are propagated and hence they are of fundamental physical importance as the paths along which disturbances move in the generation of the solution of the problem.

3.4 Boundary conditions in flood routing

The basic equations in any of the forms discussed above cover all cases to which the basic assumptions apply. The individual problems are distinguished from one another by different sets of boundary conditions. The boundary conditions include both temporal conditions which apply to all sections at a given time and terminal conditions which apply at a given section at all times. In the case of the St. Venant equations, two temporal and two terminal boundary conditions are required for the solution of any particular problem. The question of appropriate boundary conditions is best discussed in terms of the characteristic paths since the number of conditions required at any boundary is equal to the number of characteristics entering the solution region from a point on that boundary. For the case of tranquil flow, i.e. for Froude number less than unity, the characteristic direction defined by Equation (3.3.5) will be downstream and the characteristic direction defined by Equation (3.3.7) will be upstream. For the case of rapid flow, i.e. Froude number greater than unity, both of the characteristic paths will be in a downstream direction. We will confine ourselves to the case of tranquil flow.

For tranquil flow the solution region will be divided into four zones as shown on Figure 3.4.1. The solution in Zone A is affected only by the two initial conditions specified along the x-axis LR for t = 0. The solution in Zone B will be affected by the Zone A solution for the two dependent variables along the leading positive characteristic LM and by the single terminal condition given on the boundary LS for x = 0. Similarly, the solution in Zone C will be affected by the Zone A solution for the two dependent variables along the leading negative characteristic RM and the single terminal condition given on the boundary RN for x = L. Finally, the remainder of the solution in Zone D will depend on the solution in Zone C on the leading positive characteristic MN and in Zone B on the leading negative characteristic MS and on both the upstream and downstream boundary conditions.

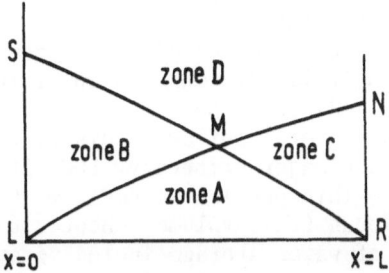

Fig. 3.4.1. Zones of solution for tranquil flow

If the downstream control (at x = L) is very far downstream of the
river reach through which the flood is routed, then this control will
have little influence. At the downstream end of this reach of interest,
the outflow hydrograph will consist of a steady segment controlled by the
initial conditions in the channel (Zone A), the bulk of the flood wave
as yet unaffected by the downstream control (Zone B), and the final
recession which will be affected to some extent by the downstream control
(Zone D). In the common methods of hydrologic flood routing (discussed
in section 3.12 below) the effect of the downstream control is neglected.

Figure 3.4.2 shows a typical set of boundary conditions for a flood
routing problem. The flow variables are prescribed along the initial
boundary t = 0. The single terminal condition at the upstream boundary
x = 0 usually prescribes the variation of some hydraulic flow parameter
(discharge, area of flow, water level, depth of flow, etc.) as a function
of time. Once the values for two independent flow parameters are known
at a point, the values of the other flow parameters of interest can be
readily calculated. The downstream boundary condition depends on the
downstream control for the channel reach and may be difficult to specify.

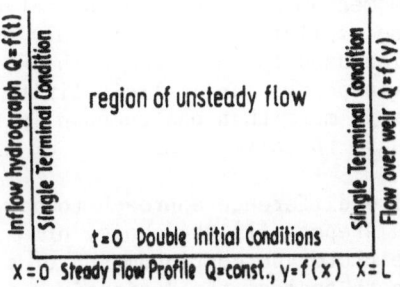

Fig. 3.4.2. Typical boundary conditions

The assumption is often made that at the terminal boundary x = L the
relation between water level and discharge is the same as for the steady

flow condition i.e. by using the steady-state rating curve. This assumption is less harmful for the downstream boundary condition at x = L than it would be for the upstream boundary condition at x = 0 since disturbances propagating upstream against the flow are more heavily damped than disturbances propagating downstream with the flow.

If a substantial quantity of water is added to, or diverted from a channel reach at a particular point, then it is advisable to split the reach into two sub-reaches at that point and route the flows through these two sub-reaches separately. If the lateral inflow is distributed along a reach rather than concentrated in a major tributary, then it is normally lumped in with upstream inflow. This problem is dealt with in Cunge, Holley and Verwey (1980) and in Young (this volume, chapter 6). If overbank flow occurs, the total area of water storage including overbank storage must be included in the continuity equation (3.1.2). The delineation of the area of water flow for use in the momentum equation is a matter of difficulty and varies from model to model. If overbank flow is to be taken into account, then a very small initial depth of water is maintained on the overbank area between flood events to facilitate computation (see Grijsen, this volume, chapter 9).

3.5 The finite difference approach

The basic equations for the solution of the channel routing problem comprise a system of non-linear partial differential equations. As such they cannot be solved analytically in their primitive form. To obtain a solution the original equations must either be solved numerically or else simplified to facilitate an analytical solution.

The numerical approach most commonly used in the solution of the complete non-linear equations for channel routing is the finite difference method. This approach was used even before the advent of digital computers (Thomas 1937). The large number of finite difference methods are often grouped into (a) characteristic methods in which the characteristic formulation of the basic equations described in section 3.3 is solved either on a characteristic network or on a rectangular network; (b) explicit methods for solving the basic equations of continuity and momentum on a rectangular network; and (c) implicit methods for solving the basic equations of continuity and momentum on a rectangular network. A finite difference scheme is said to be explicit if there is only one unknown in the finite difference equation used to approximate the original differential equation at a given point. An implicit finite difference scheme is one in which there is more than one unknown in the finite difference equation so that we have to solve a whole set of simultaneous equations.

As indicated on Figure 3.5.1 a finite difference approach to a differential equation consists of three steps: (1) the choice of a finite difference mesh to represent the prototype continuum and of a type of finite difference approximation to replace the derivatives in the original equation by finite differences; (2) the choice of a method for linearizing the resulting algebraic equations; and (3) the choice of a method for the solution of the resulting linear algebraic equations. In step (1) either regular or irregular grids of nodal points may be used but

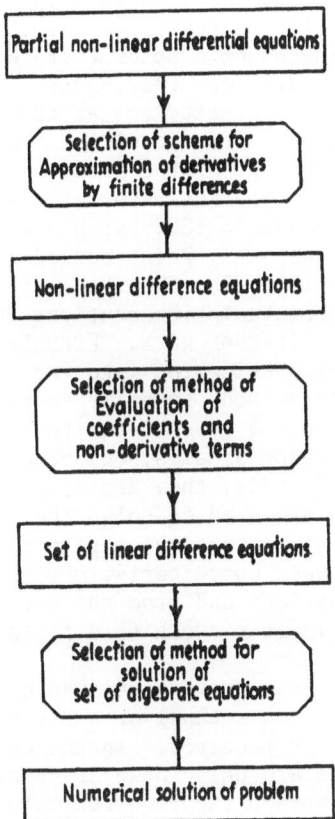

Fig. 3.5.1. Choices in explicit and implicit methods

regular grids (and in particular rectangular grids) are favoured because
of the greater ease of programming. The derivatives in differential
equations are replaced by the ratios of the finite differences between
the nodal values of the variables concerned. This produces a set of non-
linear algebraic difference equations which must then be solved.

A finite difference scheme is said to be consistent if the finite
difference equation approaches the differential equation as the mesh size
goes to zero. A finite difference equation is said to be _stable_ if the
cumulative effect of all the roundoff errors arising during the computations do not swamp out the solution. A finite difference scheme is said
to be _convergent_ if the values of the finite difference solution approach
those of the differential equation solution as the mesh size goes to
zero. A finite difference scheme is said to be _conservative_ if the
conservation principles on which the differential equations are based is
preserved in the finite difference solution.

3.6 Characteristic finite difference schemes

A number of finite difference schemes have been developed for the solution

of the characteristic form of the St. Venant equations described in
section 3.3. In these methods we seek a solution of Equations (3.3.5)
and (3.3.7) in order to determine the co-ordinates of a point (x,t) on
the characteristic network and of Equations (3.3.6) and (3.3.8) to
determine the values of the velocity (u) and the celerity (c) at that
point. Once the latter two values are known, the water level and the
discharge can be readily calculated from a knowledge of the cross-section
properties at the point x of the channel reach. The characteristic equations and their solution were discussed by Massau in 1889 but his work
was ignored and the method of characteristics was only reintroduced into
hydraulics over fifty years later via compressible flow theory.

Finite difference methods based on the characteristic equations for
open channel flow may be subdivided in three different ways. Firstly
they are divided according to the manner of discretization into (a)
characteristic grid methods and (b) fixed grid methods (usually rectangular). Secondly, they are divided according to the type of finite
difference approximation used into (a) first order schemes; (b) second
order schemes and (c) higher order schemes. Thirdly, they are divided
depending on whether or not iterative methods are used to solve the
equations, into (a) implicit characteristic methods and (b) explicit
characteristic methods. Other variations between characteristic methods
arise from the treatment of the boundary conditions and from the methods
used to interpolate from a variable characteristic network to a fixed
rectangular network.

The solution on a characteristic grid does not provide the hydrograph
of the variable of interest at a fixed location on a canal or a river
and to obtain such a hydrograph interpolation is required. In the case
of a natural channel, the sectional properties are only known at specific
locations and interpolations must be used to obtain the required values
at the node points of the characteristic network. To overcome the
disadvantage of the uneven spatial and temporal distribution of points
resulting from use of a characteristic network, methods have been
developed for the solution of the characteristic equations on a fixed
rectangular grid. The spatial interval of the grid is made to correspond
to the surveyed cross-sections so that no interpolation of section
properties is required.

3.7 Explicit finite difference schemes

Explicit finite difference schemes were the first to be applied to the
St. Venant equations and were widely used at first because of their
simplicity in programming and execution. In recent years they have
been largely displaced by implicit schemes whose additional complexity
is compensated for by their inherent numerical stability. However,
there are circumstances where an explicit scheme may be preferable to
other methods. The use of an explicit scheme is appropriate when a
computer program is required urgently for occasional rather than frequent
use or when the forecasting procedure of necessity involves a small time
step (thus ensuring stability).

When using the characteristic equations on a rectangular grid, it is
necessary to interpolate between the known values of the variables at

the initial time level to determine the starting values for the characteristics that intersect at the point of interest at the new time level.
This interpolation procedure may be a source of error but the error
should not be serious if flood hydrograph and the variations in channel
properties are relatively smooth.

Explicit finite difference schemes are usually applied to the conservative form of the basic equations, as given by equation 3.2.6. The
solution is advanced from one time level to the next for each node in
the channel reach in turn until the solution has been obtained all along
the channel at the advanced time step and then the process is repeated.
The approximation to the space derivatives is evaluated on the initial
time level. The non-derivative terms in the equations can be evaluated
as a weighted average of the node values at the original and advanced
time levels without destroying the explicit nature of the formulation.
The various steps in a finite difference scheme procedure and the choices
available are shown on Figure 3.5.1. The explicit finite difference
approximations that have been applied to flood routing include
  the simple explicit scheme
  the diffusive scheme
  the leapfrog scheme
  the Lax-Wendroff scheme.
The simple explicit scheme uses forward differencing in time and
central differencing in space. This scheme can be shown to be unconditionally consistent but to be unstable when the friction slope is
evaluated at the known time level. The diffusive scheme overcomes the
stability problem but at the cost of introducing numerical diffusion.
The leapfrog scheme is a three level scheme with centred differences in
both time and space. The Lax-Wendroff scheme is a two-step predictor-
corrector method using the diffusive scheme in the first step and leap-
frog scheme in the second step.

The time step in explicit methods is limited by the Courant-Friedrichs-
Lewy criterion

$$\frac{\Delta t}{\Delta x} < \frac{1}{(u + c)_{max}}$$

where $u(x,t)$ is the mean velocity of flow and $c(x,t)$ is the wave celerity
defined by equation (3.3.1). This criterion implies that in every case
the point at the new time level should lie within the domain of dependence of two adjacent points at the old time level. This domain of dependence is delimited by the characteristics passing through the two
adjacent points at the old time level. For many practical cases of flood
routing the CFL criterion would restrict time steps in explicit methods
to values of the order of 1 or 2 minutes. This restriction does not
apply to implicit methods because all of the values at the old time
level are taken into account simultaneously.

3.8 Implicit finite difference schemes

Implicit methods of finite difference approximations have been increasingly used because of the necessity to restrict the time step in explicit

schemes so as to avoid instability. In implicit methods the solution is
obtained by applying the implicit scheme at each of the nodes along the
channel simultaneously, combining the difference equations obtained in
this way with the boundary conditions at each end, and solving the whole
set of equations simultaneously to advance the solution through a time
step.

The simplest implicit schemes are those involving relationships
between the unknown values of the dependent variables at two adjacent
nodes in the advanced time level and the known values of the variables
at the two corresponding nodes in the original time level for which a
solution has already been found. Such a four-point implicit scheme based
on an averaging of the values at the initial and final time levels was
proposed for flood routing by Thomas in 1937 but he was unable to carry
out the required computations with the calculation aids of that time.
The general four-point implicit scheme with a weighting of Θ on the
values at the unknown time level was proposed by Preissman in 1960. It
can be shown that four-point implicit schemes are unconditionally stable
for $\Theta > 0.5$. A great number and variety of implicit difference schemes
were discussed by Liggett and Cunge (1975).

The application of an implicit finite difference scheme to the St.
Venant equations results in a set of non-linear simultaneous algebraic
equations. The number of equations, and hence the size of the matrix to
be inverted, is twice the number of step lengths used to discretize the
channel reach. The matrices will be banded but their solution remains
very difficult for the full non-linear case. Some of the difficulties
can be overcome by evaluating the non-linear terms at the original time
level or with less loss of accuracy by linearising these terms.

The efficiency of any implicit finite difference procedure depends
largely on the manner in which the set of algebraic equations obtained
by finite difference approximation is solved. All of the practical techniques for solving sets of non-linear simultaneous algebraic equations
involve the successive solution of a set of linearised equations on an
iterative basis until the change in the values of the unknown dependent
variables after an iteration is less than a prescribed tolerance. A
popular method for the solution of the linearised equations is the double
sweep method (Liggett and Cunge 1975).

The solution of the complete St. Venant equations by the six point
implicit method is illustrated in the case study of the Parana-Paraguay
model described by Grijsen in Chapter 9.

3.9 Linearisation of the St. Venant equations

The main obstacle to finding an analytical solution for the flood
routing problem is the non-linear character of the momentum equation.
Consequently an important simplification of the problem is to linearise
that equation. The highly non-linear equation can be linearised by
treating the unsteady flow conditions as perturbation or deviation from
a steady reference flow condition and neglecting second higher order
terms. The question of how well the linearised solution approximates
the complete non-linear solution in any given case can only be properly
evaluated by comparing the two solutions. Apart from its use for fore-

casting, a linearised solution may give greater insight into the nature
of the solution of the complete non-linear problem than the solution of
a simplified non-linear version of the original equations.

The linearised equation for unsteady flow in an open channel without
lateral inflow is given by

$$g\bar{y}_0(1 - F_0^2) \frac{\partial^2 Q'}{\partial x^2} - 2u_0 \frac{\partial^2 Q'}{\partial x \partial t} - \frac{\partial^2 Q'}{\partial t^2} = gA\left(\frac{\partial S_f}{\partial Q_0} \frac{\partial Q'}{\partial t} - \frac{\partial S_f}{\partial A_0} \frac{\partial Q'}{\partial x}\right) \quad (3.9.1)$$

where $Q'(x,t)$ is the perturbation of the flow from its reference value Q_0.
The classical problem of flood routing in a prismatic channel can be
approximated by solving Equation (3.9.1) subject to the appropriate
boundary conditions. It can be shown that the effect of the downstream
boundary condition dies out rapidly as we move up the channel reach.
Accordingly this discussion can be confined to the effect on $Q_B(x,t)$ of
an upstream boundary condition $Q_A(t)$. Since the system is linear it is
only necessary to determine the linear channel response $h(x,t)$ due to a
delta function input. The downstream flow $Q_B(x,t)$ is given by:

$$Q_B(x,t) = \int_0^t Q_A(\tau).h(x,t-\tau)\,d\tau \quad (3.9.2)$$

which is the convolution of the upstream inflow $Q_A(t)$ with the linear
channel response $h(x,t)$.

The solution of Equation (3.9.1) for the response to an upstream
disturbance only (Lighthill and Whitham 1955, Dooge and Harley 1967) is
found to consist of two parts, one representing the head of the wave and
the other representing the slower moving body of the wave. If we represent this by writing

$$h(x,t) = h_1(x,t) + h_2(x,t) \quad (3.9.3)$$

then the solution is

$$h_1(x,t) = \delta(t - x/c_1) \exp(-ax) \quad (3.9.4)$$

and

$$h_2(x,t) = \exp(-bt + dx)e^2(x/c_1 - x/c_2) \sum_{k=0}^{\infty} \frac{\left[e^2(t-x/c_1)(t-x/c_2)\right]^k}{k!(k+1)!}$$

$$(3.9.5)$$

where c_1 and c_2 are the positive and negative characteristic velocities
and a, b, d, and e are parameters depending on the hydraulic properties
of the channel.

The $h_1(x,t)$ part of the solution for the impulse response, which is
given by equation (3.9.4), represents a delta function travelling downstream at the dynamic wave speed given by equation (3.3.5) above but
rapidly decreasing in volume. The $h_2(x,t)$ part of the impulse response
given by equation (3.9.5) represents the main body of the wave which
moves more slowly at a speed close to the kinematic wave speed given by
equation (3.10.9) below. For short lengths of channel this body of the
wave consists of a finite depth at the head of the wave followed by a
monotonically declining tail. For long lengths of channel, the flood
wave consists of a rising limb, a crest segment, and a falling limb.

3.10 Simplification of the St. Venant equations

There have been many proposals for the simplification of the St. Venant
momentum equation as a means of making the channel routing problem more
amenable to solution. The momentum equation described in section 3.2
represents the balance between the force of gravity represented by gS_0,
the resistance of boundary friction represented by gS_f, the unbalanced
pressure force represented by $g\partial y/\partial x$, the advective acceleration represented by $u\partial u/\partial x$, the local acceleration represented by $\partial u/\partial t$ and the
momentum drag of any lateral inflow. All of the terms mentioned above
have the dimension of force per unit mass. The equation representing
this balance is more usually written in terms of force per unit weight
as:

$$S_0 = S_f + \frac{\partial y}{\partial x} + \frac{u}{g} \frac{\partial u}{\partial x} + \frac{1}{g} \frac{\partial u}{\partial t} + \frac{ur}{gA} \qquad (3.10.1)$$

For the conditions usually encountered in channel routing, the third,
fourth and fifth terms on the right hand side of Equation (3.10.1) i.e.
the acceleration terms, are usually two orders of magnitude smaller than
S_0 and S_f and one or two orders of magnitude less than $\partial y/\partial x$ (Henderson
1966, Kutchment 1972, Cunge et al. 1980). This suggests that the solution
of the simplified equation obtained by dropping these three terms may
provide a good approximation to the solution based on the full equations.
Since this simplification replaces the original system of hyperbolic
equations by a system of parabolic equations the solution is often referred to as the parabolic solution of the St. Venant equations. Because
the resulting equation is the same as that obtained by replacing the
momentum equation by an equation representing convection and diffusion
in the form:

$$Q(x,t) = c(Q) A(x,t) - D(Q) \partial A/\partial x \qquad (3.10.2)$$

it is frequently referred to as the method of diffusion analogy.
    If in the linear parabolic formulation instead of neglecting the
acceleration terms completely, they are expressed in terms of $\partial y/\partial x$ by
using the linear kinematic solution then we have (Dooge and Harley 1967)

$$\frac{\partial y}{\partial x} \left(1 - \frac{F_0^2}{4}\right) = S_0 - S_f \tag{3.10.3}$$

and the diffusivity of Equation (3.10.2) is then given by

$$D = \frac{Q_0}{2S_0 T_0} \left(1 - \frac{F_0^2}{4}\right) \tag{3.10.4}$$

It can be shown that the first two moments with respect to time of this solution are identical to the first two moments of the complete linear solution. The linear diffusion analogy is applied by Moll in Chapter 11 and the non-linear version applied to the Upper Dee by O'Connell in Chapter 8.

If the term representing the unbalanced pressure force $\partial y/\partial x$ is also neglected then the momentum equation is reduced to

$$S_0 = S_f \tag{3.10.5}$$

which corresponds to steady uniform flow and for any given cross-section can be written as

$$Q(x,t) = f[A(x,t)] \tag{3.10.6}$$

where the function $f[A]$ depends on the shape of the cross-section, the friction law, and the roughness parameter. For a prismatic channel the same function $f[A]$ will apply all along the channel but for a natural channel the function will vary from section to section. The equation of continuity for the case of no lateral inflow is given by

$$\frac{\partial A}{\partial t} = -\frac{\partial Q}{\partial x} \tag{3.10.7}$$

Combination of equations (3.10.6) and (3.10.7) gives the relationship

$$\frac{\partial A}{\partial t} = -c(A) \frac{\partial A}{\partial x} \tag{3.10.8}$$

where $c(A)$ has the dimension of velocity and is given by

$$c(A) = \frac{dQ}{dA} = f'[A] \tag{3.10.9}$$

It can be shown that if $A(x,t)$ is eliminated from (3.10.6) and (3.10.7) we get

$$\frac{\partial Q}{\partial t} = -c(A) \frac{\partial Q}{\partial x} \tag{3.10.10}$$

The solution of equation (3.10.8) is clearly given by

$$A = \phi\left[t - \frac{x}{c}\right] \tag{3.10.11}$$

where the form of the function ϕ is determined by the boundary condition
for x = 0. This solution is known as the kinematic wave solution
(Lighthill and Whitham 1955) because the dynamic relationship of equation
(3.10.1) has been replaced by the kinematic relationship of equation
(3.10.6).

If c is taken as constant we have the linear kinematic wave which
takes the form of a flood wave moving downstream at the kinematic wave
speed c given by equation (3.10.9) without change of shape. Comparison
with the complete linear solution discussed in section 3.9 indicates that
the linear kinematic wave gives the same first moment of lag of the flood
wave and thus represents the general downstream movement of the wave
fairly well. However, it fails to predict any subsidence of the flood
peak and thus would introduce errors in flood routing except for the
limiting case where F = 2. In contrast, the diffusion analogy is exact
as the Froude number F tends to zero and reproduces both the first and
second moments of the complete linear solution for higher Froude numbers.

If the kinematic equation given by equation (3.10.8) is solved by
finite differences, then numerical diffusion will be introduced into the
solution (see section 9.3 of Grijsen in this volume). If the values of
c and α are chosen so that this numerical diffusion is equal to the
hydraulic diffusion of the channel reach, then the subsidence of the
flood wave will be accurately modelled. This is the basis of the so-
called Muskingum–Cunge model (Cunge 1969, Cunge, Holley and Verwey, 1980).

For the non-linear kinematic wave approach we have a variable kinematic wave speed so that the higher parts of the flood wave will travel
more quickly than the lower parts. As a result, the shape of the flood
wave will change but there will still be no subsidence of peak value of
discharge. Due to the varying kinematic wave speed the front of the
flood wave will steepen and the tail of the wave will become flatter.
This steepening of the front of the flood wave will increase the value
of $\partial y/\partial x$ as the wave travels downstream and this renders less and less
plausible the neglect of this term in the kinematic simplification of
the momentum equation. When the kinematic wave solution is used in
longer reaches it will predict the formation of shock waves at the front
of the flood wave under conditions where the complete equations would
not predict any such phenomenon. Another feature of the kinematic wave
approach is that it cannot take account of the effect of variation in a
downstream boundary condition on flow conditions in a channel reach.

The non-linear kinematic wave forms the flood-routing-element of
the catchment model applied by Fleming to the Santa Ynez, the Derwent
and the Orchy (Chapter 13).

3.11 Comparison of hydraulic solutions

The wide variety of hydraulic methods of channel routing provide a
bewildering choice for a potential user. Numerical experiments can,

however, give some information on their relative merits. The present section describes the results of such numerical experiments (Maczuga 1979). The experiments were based on a wide rectangular channel of constant bed slope (S_0) and Chézy friction with $C = 31.3 \, m^{\frac{1}{2}}s^{-1}$.

The following computational methods:
(a) the characteristic grid method
(b) the fixed grid characteristic method
(c) the simple explicit scheme
(d) the diffusive explicit scheme
(e) the leapfrog explicit scheme
(f) the four-point implicit scheme
(g) the six-point implicit scheme
were compared on the basis of their ability to predict the downstream hydrograph at 100 km. In the first few series of tests this was done on the basis of the linearised equations for which an analytical solution is available and can be used to determine the error for any particular set of computations. Further series dealt with the non-linear forms of the St. Venant equations. The general conclusions from the series of experiments were:
(1) the best numerical scheme for a slow-flowing river $(F_0 = 0.2)$ would appear to be the four-point implicit method with a mesh ratio based on the kinematic wave speed;
(2) the best numerical scheme for faster streams $(F_0 = 0.8)$ would appear to be one based on the solution of the characteristic equations with parabolic interpolation on a fixed rectangular grid with a mesh ratio corresponding to the dynamic wave speed;
(3) if a very high accuracy is not required, analytical solutions based on either complete linearised equations or on non-linear kinematic equations are more economical of CPU time than numerical finite difference methods.

It must be appreciated that the above results apply to a particular shape of prismatic channel and a particular inflow. It is not anticipated that the result would be widely different for other cases but further numerical experiments over a wide range of conditions would be desirable.

3.12 Nature of hydrologic methods

The term hydrologic methods is applied to all methods of channel routing in which a channel reach is treated as a lumped system and the momentum equation replaced either by a black-box representation or by a conceptual model. Such methods require only data from past flood events for the upstream and downstream end of the reach for their calibration and can be used even when nothing is known about the topographical or roughness characteristics of the intermediate channel sections.

The distributed equation of continuity given by Equation (3.1.2) can be integrated along the channel reach to give

$$\frac{d}{dt} \int_A^B A(x,t) \, dx = - \left[Q(x,t)\right]_A^B + R_{AB}(t) \tag{3.12.1}$$

which in turn gives

$$dS_{AB}/dt = Q_A(t) - Q_B(t) + R_{AB}(t) \tag{3.12.2}$$

where $S_{AB}(t)$ is the storage in the reach, $Q_A(t)$ is the upstream inflow, $Q_B(t)$ is the downstream outflow, and $R_{AB}(t)$ is the total lateral inflow in the reach AB. Equation (3.12.2) is often referred to as the hydrologic storage equation.

The lumping of the St. Venant momentum equation discussed in section 3.2 is not so readily accomplished. In the hydrological approach the momentum equation is abandoned and replaced by a postulated relationship between the variables in the hydrologic storage equation.

$$S_{AB}(t) = F[Q_A(t), Q_B(t), R_{AB}(t)] \tag{3.12.3}$$

If the function F[ ] is known or assumed then the downstream outflow $Q_B(t)$ can be predicted for any given upstream inflow $Q_A(t)$ and any given lateral inflow $R_{AB}(t)$.

By substituting for $S_{AB}(t)$ from Equation (3.12.3) into Equation (3.12.2) one can obtain the input-output formulation

$$P_B[Q_B(t)] = P_A[Q_A(t), R_{AB}(t)] \tag{3.12.4}$$

for any particular hydrologic model. Similarly by eliminating $Q_B(t)$ from Equation (3.12.2) and (3.12.3) one can obtain the equivalent state-transition equation

$$dS_{AB}/dt = G[S_{AB}(t), Q_A(t), R_{AB}(t)] \tag{3.12.5}$$

for the given hydrologic model.

One method of determining the input-output function F[ ], or the equivalent functions $P_A[\ ]$ and $P_B[\ ]$, is by black box analysis which depends on the input-output records without assuming any structure for the model. In linear black-box analysis, the unit hydrograph approach and techniques can be adapted to find the linear channel response (Dooge and Harley 1967, Sauer 1973). It is surprising, in view of the popularity of the unit hydrograph approach to finding the catchment response, that this approach has not been used more widely in channel routing. Black-box analysis can be extended to the non-linear case through the use of the Volterra series (Amorocho 1973).

3.13 Linear conceptual models

All of the conceptual models widely used in channel routing were originally proposed in linear form and in many applications are still used as
linear models. Accordingly, it is appropriate to deal first with linear
lumped conceptual models. The simplest form of the storage function of
Equation (3.12.3) is

$$S_{AB}(t) = K.Q_B(t) \tag{3.13.1}$$

which represents the elementary model of a linear reservoir with a constant storage delay time K. Substitution of this relationship in the
storage relationship of Equation (3.12.2) gives, for $R_{AB}(t) = 0$

$$Q_B(t) + K.dQ_B/dt = Q_A(t) \tag{3.13.2}$$

as the input-output relationship. The solution of the differential
equation for relaxed initial conditions, i.e. $Q_B(0) = 0$ is

$$Q_B(t) = \exp(- t/K)/K \int_0^t \exp(\tau/K).Q_A(\tau) \, d\tau \tag{3.13.3}$$

so that the linear channel response for a delta function input is given
by

$$h(t) = \exp(- t/K)/K \tag{3.13.4}$$

which can be convoluted with any input $Q_A(t)$ to find the corresponding
predicted output. This one-parameter model is not adequate for channel
routing but the linear reservoir is an important building block of many
widely used conceptual models in hydrology.
  If the storage function is taken as a linear function of both inflow
and outflow we have

$$S_{AB}(t) = a \, Q_A(t) + b \, Q_B(t) \tag{3.13.5}$$

which is the basic structure of the Muskingum model

$$S_{AB}(t) = K[xQ_A(t) + (1 - x)Q_B(t)] \tag{3.13.6}$$

proposed by McCarthy (1939).
  Equations (3.12.2) and (3.13.5) can be combined to give for the case
of no lateral inflow:

$$Q_B(t) + b \frac{dQ_B}{dt} = Q_A(t) - a \frac{dQ_A}{dt} \tag{3.13.7}$$

The linear channel response $h(t)$ for this model can be obtained by setting $Q_A(t) = \delta(t)$. The solution is

$$h(t) = \left[1 + \frac{a}{b}\right] \frac{\exp(-t/b)}{b} - \left[\frac{a}{b}\right] \cdot \delta(t) \qquad (3.13.8)$$

It is clear from this solution that this model may predict negative ordinates during the early part of a flood event if a and b are both positive. If the time-derivatives in equation (3.13.7) are replaced by first order finite differences we get a linear relationship between the four values $Q_A(t)$, $Q_A(t + t)$, $Q_B(t)$ and $Q_B(t + t)$. Since the application of first order finite differences to the kinematic wave equation for (x,t) given by equation (3.10.10), also results in a linear relationship between the same four values, this numerically diffusive finite difference solution of the kinematic wave solution can be linked to the Muskingum model (Cunge, 1969).

If a number of equal linear reservoirs are taken in series then the linear channel response is

$$h(t) = (t/K)^{n-1} \exp(-t/K)/[K(n-1)!] \qquad (3.13.9)$$

Equation (3.13.8) represents the model applied to channel routing by Kalinin and Milyukov (1957) and to catchment response by Nash (1959). It was shown by Nash (1960) that any cascade leads to a shape intermediate between, and closely bounded by, (1) the cascade of equal linear reservoirs represented by Equation (3.13.8) and (2) the cascade of a single linear reservoir and a linear channel represented by

$$Q_B(t) + K \, dQ_B/dt = Q_A(t - \tau) \qquad (3.13.10)$$

where τ is the delay time of the linear channel.

The Muskingum model and the Kalinin-Milyukov model (which are both two-parameter models) can be related to one another. In particular the Muskingum model for $x = 0$ and the Kalinin-Milyukov model for $n = 1$ are identical since each of these special cases is a single linear reservoir. The first and second moments about the origin of the complete linear solution of the St. Venant equation for a wide rectangular channel with Chezy friction, are given by Dooge (1973)

$$u_1' = L/1.5u_0 \qquad (3.13.11)$$

and

$$u_2 = \frac{2}{3}\left[1 - \frac{F_0^2}{4}\right]\left[\frac{\bar{y}_0}{S_0 L}\right]\left[\frac{L}{1.5u_0}\right]^2 \qquad (3.13.12)$$

The first and second moments of the linear channel response for the Muskingum model are given by Dooge (1973):

$$u_1' = K \tag{3.13.13}$$

$$u_2 = (1 - 2x)K^2 \tag{3.13.14}$$

Equating the respective moments we get

$$(1 - 2x) = \frac{2}{3}\left(1 - \frac{F_0^2}{4}\right)\frac{\bar{y}_0}{S_0 L} \tag{3.13.15}$$

It is clear from equation (3.13.15) that x cannot exceed 0.5 unless $F_0 >$ 2 at which value the flow becomes unstable. If x = 0 we have a single linear reservoir and hence the expression

$$L = \frac{2}{3}\left(1 - \frac{F_0^2}{4}\right)\frac{\bar{y}_0}{S_0} \tag{3.13.16}$$

for the Kalinin–Milyukov characteristic length. If the length of the channel reach is less than this characteristic length, then the second moment of the complete linear solution can only be reproduced by the Muskingum model if x < 0.

The hydrologic models can accommodate lateral inflow or outflow. In the case of the single-reach Muskingum model, the lateral inflow (or outflow) is lumped in with the upstream inflow. In this case the Muskingum model is no longer a closed system and shows a system gain (or loss). A field example of such a system gain is shown by Young in section 6.2 of Chapter 6 of this volume. In the case of the Kalinin-Milyukov model any known inflow or outflow can be lumped as lateral inflow in one of the reservoirs of the cascade. If the amount and location of the lateral inflow is unknown it can be lumped with the upstream inflow and the system gain computed from the ratio of the volume of measured outflow to the volume of measured inflow.

3.14 Comparison of linear hydrologic models

The channel response of the linear conceptual models described above can be compared with the channel response for the linearised St. Venant equations. Any channel response can conveniently be characterised by its moments (or cumulants) with respect to time (Dooge 1973). The scale of the channel response is represented by the first moment about the origin which is the first cumulant. The shape of the channel response can be characterised by means of dimensionless higher order moments (or cumulants). Though moments will be more familiar to most readers, the use of cumulants which are related to moments (Dooge 1973) enables us to present results for linear models in a much more compact form. Thus we can define an R-th order shape factor (Nash 1959, Dooge and Harley 1967, Dooge 1973) as

$$S_R = k_R/(k_1)^R \tag{3.14.1}$$

Figure 3.14.1 shows a comparison of various 2-parameter hydrologic models on a $(S_2 - S_3)$ shape factor diagram. Three models are shown

    Muskingum model
    Lag and single reservoir
    Cascade of reservoirs (Kalinin-Milyukov)

together with the solution of the linearised St. Venant equations for a rectangular channel for F = 0 and F = 1 and for both Chézy and Manning friction. The three conceptual models have a common point ($S_2 = 1$, $S_3 = 2$) which corresponds to a single linear reservoir (SLR). For higher values of S_2 (i.e. for shorter reaches) the Muskingum parameter x is

negative. A comparison with the plottings for the linearised St. Venant equations gives an indication of the ability of the hydrologic models to represent the linearised St. Venant equations. It is seen, for example, that the Muskingum method will give a poor representation for low values of S_2 i.e. for long lengths of channel. The diffusion analogy gives an exact result for F = 0.

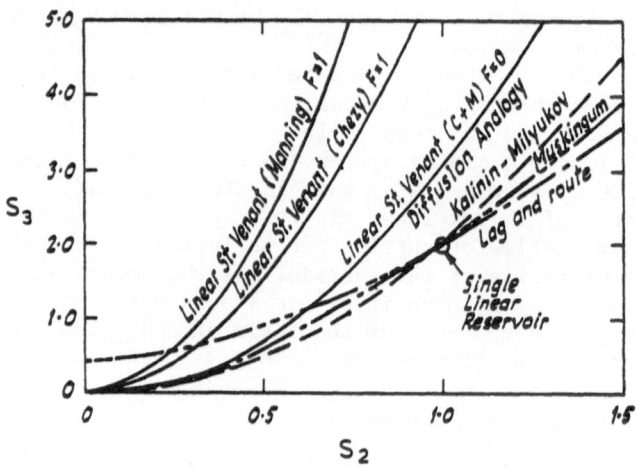

Fig. 3.14.1. Comparison of 2-parameter conceptual models

    If the two-parameter conceptual models are combined with a pure lag component, to give lagged conceptual models, then three cumulants of the natural channel are required in order to determine the three parameters of the model. The shapes can no longer be compared by dimensionless shape factors based on the first moment. Instead we can use shape factors based on the second moment (Strupczewski and Kundzewicz 1980) and defined by

$$f_R = k_R/(k_2)^{R/2} \qquad\qquad (3.14.2)$$

Figure 3.14.2 shows the $(f_4 - f_3)$ shape factor diagram for the linearised
St. Venant solution for a wide rectangular channel. The solution is seen
to be relatively insensitive to the friction law and to the Froude
number. The solution for the combination of a pure lag with a cascade
of equal linear reservoirs is seen to correspond closely to the solution
of the complete equation.

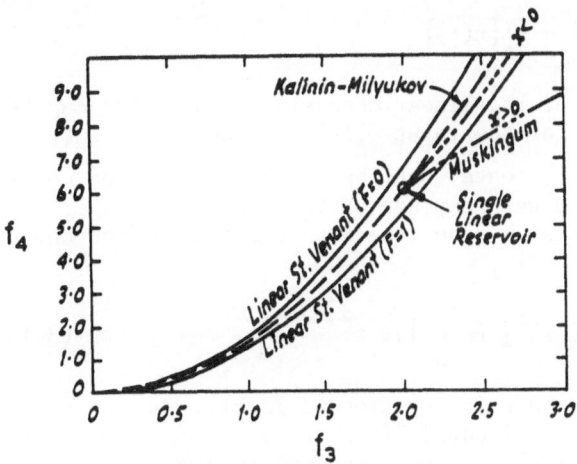

Fig. 3.14.2.   Comparison of lagged conceptual models

From the above comparison of hydrologic models one can conclude that
for slow flowing rivers (i.e. lower values of the Froude number) the
most suitable model is the diffusion analogy and that for swifter flowing rivers the most suitable model is a cascade of equal reservoirs coupled with a pure lag. It can also be concluded that the Muskingum model
is only suitable for short reaches of channel for which optimal value of
the parameter x is negative.

While the above comparisons of hydrologic models (as in the case of
hydraulic models) are confined to a prismatic channel of given shape the
conclusions should not lose all validity when applied to natural channels.

We have discussed hydraulic models and conceptual hydrologic models
for flood routing. The black box approach used widely in rainfall runoff studies, described by Schultz in Chapter 1 and O'Donnell in Chapter
2 of this volume, could equally be applied to linear flood routing. A
suggestion that this be done (Dooge and Harley 1967) was partially
taken up by Sauer (1973) but did not find wide acceptance. Such an
approach would be equivalent to the use of a moving average model and the
estimation of its parameters. It is clear from Chapter 6 by Young in
this volume that the information available in the data can be used more
efficiently to estimate the parameters of a more parsimonious model.

3.15 Calibration of linear models

In applying any of the above hydrologic models to a given channel reach
it is necessary to determine the parameters of the model from past flood
events. Many methods have been proposed for this. One method that is
objective and is applicable to all models is the method of moment or
cumulant matching. Moment matching and hence cumulant matching becomes
subject to error if the inflow and outflow have long recessions but this
difficulty can be overcome by using moments in which the ordinates are
weighted by a negative exponential. For any linear model of channel
routing we have (Dooge 1973)

$$k_R[Q_B(t)] = k_R[Q_A(t)] + k_R[h(t)] \qquad (3.15.1)$$

so that the R-th cumulants of the linear channel response h(t) can be
got from the R-th cumulants of the input $Q_A(t)$ and the output $Q_B(t)$.

These derived cumulants (or moments) can then be equated to the corresponding cumulants of the channel response of the chosen model.
    In the case of the Muskingum model the R-th cumulant of the channel
response is given by

$$k_R(M) = [(1 - x)^R - (-x)^R]\ (R - 1)!\ K^R \qquad (3.15.2)$$

where x and K are the parameters in Equation (3.13.6). For the Kalinin-
Milyukov model of the cascade of equal linear reservoirs, each of delay
time K, the R-th cumulant of the channel response is given by

$$k_R(KM) = n\ (R - 1)!\ K^R \qquad (3.15.3)$$

If a lag is added to a 2-parameter model, the first cumulant will be
increased by the amount of the lag but the remainder will be unchanged.
The derivation of the parameters of the new 3-parameter model gives some
added difficulty and possible imprecision but in forecasting this may be
compensated for by the lead time provided by the lag component.
    The Muskingum model is usually used in the design mode i.e. it is
assumed that inflow and outflow are both known at the end of the time
step as well as at the beginning. The lumped equation of continuity
given by the equation (3.10.2) reduces for the case of no lateral inflow
to

$$\frac{dS_{AB}}{dt} = Q_A(t) - Q_B(t) \qquad (3.15.4)$$

This can be expressed in finite difference form as

$$S_{AB}(t + \Delta t) - S(t) =$$

$$\Delta t \left[\frac{Q_A(t + \Delta t) + Q_A(t)}{2} - \frac{Q_B(t + \Delta t) + Q_B(t)}{2} \right] \qquad (3.15.5)$$

But from equation (3.13.5) we can write

$$S_{AB}(t + \Delta t) - S(t) =$$

$$a \left[Q_A(t + \Delta t) - Q_A(t) \right] + b \left[Q_B(t + \Delta t) - Q_B(t) \right] \qquad (3.15.6)$$

Since the left hand sides of equations (3.15.5) and (3.15.6) are equal we can equate the right hand sides and re-arrange the terms to express $Q_B(t + \Delta t)$ in terms of $Q_A(t + \Delta t)$, $Q_A(t)$, and $Q_B(t)$. The result of such an operation is

$$Q_B(t + \Delta t) = C_1 Q_A(t + \Delta t) + C_2 Q_A(t) + C_3 \cdot Q_B(t) \qquad (3.15.7)$$

where

$$C_1 = \frac{\Delta t - 2a}{\Delta t + 2b} \qquad (3.15.8)$$

$$C_2 = \frac{\Delta t + 2a}{\Delta t + 2b} \qquad (3.15.9)$$

$$C_3 = \frac{2b - \Delta t}{\Delta t + 2b} \qquad (3.15.10)$$

which are the classical Muskingum coefficients.

When operating in the forecast mode the value of $Q_A(t + \Delta t)$ is not known and must be estimated in some fashion. If the inflow is assumed to remain constant during the interval from time t to time $(t + \Delta t)$ then

$$Q_B(t + \Delta t) = (C_1 + C_2)Q_A(t) + C_3 \cdot Q_B(t) \qquad (3.15.11)$$

If on the other hand the inflow at time $t + \Delta t$ is taken as a linear extrapolation of the values at times $t - \Delta t$ and t then we have:

$$Q_A(t + \Delta t) = 2Q_A(t) - Q_A(t - \Delta t) \qquad (3.15.12)$$

Substitution of equation 3.15.12 into equation 3.15.7 gives

$$Q_B(t + \Delta t) = (2C_1 + C_2)Q_A(t) - C_1 Q_A(t - \Delta t) + C_3 Q_B(t) \qquad (3.15.13)$$

In the case of either equation (3.15.11) or (3.15.13) there is only one unknown value at the end of the time step.

While the coefficients in equations (3.15.7), (3.15.11) and (3.15.13) can be related to the Muskingum parameters a = xK and b = (1 - x)K as shown above, there seems little sense in deriving these coefficients in an indirect manner. It seems more sensible to derive the coefficients directly by time series analysis as described by Young in Chapter 6 of this volume.

3.16 Non-linear hydrologic models

All the classical methods of flood routing have been extended to cover non-linear flood routing. Thus the basic conceptual model of a linear reservoir given by Equation (3.13.1) can be generalised to

$$S_{AB}(t) = K\left[Q_B(t)\right]^c \qquad (3.16.1)$$

which represents a non-linear reservoir, and has been used as a routing element in a number of models of catchment response. In Chapter 8, O'Connell describes the application of a single non-linear reservoir on the lower Dee.

The Muskingum method is often presented in the text books in a non-linear form and then linearised. This is done by expressing the storage in the reach in terms of the end areas.

$$S_{AB}(t) = xA_A(t) + (1 - x)A_B(t) \qquad (3.16.2)$$

and then using the relationship between area and discharge to express storage in terms of discharge as

$$S_{AB}(t) = K\left[xQ_A^c(t) + (1 - x)Q_B^c(t)\right] \qquad (3.16.3)$$

The dependence of the parameters c, K and x on $Q_B(t)$ is determined from the records of a range of flood events (Kohler 1958).

Since the cascade of equal linear reservoirs is widely used in hydrology as a model for linear channel response and linear catchment response, it is natural to consider a cascade of equal non-linear reservoirs as a model of non-linear response (Kalinin and Milyukov 1957, Dooge 1969, Napiórkowski and Strupczewski 1979). O'Connell describes in Chapter 8 the use of a model of a cascade of non-linear reservoirs on the Bedford Ouse. Figure 3.16.1 shows the contributions of the linear and quadratic terms to the predicted downstream outflow based on a cascade of non-linear reservoirs each with the storage function

$$Q_B(t) = a\left[S_{AB}(t)\right] + b\left[S_{AB}(t)\right]^2 \qquad (3.16.4)$$

compared with the complete non-linear solution of the St. Venant equations due to a rectangular input. Figure 3.16.2 shows the prediction, by means of such quadratic Volterra series, of the downstream outflow due

to a bimodal smooth inflow using parameters optimised on the output due
to the rectangular pulse input (Napiórkowski et al. 1985).

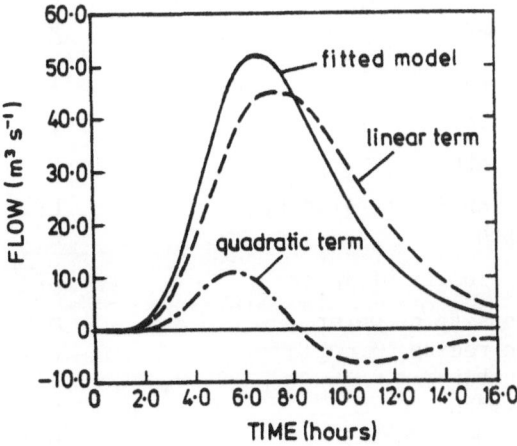

Fig. 3.16.1. Fitted response for a pulsed input

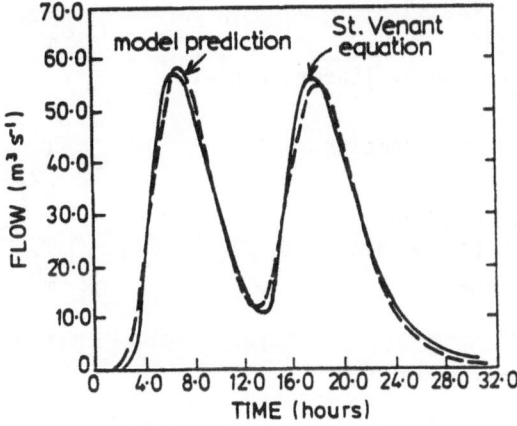

Fig. 3.16.2. Predicted response to bimodal input

It is also possible to generalise the convolution integral described
by O'Donnell in Section 2.5 of Chapter 2 to a Volterra series so as to
include non-linear effects (Amorocho 1973) in the black box approach but
the derivation of even the quadratic kernel is extremely onerous. The
use of a conceptual non-linear model greatly simplifies the derivation
of the non-linear kernels (Napiórkowski 1985).

SYMBOLS

Q(x,t)   flow $L^3 T^{-1}$

$A(x,t)$	wetted area	L^2
$r(x,t)$	lateral inflow	L^2T^{-1}
z	elevation of the water surface	L
τ_0	shear stress	$M\,L^{-1}T^{-2}$
S_0	bottom slope	1
S_f	friction slope	1
B	channel width	L
g	acceleration due to gravity	LT^{-2}
$y(x,t)$	water depth	L
$u(x,t)$	average velocity of water	LT^{-1}
γ	weight density of water	$M\,L^{-2}T^{-2}$
$R(x,t)$	hydraulic radius	L
$F(x,t)$	Froude number	1
$\underline{T}(x,t)$	surface width	L
$\bar{y}$	hydraulic mean depth	L
c	wave celerity	LT^{-1}
$R_{AB}(t)$	total lateral inflow in reach AB	L^3T^{-1}
$S_{AB}(t)$	total storage in reach AB	L^3
τ	delay time	T
$h(x,t)$	impulse response, linear channel response	T^{-1}
$\delta(t)$	Dirac's delta function	T^{-1}
D	diffusivity	L^2T^{-1}
K	reservoir constant	T
K,x	Muskingum parameters	$T, 1$
k_R	Rth cumulant	T^R
S_R	Rth shape factor	1
f_R	Rth shape factor	1

REFERENCES

Abbott, M.B. 1966 "An introduction to the method of characteristics", American Elsevier Publishing Co., New York.

Amorocho, J. 1973 <u>Advances in Hydroscience</u>, vol. 9, edited by V.T. Chow, Academic Press, New York. Nonlinear hydrologic analysis.

Cunge, J.A., Holly, F.M. and Verwey, A. 1980 "Practical Aspects of Computational River Hydraulics", Pitman Advanced Publishing Program, Boston.

Dooge, J.C.I. and Harley, B.M. 1967 <u>Proc. Int. Hydrol. Symp.</u>, Fort Collins, Colo., Pap. No. 8, 1:57-63. Linear routing in uniform open channels.

Dooge, J.C.I. 1969 <u>IAHS Publication 76</u>, 409, A new approach to non-linear problems in surface water hydrology.

Dooge, J.C.I. 1973 "Linear Theory of Hydrologic Systems", USDA Agricultural Research Service, Tech. Bull. no. 1468, Washington, D.C.

Henderson, F.M. 1966 "Open Channel Flow", Macmillan, New York.

Kalinin, G.P. and Milyukov, P.I. 1957 <u>Met. i. Gidrologiya Zhurnal</u>, 10, On raschete neustanovivshegosya dvizheniya vody v otkeretykh ruslakh.

Kohler, M.A. 1958 <u>Proc. ASCE</u>, <u>84</u>, HY2, 1, Mechanical analogs aid graphical flood routing.

Kuchtment, L.S. 1972 "Matematicheskoe modelirovanie rechnogo stoka", Gidrometeoizdat, Leningrad.

Liggett, J.A. and Cunge, J.A. 1975 Chapter 4 of "Unsteady Flow in Open Channels", edited by K. Mahomaad and V. Yevjevich, Water Resources Publication. Numerical methods of solution of the unsteady flow equations.

Lighthill, M.J. and Whitham, G.P. 1955 <u>Roy. Soc. London Proc.</u>, <u>A229</u>, 281, On kinematic waves I, Flood movements in long rivers.

Maczuga, T. 1979 "Unsteady Flow in Open Channels", Ph.D. Thesis, National University of Ireland.

Massau, J. 1889 "L'integration graphique des équations aux derivées partielles (graphical integration of partial differential equations)", Assoc. Ingénieurs sortis des Ecoles Spéciales de Gand, Annales, Vol. 12.

McCarthy, G.T. 1939 "The Unit Hydrograph and Flood Routing", U.S. Corps of Engineers.

Napiórkowski, J.J. and Strupczewski, W.G. 1979 <u>J. Hydrol. Sciences</u>, <u>6</u>, 121, The analytical determination of the kernels of the Volterra series describing the cascade of nonlinear reservoirs.

Napiórkowski, J.J., Dooge, J.C.I. and Strupczewski, W.G. 1985 submitted to <u>Wat. Res. Research.</u>, Identification of the parameters of the kernels of the Volterra series describing open channel flow.

Nash, J.E. 1959 <u>J. Geophys. Res.</u>, <u>64</u>, 111, Systematic Determination of Unit Hydrograph Parameters.

Nash, J.E. 1960 <u>Inst. Civ. Ing. Proc.</u>, <u>17</u>, 249. Unit Hydrographs with particular reference to British catchments.

Preismann, A. 1960 <u>Proc. A.F.C.A.L.</u>, p. 433. Propagation des intumescences dans les canaux et rivières (Flood wave propagation in channels and rivers).

Saint-Venant (Barre de) 1871 C.R. <u>Acad. Sci. Paris</u>, <u>73</u>, 148-154; 237-240. Theory of unsteady water flow with application to river floods and to propagation of tides in river channels.

Sauer, V.B. 1973 <u>Proc. ASCE</u>, <u>99</u>, HYI, 179, Unit-response of channel flow routing.

Strupczewski, W.G. and Kunzewicz, Z.W. 1980 <u>Acta Geophys. Polonica</u>, <u>28</u>, 129, Choice of a linear three-parametric conceptual flood routing model and evaluation of its parameters.

Thomas, H.A. 1937 "The Hydraulics of Flood Movements in Rivers", Carnegie Institute of Technology.

4. LOW FLOW SUSTAINED BY GROUND WATER

R. Mull

University of Hannover
Callinstrasse 32
3000 Hannover 1
F.R.G.

4.1 INTRODUCTION

4.1.1 Phenomenon of "low flow"

The term "low flow" in natural rivers is qualitatively indicated by a low water level. This low water level is obvious. The quantitative relation between water level and discharge can only be revealed by measurements.
Table 1 shows daily values of discharges at a gauging station of the Leine River at Hannover, West Germany (area of the catchment: 5329 km^2). The lowest discharge in every month is indicated and also the lowest value of the given year (17.2 m^3/s). The lowest discharge ever measured at the station is 8.9 m^3/s, which is far below the value found in 1980. This example shows that low flow is not well defined. It only indicates that less water is discharged than normal, and normal is the given discharge in a given time interval. The time interval is a week, a month, a year; or the time a gauging station exists. In 1980 the medium discharge was found to be 50 m^3/s, the lowest discharge in this year was about one third of the medium (Deutsches Gewässerkundliches Jahrbuch 1981).

4.1.2 Importance of low flow

Low flow periods in rivers are important for various aspects of economy and ecology.

Quantitative aspects: – Water supply for domestic, industrial and agricultural purposes
– Hydropower generation
– Navigation
Qualitative aspects: – Chemistry and biology in the water courses
– Ecosystems (fish)

In Fig. 4.1.1 an example is given related to the Neckar River in West Germany, concerning the quantity of treated water which flows into

D.A. Kraijenhoff and J.R. Moll (eds.), River Flow Modelling and Forecasting, 67-97
© 1986 by D. Reidel Publishing Company.

TABLE 1. Mean daily discharge (m^3/s) in the Leine River at Hannover, West Germany, 1980

River: Leine
Gauging station: Hannover
Catchment area: 5 329 km^2

Day	Nov	Dec	Jan	Feb	Ma	Ap	May	June	July	Aug	Sept	Oct
1	17.2	22.5	42.2	94.8	44.9	49.7	91.4	38.0	74.2	50.0	36.7	31.9
2	17.8	23.1	40.6	81.2	44.9	71.9	82.0	33.2	84.4	45.6	42.5	28.2
3	17.5	24.0	36.4	73.4	44.5	130.0	74.2	31.0	98.9	42.5	36.4	29.1
4	17.2	24.8	34.5	103.0	42.9	122.0	67.3	29.7	129.0	42.5	32.6	28.5
5	18.6	24.0	34.8	125.0	42.2	100.0	60.9	28.5	109.0	44.5	31.0	27.9
6	30.7	24.0	35.1	171.0	41.9	85.2	58.3	27.9	95.6	41.9	30.7	27.6
7	31.9	25.5	36.4	194.0	43.2	75.0	56.5	27.3	82.4	38.7	33.9	29.7
8	31.6	26.7	38.3	205.0	45.9	66.6	55.4	27.3	83.2	43.9	32.6	35.8
9	35.5	28.5	40.6	187.0	44.9	71.1	54.7	35.5	81.6	37.4	52.9	34.5
10	48.3	31.6	39.9	156.0	41.6	69.2	51.5	39.9	73.8		75.7	30.4
11	34.5	44.5	37.7	133.0	40.9	65.0	46.2	31.9	75.0	34.8	59.0	29.1
12	29.7	107.0	35.1	119.0	40.9	63.1	43.9	43.9	72.6	38.7	63.5	28.8
13	28.0	97.3	31.0	112.0	41.6	58.7	43.9	37.7	71.5	38.3	56.5	27.9
14	27.0	75.7	28.5	103.0	42.9	54.7	40.9	31.6	78.1	35.5	51.1	27.0
15	27.9	79.2	30.4	95.2	40.9	52.9	38.3	64.6	91.0	33.5	49.7	27.9
16	30.1	92.7	30.7	88.9	39.3	51.5	37.4	85.2	88.9	32.6	52.9	26.4
17	26.7	104.0	30.7	82.8	37.7	49.3	35.8	49.7	82.0	30.7	52.2	27.6
18	24.6	99.0	29.1	75.3	37.7	49.3	35.5	45.6	72.3	31.9	54.0	28.8
19	23.1	84.8	28.5	70.7	37.7	47.3	33.9	45.6	66.6	40.3	51.1	28.5
20	23.1	80.4	27.0	66.6	37.7	47.3	33.9	46.2	69.2	39.0	45.9	26.7
21	22.8	71.1	26.7	61.3	36.7	46.2	32.9	55.8	96.0	37.7	42.2	27.3
22	22.5	63.5	26.4	56.5	35.8	44.9	31.9	49.0	134.0	39.6	38.0	27.0
23	21.6	56.8	27.3	54.3	35.5	42.9	31.3	44.6	139.0	35.8	36.1	27.9
24	21.6	50.0	28.2	50.7	34.5	40.1	31.0	43.9	125.0	32.9	35.5	25.8
25	21.0	47.3	28.2	47.3	35.1	57.6	30.4	45.9	107.0	29.7	35.1	27.3
26	20.7	43.2	30.1	48.9	35.1	82.8	30.1	43.2	89.7	30.7	35.1	26.4
27	21.0	40.6	28.5	45.9	35.5	80.4	30.4	44.9	76.9	30.7	33.9	27.6
28	21.8	40.8	26.4	44.9	44.2	83.5	30.1	46.6	68.1	30.1	32.3	27.6
29	21.8	40.8	24.3	51.8	54.0	122.0	30.1	45.9	66.2	30.1	31.0	26.4
30	22.2	40.6	26.7		52.9	105.0	38.0	56.5	59.0	37.5	30.7	29.4
31		42.5	43.5				54.0		58.3			30.4

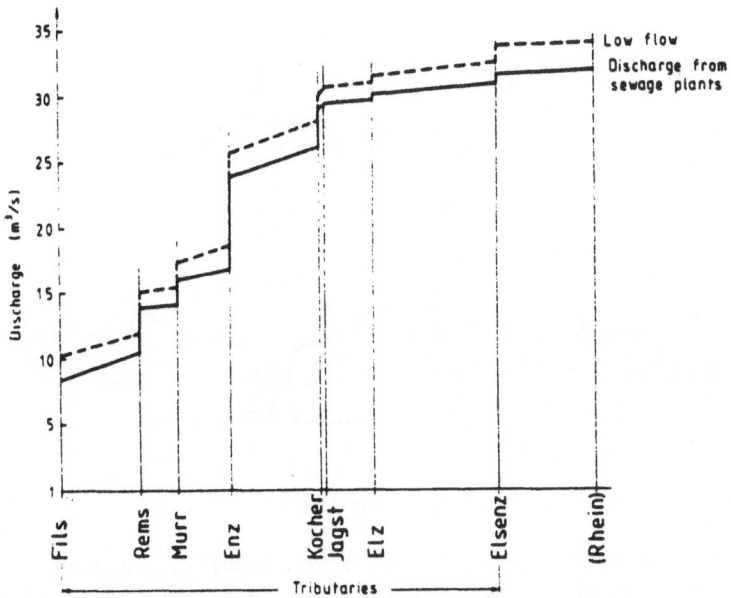

Fig. 4.1.1. Low flow and cumulative discharge from sewage plants
along the Neckar River

this river. The medium low flow is also given. The graphs show that the
volume of treated water running into the river almost equals the medium
low flow. This example indicates the relation between low flow and
water quality. An increasing discharge is generally coupled with an
increasing flow velocity. An increasing flow velocity means (Lehmann
and Rubach 1982):

- decrease of bacteria concentration at a certain location;
- dislocation of degradation processes downstream;
- less intensive degradation at the inlet of treated sewage water;
- more intensive oxygen intake;
- reduction of sedimentation;
- erosion of mud and an increasing oxygen demand;
- reduction of oxygen generation by photosynthesis because of increasing
  water depths.

4.1.3 Phenomena affecting low flow

Geometry and hydraulics of subsurface water systems

In the areas under investigation, low flow results from interaction
processes between ground- and surface-water systems. Under effluent
conditions ground water exfiltrates into surface-water courses. Influent
conditions exist if surface water infiltrates into the subsurface (Fig.
4.1.2).
Before giving the mathematical description of ground-water flow the

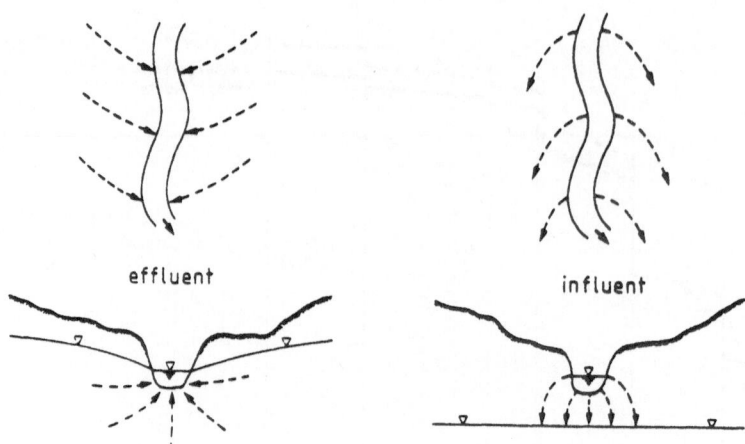

Fig. 4.1.2. Effluent and influent conditions

subsurface water systems will be described schematically. Looking
vertically downward from the ground surface, the subsurface water system
is divided into two zones: (a) unsaturated or aerated zone, (b) saturated zone (ground-water zone).

In the unsaturated zone water and air fill the voids (pores, fissures
and holes) within the consolidated and unconsolidated rocks. In the
saturated zone all pores are filled with water. In the unsaturated zone
there is a preferred vertical movement of water. Water originating from
infiltrating rain water seeps to the ground water or ground water rises
by capillary forces into the root zone. Within the saturated zone there
is a preferred horizontal movement of ground water.

The ground-water regime can be divided schematically into the three
zones indicated in Fig. 4.1.3: recharge zone, transition zone, and
discharge zone.

Within the recharge zone water originating from precipitation or
surface-water systems recharges the ground water. In the transition zone
impermeable layers near the surface or a sealing of the surface by all
kinds of buildings can prevent a recharge.

In the discharge zone ground water comes up and exfiltrates into
drainage systems within the flood plain or into the river itself. In
flat areas in Central Europe effluent conditions are dominant. In
mountainous areas effluent and influent conditions can occur alternately.
Water is collected in the rivers within the mountains. When the river
enters the plains the river bed is located above the ground-water level.
Water infiltrates into the ground and recharges the ground water. So,
many rivers disappear along two maintain ranges in West Germany (the
Odenwald and the Black Forest). Further downstream this water comes up
to the surface. The surface water-course is still existing as a tributary of the Rhine River. Fig. 4.1.4 shows parts of a river course with
alternating influent and effluent conditions. For effluent conditions
the ground-water system can be compared with a reservoir from which water
flows into surface water systems. During droughts the outflow is greater

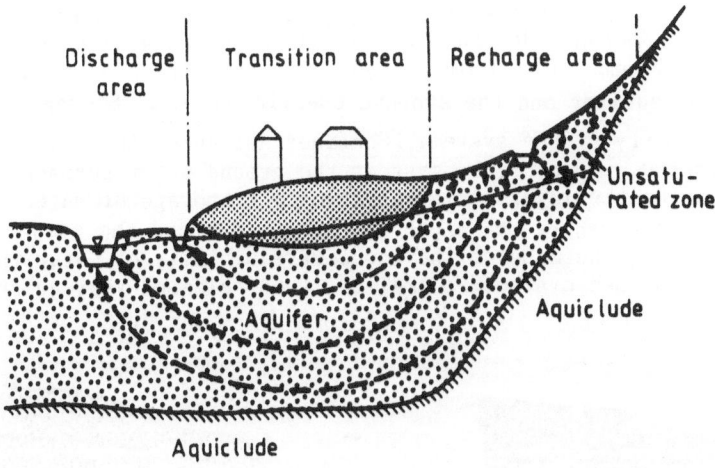

Fig. 4.1.3.   Cross section of a ground-water regime

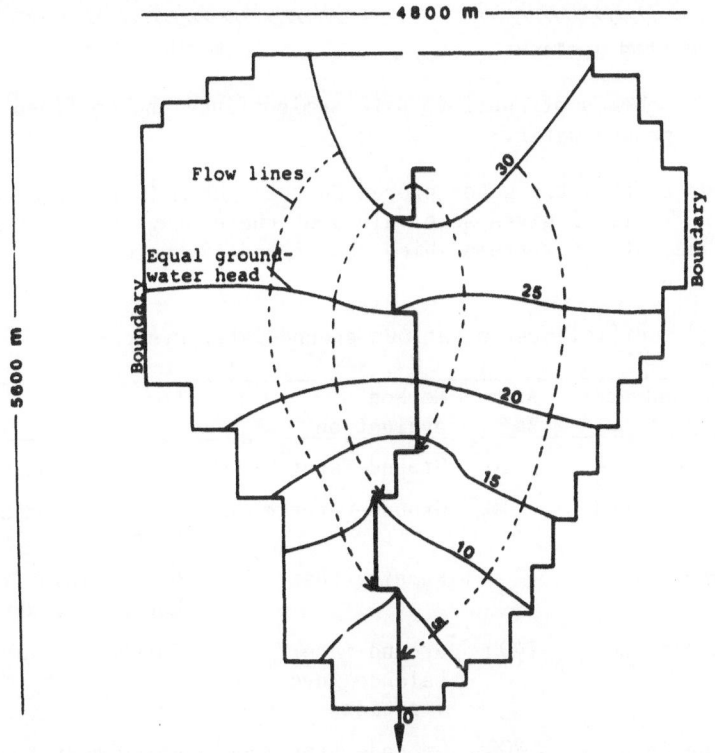

Fig. 4.1.4.   Flow lines and lines of equal ground-water head for
             influent and effluent conditions

than the recharge. The discharge in the river decreases continuously.
The velocity of the decrease depends on the hydraulic and geometric
parameters of the system. The dominant hydraulic parameters are transmissivity T of the aquifer and the storage coefficient C_s. Storage is
related to the geometry of the system. Confined and unconfined systems
have to be considered. In confined systems the ground water surface
lies within low permeable layers (Fig. 4.1.5). The storage of water
within the system is given by the depth of the aquifer and the compressibility of the water. Because of the low compressibility of water the
volume stored in confined systems is small. It is released very quickly
from the system.

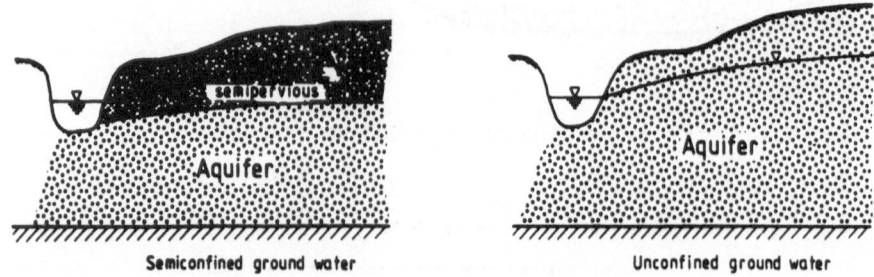

Semiconfined ground water   Unconfined ground water

Fig. 4.1.5   Schemes of aquifers with semiconfined and confined
              ground water.

In unconfined systems the water stored in the system is related to
the pore volume. Table II gives some values of the storage coefficient
in confined and unconfined systems which show the difference.

Table II Storage coefficients in various ground-water systems

Storage coefficient	Groundwater condition	Area km^2	Method of evaluation	Location
10^{-1}	unconfined	10	Ground-water model	City of Hannover
10^{-1}	unconfined	10	Ground-water model	Valley of Leine River
8×10^{-2}	unconfined	5	Pumping test	Flood plain of Lower Elbe River
2×10^{-1}	unconfined	1400	Ground-water balance inves-tigation	Northern Germany
5×10^{-5}	confined	200	Ground-water model	Rhine near Bonn 2. aquifer
3×10^{-5}	confined	5	Ground-water model	Valley of Leine River 1. aquifer

Hydrological aspects

Droughts occur because of various reasons. Generally a drought is
defined as a lack of rainfall. But the term "drought" has different
connotations in various parts of the world as pointed out by Hudson and
Hazen (1964). In humid areas a period of several days without rain is
a drought. In semiarid areas, droughts are recognized after several
years without rain. In the alpine region in Europe the lowest discharge
in rivers generally occurs during the winter season, when precipitation
is stored as snow and ice.

In the flat lands in Central Europe long periods of permanent snow
and frost are not very frequent. Low flow periods in the summer season
are more frequent. High evapotranspiration will reduce the recharge of
the aquifer. The difference between precipitation and potential evapotranspiration is given for a weather station in the northern part of
West Germany in Fig. 4.1.6. Generally 20% of the total annual recharge
occurs in the summer season. Recharge is also dependent on soil characteristics (permeability, storage capacity), land use and morphology.

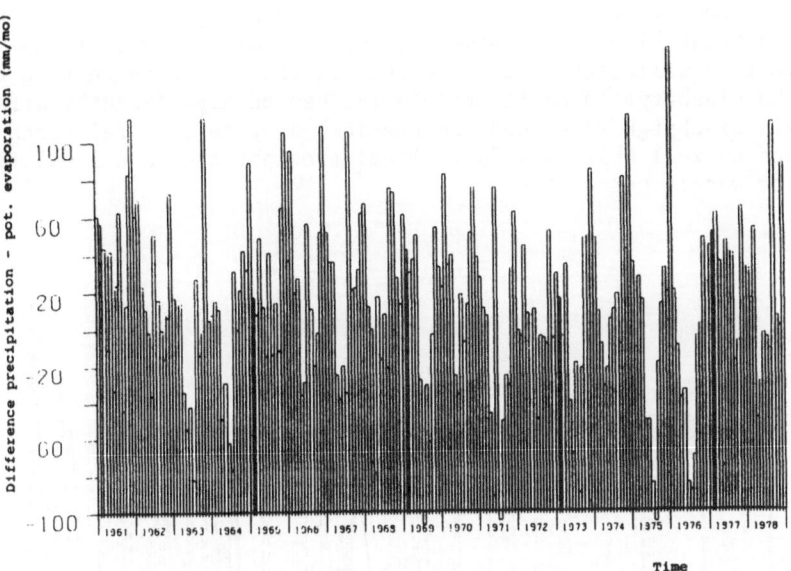

Fig. 4.1.6. Difference between precipitation and potential
            evaporation (North Germany)

Man made effects

Low flow is first of all dependent on the natural parameters discussed
before. Man made effects can influence the situation in one or the other
direction. The storage of water in storage basins can cause low flow,
the release of water from reservoirs can increase the discharge and
moderate critical situations. River regulation can maintain high water

levels in spite of low flow conditions but it also reduces the storage
capacity of the system. In this case flood water is more quickly
released from the system. During droughts the discharge is lower than
under natural conditions.

Water use has to be considered as an influence on low flow. First
of all, power plants and irrigation areas are big consumers because an
important part of the water extracted from the system evaporates. There
is a direct withdrawal of water from the river and an indirect one. It
is well known that water is pumped from wells which are located near
rivers. Because of the drawdown of the ground-water table by pumping,
a slope from the river bed to the wells is generated. Water infiltrates
from the river underground, and exfiltration of ground water into the
river is prevented. Both effects have to be considered regarding low
flow: the underground infiltration of surface water and the prevention
of the flow of ground water into the river.

4.1.4 Rainfall discharge relations.

As has already been pointed out, low flow results from lack of rainfall.
But if the discharge of the Nile River in Central Egypt is considered,
there is a discharge without any rainfall in this region. On the other
hand if the Rhine River is considered, heavy local rainfall of the
convective type will cause a minor effect on the discharge in this river,
whereas the discharge in small tributaries can be significantly affected.
If rainfall discharge relations are considered space-time relations must
be regarded as well (Fig. 4.1.7a): local precipitation and local flow
may only be weakly correlated.

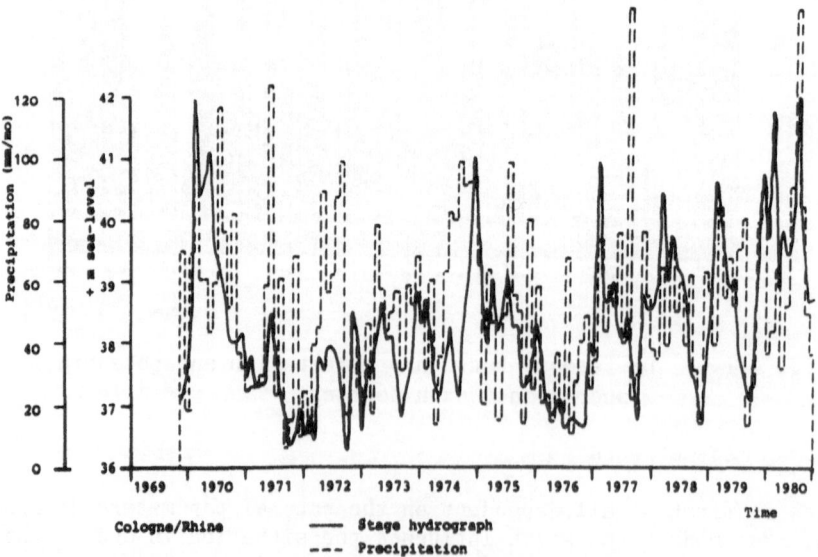

Fig. 4.1.7(a). Precipitation and discharge of the Rhine River

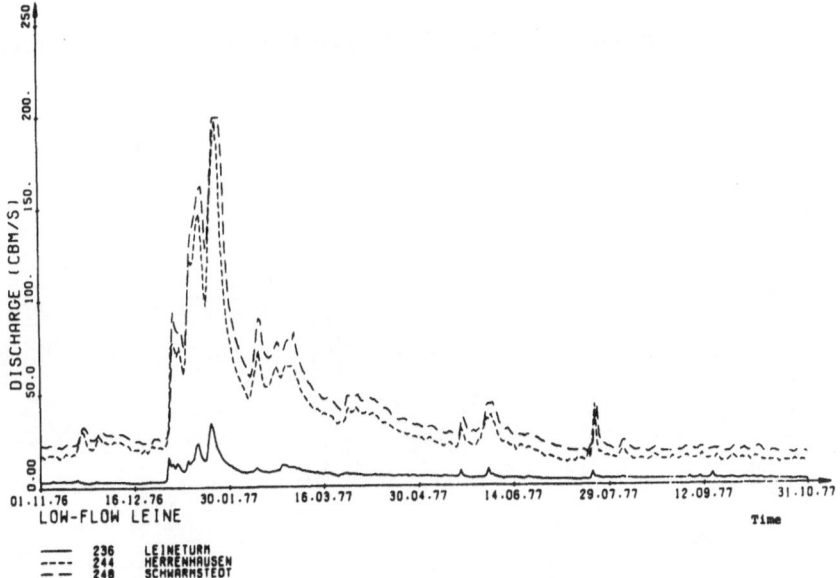

Fig. 4.1.7(b). Discharge at three gauging stations of the Leine
River

In small catchments the low flow can be related to the weather
conditions in the area. Droughts of several days will influence low flow
conditions (Fig. 4.1.7(b)).

As has already been stated, seasonal effects are obvious in all
river streams. During spring-time melting, water goes down the rivers.
During this time low evapotranspiration and high recharge of the ground
water are responsible for a high recharge in the rivers. In the alpine
region melting water dominates discharge towards the end of springtime
and the beginning of the summer (Fig. 4.1.7(c)).

Besides these seasonal effects long-term variations in precipitation
contribute to a variation of low flow. In Central Europe the time
between 1971 and 1977 was dry, so the discharge in large, as well as in
small, rivers was relatively low. Regarding rainfall discharge relations
we have to distinguish between short, medium and long-term effects.

4.1.5 Clients interested in low flow forecasting

Both the demand for information concerning low flow and the need for a
given accuracy of prediction may vary. Generally the accuracy of a
forecast decreases as its lead-time increases.

Clients interested in short-term prediction are working in the
following branches:
  Navigation
  Hydropower-plants
  Power-plants using cooling water and discharging thermal water
  Industry discharging waste water
  Fishery (oxygen content)
  Authorities managing multipurpose reservoirs

Besides the above mentioned groups the water supply industry and farmers concerned with irrigation are primarily interested in medium-term prediction.

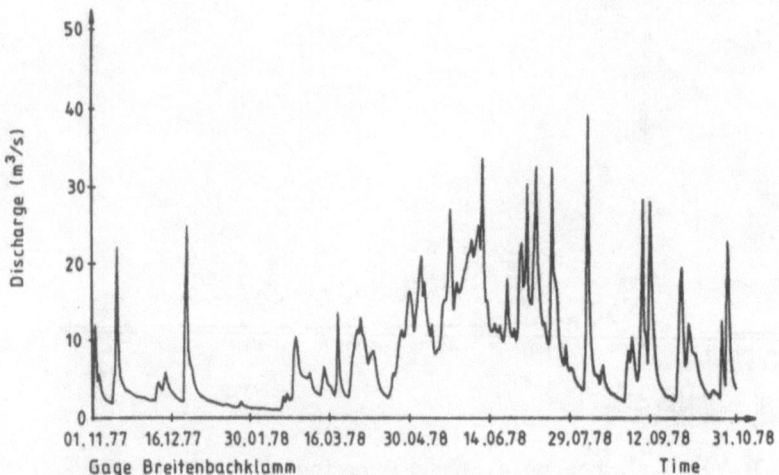

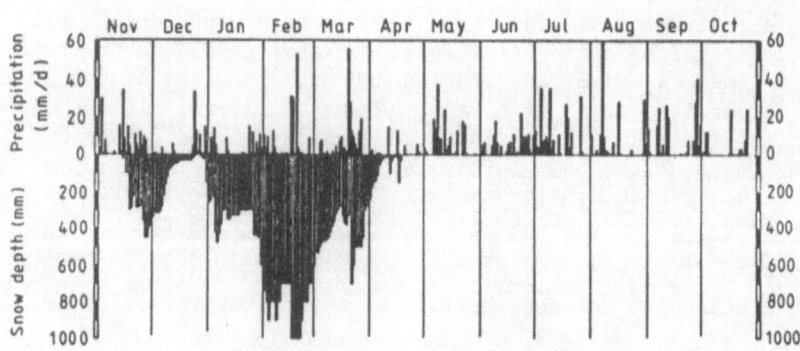

Fig. 4.1.7(c).   Precipitation, snow and discharge of an alpine
                 river

In those areas where ground water contributes to the water supply of the vegetation, farmers and foresters are also interested in long-term predictions of low flow because low flow is related to the ground water level. Declining ground water levels may necessitate the construction of irrigation systems.

These decisions do not depend on short-term variations of the weather

but on long-term trends in the ground water regime. Weather, ground
water reservoir parameters and man-made effects are involved.

Ecologists have similar interests if they are concerned with wet-
lands and aquatic biotops. Also in this case long-term trends are
helpful for decisions to take measures for preventing damage to important
biotops.

4.1.6 Concept of forecasting

Short-term

The ground water system can be considered as a reservoir. This reservoir
gains water from recharge and loses water by discharge to surface water
systems.

Short-term forecasting implies the extrapolation of the depletion
curve of a reservoir under the assumption that rainfall and surface
runoff do not change the flow in a river.

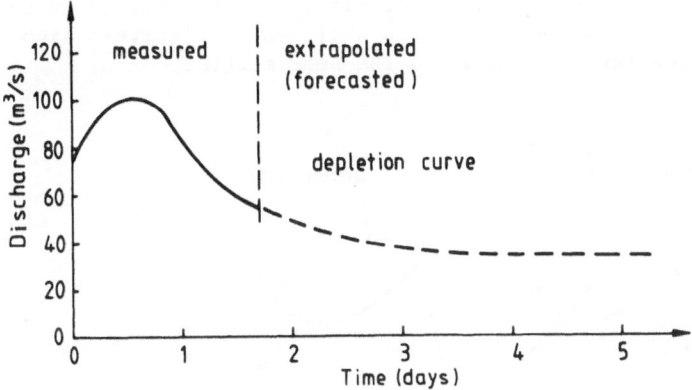

Fig. 4.1.8. Hydrograph, depletion of the ground water reservoir

The depletion curve as it is shown in Fig. 4.1.8 can be described
by Equation 4.2.3 of section 4.2.1. Values of the parameters Q(o) and k
are found in a calibration procedure. The forecast can start at every
moment for which the actual river discharge can be defined.

The interval in which forecasts are valid depends on the time for
which droughts can be predicted. The length of these intervals is
different in various climatic zones and seasons within these zones.

Medium-term

It has already been stated that the accuracy of predictions concerning
precipitation decreases with increasing time. Especially in the humid
belt of the moderate climate the accuracy decreases very rapidly.

Some phenomena counterbalance this tendency:
evapotranspiration

storage of precipitation
persistency of the system

Evapotranspiration depends on the energy available for the transfer of
water into vapor. The availability of energy varies periodically within
a year.

Precipitation can be stored as snow for longer periods in the winter
season. Changes of temperature can be predicted earlier and with
greater accuracy than precipitation.

After the summer season generally less water is stored within the
rootzone than in springtime. In autumn rainfall fills the pores up to
a certain extent before the infiltrated water percolates down into ground
water storage.

For medium-term forecasting of low flow the fluctuation of recharge
must be considered. In central Europe low flow will generally decrease
in summer and increase in springtime. A mean amplitude of the oscilla-
ting curve can be taken as a first approximation (Fig. 4.1.9). For the
second step the condition of the system has to be considered. If the
reservoir is emptied to a certain degree more time will be required for
surplus recharge to raise the ground water surface to a mean or high
level. On the other hand, if the reservoir is already filled up to a
high level, low flow will persist for a while on a relatively high level
even if little recharge will occur in the near future.

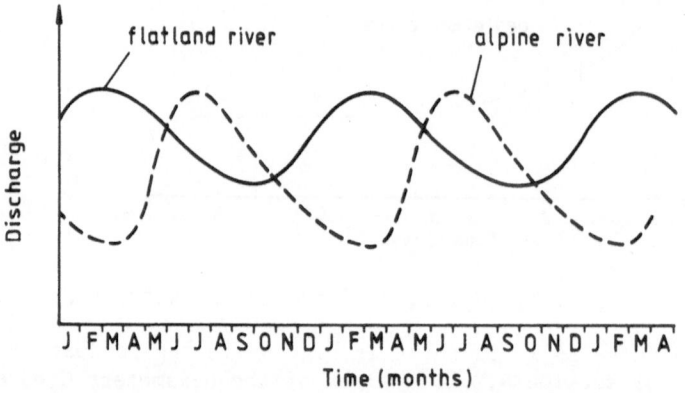

Fig. 4.1.9. Mean annual low flow hydrographs

These investigations can be supported by ground water model studies,
as will be explained in section 4.2, provided that the characteristics
of the considered system can indeed be transformed into such a model.
This is possible if the aquifers within the catchment are more or less
homogeneous. If there is a mixture of impervious, consolidated and
permeable sedimentary aquifers the application of ground water models
is difficult and even impossible for greater areas. So the use of
ground water models as a tool for low flow predictions is restricted to
flat land river systems. Otherwise these investigations become too
difficult and too costly.

Long-term

Wemelsfelder (1963) has shown a procedure to characterize persistency of a
ground water system related to low flow. He has analysed low flow
periods of the River Rhine on monthly mean values. A certain tendency
of the system was shown to persist in a once established state.
Wemelsfelder evaluated periods in which flow persists below and above a
long-term mean value (Fig. 4.1.10). For large reservoirs and under the
presumption that data are available to apply Wemelsfelder's model can
be used for long-term forecasts of low flow in a certain range of accuracy. A short description of the model concept will be given by de Ronde
in chapter 10 of this volume.

4.2 DISCUSSION OF RAINFALL-DISCHARGE RELATIONS

4.2.1 Black box method

Groundwater systems which contribute to low flow in surface watercourses
can be considered as black boxes. There is an inflow I into the box,
which is the recharge, and there is an outflow Q of groundwater from
the reservoir into surface watercourses. The difference between inflow
and outflow is the change of the water volume V_w stored in the system.
The equation of continuity is:

$$I - Q = \frac{\partial V_w}{\partial t} \qquad (4.2.1)$$

If a linear reservoir is assumed, the volume stored is directly proportional to the outflow

$$V_w = KQ, \qquad (4.2.2)$$

where K is the reservoir constant.
   If $I = 0$, then

$$Q(t) = Q(o)\, e^{-t/K}, \qquad (4.2.3)$$

where $Q(o)$ = outflow, for $t = 0$.
   This is the depletion curve of the reservoir if there is no recharge.
If a continuous recharge has to be taken into consideration, then

$$Q(t) = (Q(o) - I)\, e^{-t/K} + I \qquad (4.2.4)$$

$$Q(t) = I, \quad \text{for } t = \infty. \qquad (4.2.5)$$

The depletion of the reservoir ultimately equals the recharge I. The
parameters K and $Q(o)$ are found in a calibration procedure, which is
related to the depletion curve of a hydrograph (Fig. 4.1.8). If ground-
water is withdrawn from the system or if influent conditions occur, the

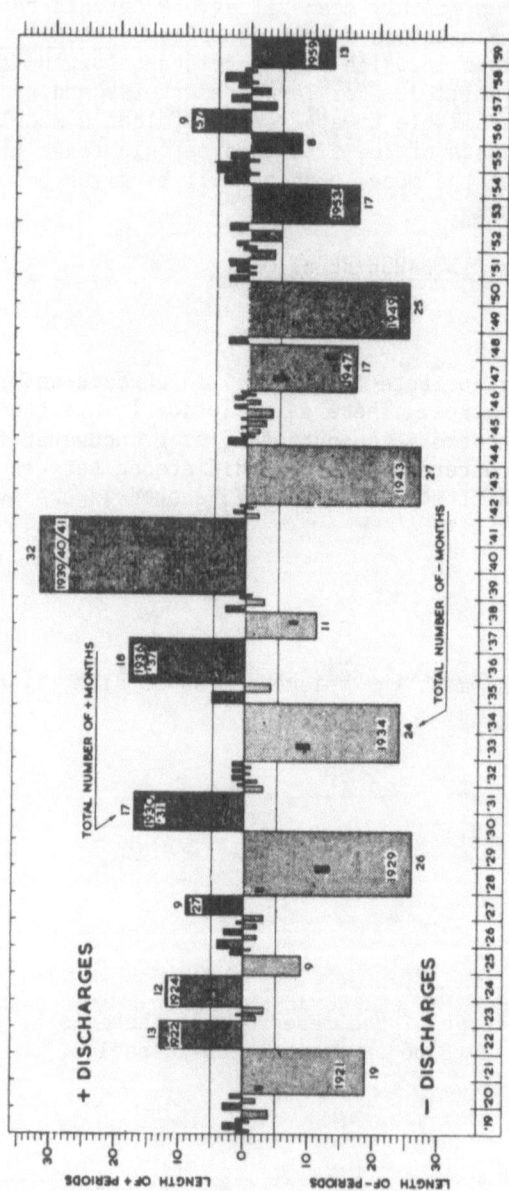

Fig. 4.1.10. Persistencies in the runoff of the Rhine, as deduced from mean monthly values. (Wemelsfelder 1963).

rate Q_a of abstraction has to be taken into consideration and added to
Equation (4.2.4):

$$Q(t) = \left[Q_0 - I\right]e^{-t/K} + I - Q_a. \tag{4.2.6}$$

The problem is reduced to finding the recharge and relating the recharge
to rainfall in order to establish the rainfall-runoff relation.

4.2.2 Rainfall-recharge

While rainwater infiltrates the soil, a certain percentage will flow
across the surface into the river system. Another quantity is lost by
interception. In this case the infiltration rate is smaller than the
rainfall rate. But also the groundwater recharge rate is smaller than the
infiltration rate. The difference will evaporate directly from the soil
or via the vegetation (transpiration).
    Table III shows some long-term data concerning the percentage of surface
runoff in four different catchments in the northern part of West Germany.

Table III   Low flow and surface runoff in four catchments,
            Northern Germany

River	Area of the catchment (km^2)	Mean surface runoff (mm/a)	Mean low flow (mm/a)
Ilmenau	1457	50	154
Luhe	480	74	256
Seeve	416	67	230
Este	184	78	222

    The flow of moisture through the unsaturated zone can be calculated
solving the following Equation (4.2.7)(Philip, 1957).

$$n\,\frac{\partial \Theta}{\partial t} = \frac{\partial}{\partial z}\left[D\,\frac{\partial \Theta}{\partial z}\right] - \frac{\partial k_a}{\partial z} - q_u, \tag{4.2.7}$$

where
n	= Porosity	1
Θ	= Moisture content	1
D	= Diffusion coefficient	$L^2 T^{-1}$
k_a	= Effective permeability	$L\,T^{-1}$
q_u	= Exchange rate	$L^3 T^{-1} L^{-3}$
z	= Space variable, positive downwards	L
t	= Time variable	T.

q_u is the volume of water per unit volume, which is extracted by the

roots of the plants per unit of time and which is transformed into vapor
and evaporates into the atmosphere. To solve this equation finite
difference or finite element techniques can be used. Fig. 4.2.1 shows
D and Fig. 4.2.2 k_a/k_f as a function of the moisture content for different soils. (k_f is the permeability for $\Theta = 100\%$). These parameters have
to be used to solve Equation (4.2.1).

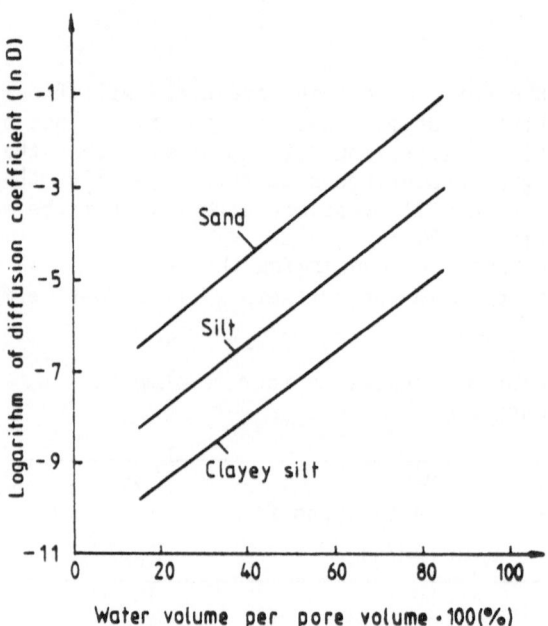

Fig. 4.2.1. Logarithm of the diffusion coefficient versus soil
             moisture

The spatial variation of depth below the soil-surface of the ground-
water table causes problems in the calculation of the effect of rainfall
on low flow in rivers. Fig. 4.2.3 shows some records concerning ground-
water levels for various depths of the ground-water table below the
surface. The greater the distance between ground-water and soil surface
the longer the travel time of the water, the smoother are these records.
For great depths the recharge is more or less uniform. Also several
days and weeks after rainfall, water seeps to the ground-water table
and recharges the system. For small distances between the soil surface
and the water table it is different. During the summer season ground
water recharge can be interrupted during droughts because in the root
zone the water content is so low that no water seeps down to the saturated zone.

Within a catchment a time- and space-dependent recharge has to be
considered related to rainfall. This complicates the situation.

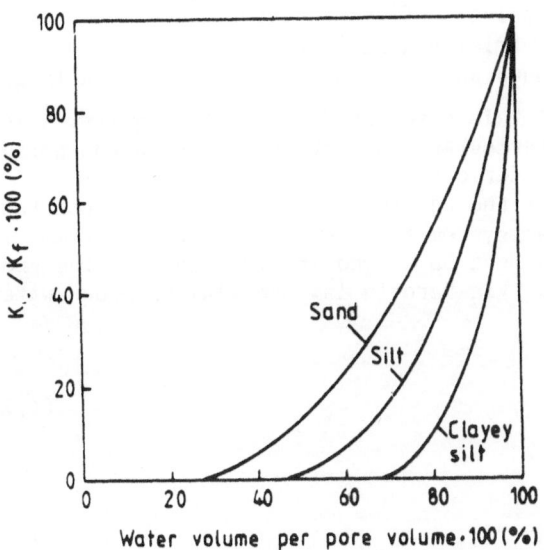

Fig. 4.2.2. Relative permeability versus soil moisture

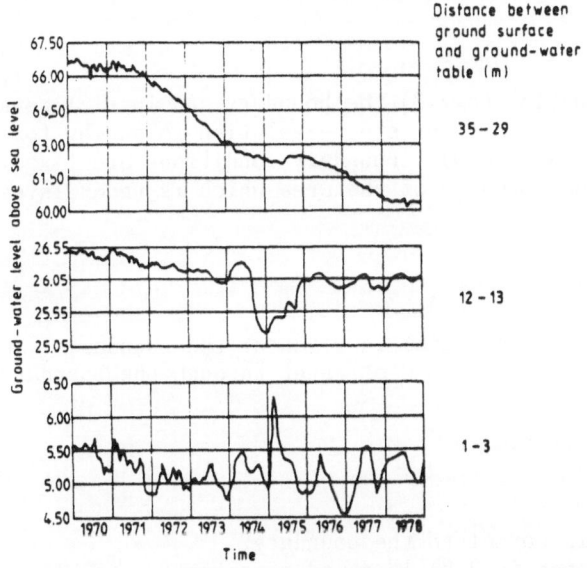

Fig. 4.2.3. Groundwater level records for various depths of the ground water below ground surface

4.2.3 Ground-water model investigations

Ground-water flow in river basins is generally analysed by ground-water models. It is beyond the scope of this paper to describe all methods

and techniques used for the analysis of ground-water flow with numerical
models. Only one approach is briefly described.

Low flow sustained by ground water is generally regarded in larger
areas (more than 20 km^2). In this case the depth of the aquifer which
releases water to surface watercourses is much smaller than the horizontal extension of the system under observation. The velocity of the
ground water over the depth of the aquifer can be regarded as constant,
which reduces the ground-water system to a two-dimensional system. The
ground-water equation (4.2.8) must be solved in order to find the ground-
water head in space and time. Via Darcy's law the flux of groundwater
is found using

$$C_s \frac{\partial h}{\partial t} = T\left(\frac{\partial^2 h}{\partial x^2} + \frac{\partial^2 h}{\partial y^2}\right) \pm q, \qquad\qquad (4.2.8)$$

where

C_s = Storage coefficient		1
h = Groundwater head		L
T = Transmissivity		$L^2 T^{-1}$
q = Volume of water exchange via the ground-water surface per unit of area and time		$L\ T^{-1}$
x,y = Space variable		L
t = Time variables		T.

In order to solve Equation (4.2.8) the boundary and initial conditions have to be defined. In a first step a solution is sought for
steady state conditions ($\partial h/\partial t = 0$). Boundary conditions are fixed
ground-water levels at the boundary of the area which is under investigation,

$$h = h\ (x_B,\ y_B), \qquad\qquad (4.2.9)$$

or the gradient $\partial h/\partial n$ of the potential is given at the boundary. This
gradient is proportional to the flow Q of water through the boundary
according to Darcy's law.

$$Q \propto \frac{\partial h}{\partial n} = f\ (x_B,\ y_B) \qquad\qquad (4.2.10)$$

means that the gradient is normal to the boundary.

The solution of Equation (4.2.8) is found numerically applying a
finite difference or a finite element procedure. In Fig. 4.2.4 a model
area is shown with a grid indicating finite differences. The solution
of Equation (4.2.8) for steady state conditions are ground water levels
at every grid point of the net. Interpolation procedures are used to
find lines of equal ground-water head.

The results of the calculation have to be compared with results
found by measurements. A calibration procedure has to achieve agreement

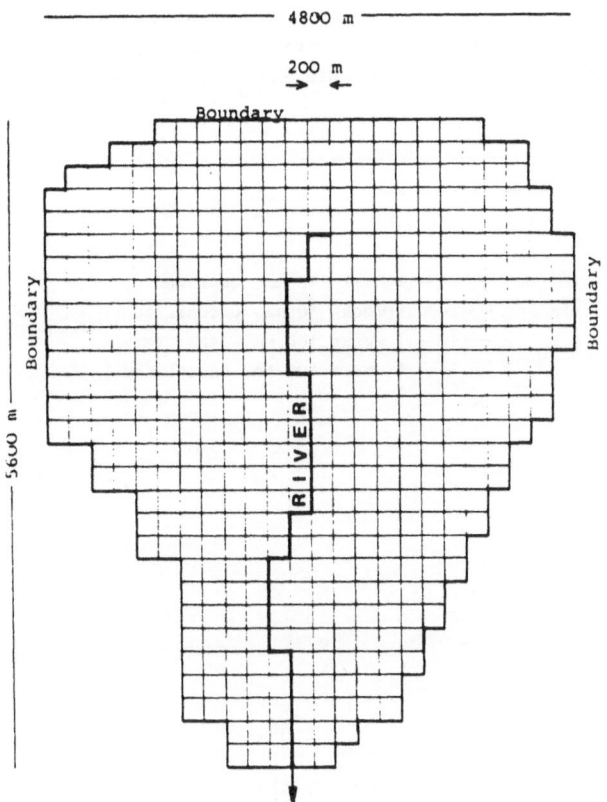

Fig. 4.2.4. Area of a ground-water regime investigated with a
            ground-water model

between calculated results and measurements in the field. In a first
step transmissivity, ground-water recharge and the exchange of water
between surface watercourses and the ground-water system have to be
estimated on the basis of existing data. When the results of the calculation during the calibration procedure come closer and closer to the
measured ground-water heads, the established values of transmissivity,
recharge and water exchange come closer and closer to reality.

    The lines of equal ground-water head found by steady state calibra-
tion are the initial condition for the non-steady state calibration.
Time dependent recharge and water exchange are then taken into consideration. Calculated fluctuations of the ground-water level are made to
agree with records of measured water levels. This calibration procedure
implies the finding of the storage coefficient. Values of C_s in Table

II (section 4.1.3) were found by such model calculations for various
areas (Mull 1982).

    These calibrated models allow calculation of the runoff in rivers at
different places. In Fig. 4.2.5 the discharge of the river in the model
area (Fig. 4.2.4) is shown. It is assumed that the groundwater is either
unconfined or confined. The difference in the reaction of the system on

a certain recharge is quite pronounced, there is a much quicker reaction
from the confined than from the unconfined system. Generally confined
and unconfined ground water exist in the same area.

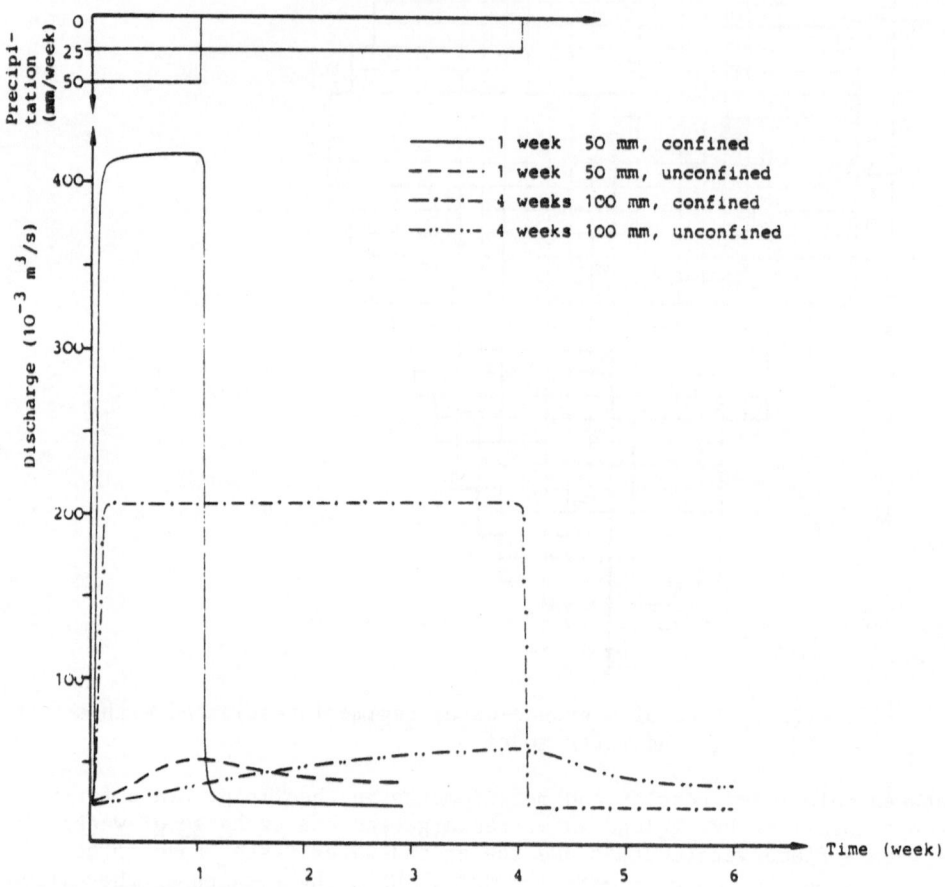

Fig. 4.2.5.   Discharge from the model area for confined or unconfined
              ground water

    The black box model permits prediction of low flow in extrapolating
the depletion curve of the reservoir if droughts reduce recharge to a
negligible quantity. Ground-water models give the total outflow of
ground water to surface water courses related to the recharge. If low
flow has to be related to rainfall then the difference between rainfall
and recharge has to be analysed. Beside this, the influence on low flow
of ground-water pumping and all other anthropogenic impacts on the system
can be better analysed with models of this type. On the other hand
ground-water extraction for industrial and domestic use is regarded in
the model as a loss of the system. In reality a certain percentage of
this water comes back to the river via sewage systems. This quantity has

to be added to the discharge in the river found by the model investigations.

4.3 EXAMPLES

4.3.1 Models for forecasting

One example of flood and low flow forecasting was carried out in the catchment of the Leine River. This river flows through Hannover. It is a tributary of the river Aller in the catchment of the Weser River.

Fig. 4.3.1 shows the investigated system with all tributaries. Some tributaries are regulated by storage basins which first of all serve as reservoirs for supply of drinking water.

Fig. 4.3.2 shows the system as it was used for model investigations concerning flood propagation as well as low flow analysis and forecasting. Calculated and measured hydrographs were compared at gauging stations and the system parameters k and $Q(o)$ evaluated by a calibration procedure. Fig. 4.3.3 shows some hydrographs used for calibration (Ludwig 1983).

The model allows the interpolation of the results between the gauging stations by flood routing methods. The discharge at a given moment in the future is found by extrapolating the depletion curve of the reservoir. Such a forecast of low flow is accurate as long as the drought continues. Interruptions of droughts by rainfall puts an end to the validity of this forecast. By analysing the history of rainfall-runoff relations it is possible to forecast low flow when the rainfall – duration and intensity – can be estimated (de Ronde, section 10.6 of this volume).

4.3.2 Seepage and ground-water flow related to low flow

By solving Equation 4.2.7 the outflow from the unsaturated zone was calculated for two different soils. The outflow equals the recharge of the ground water. The ground-water table was assumed to be 4 m below the ground surface.

The calculation period started on the first of April and ended at the beginning of the winter season. Rainfall was applied in pulses twice a month as shown in Fig. 4.3.4. The reduction of the percolation rate is due to increasing evapotranspiration. Less spectacular is the fluctuation of the recharge due to rainfall, especially in soils with low permeability and great storage capacity (field capacity).

The greater the depth of the ground-water table below the ground surface the smaller the amplitude of the seasonal fluctuation of the seepage rate. In depths greater than 10 m the seasonal fluctuations are also more or less negligible. The steady flow of seepage water down to the ground-water table equals the mean annual recharge. Mean velocities of the seepage water as related to the recharge rates are given in Table IV.

The coarser the material in the seepage zone the greater the velocity.

The oscillation of the ground-water table results from a fluctuating recharge rate.

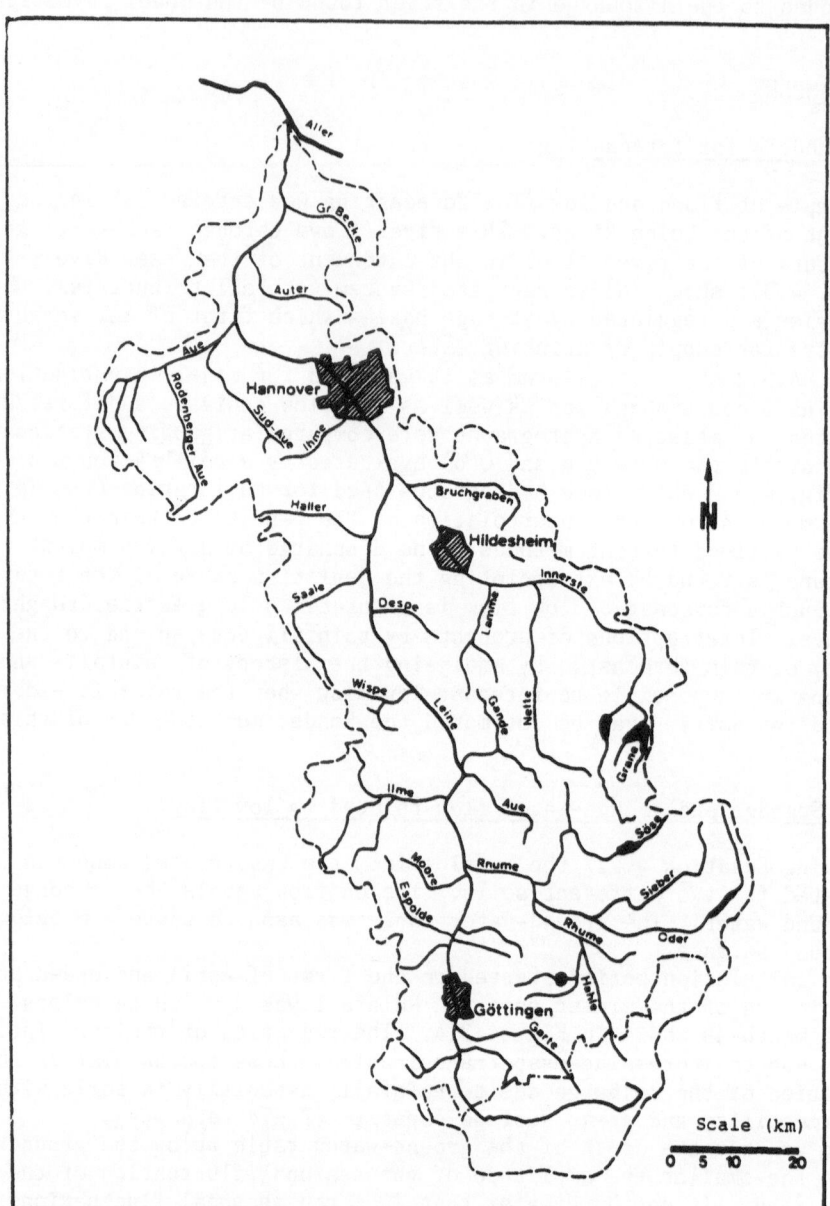

Fig. 4.3.1. Catchment of the Leine River

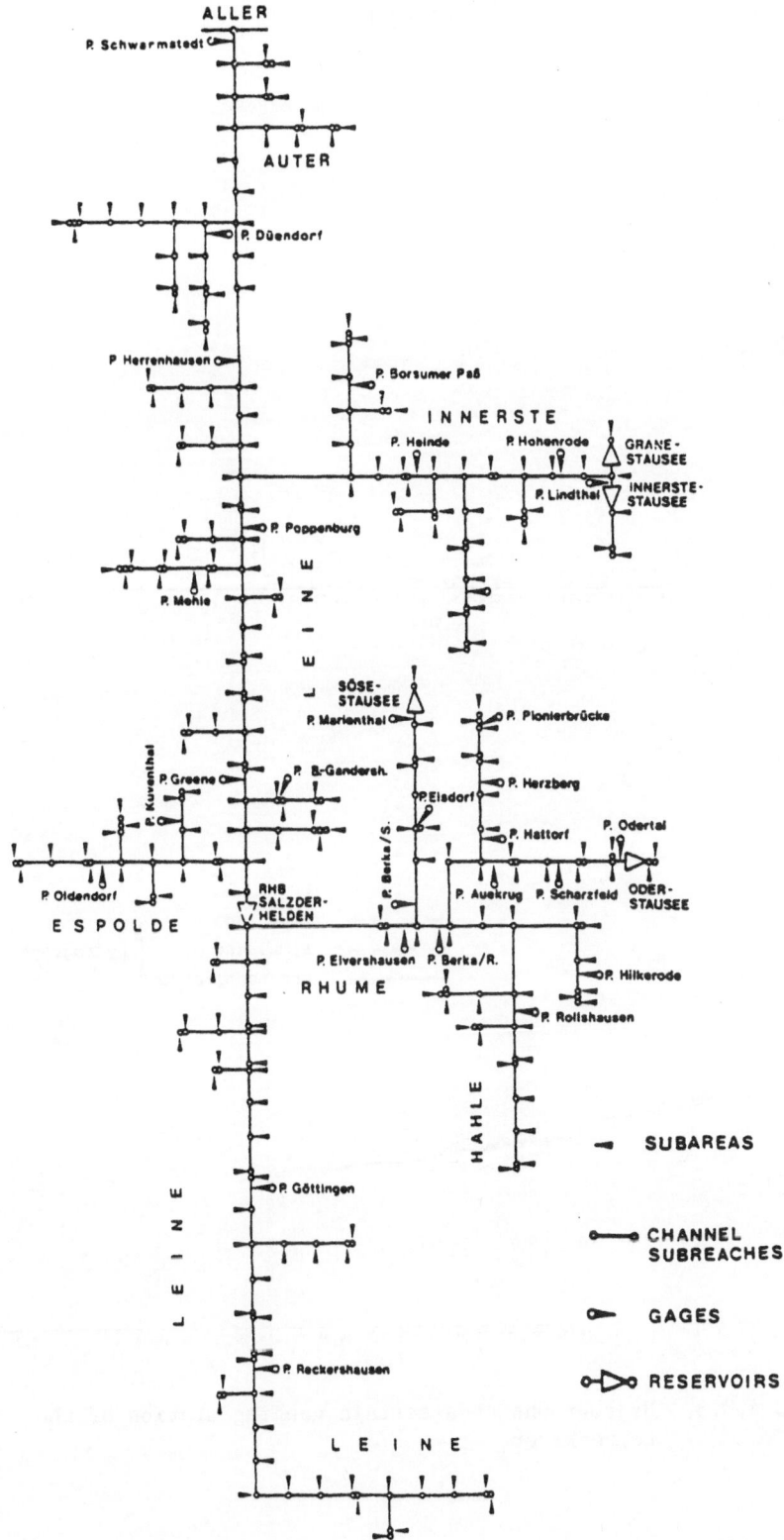

Fig. 4.3.2. Scheme of the model area used for low flow analysis and forecasting

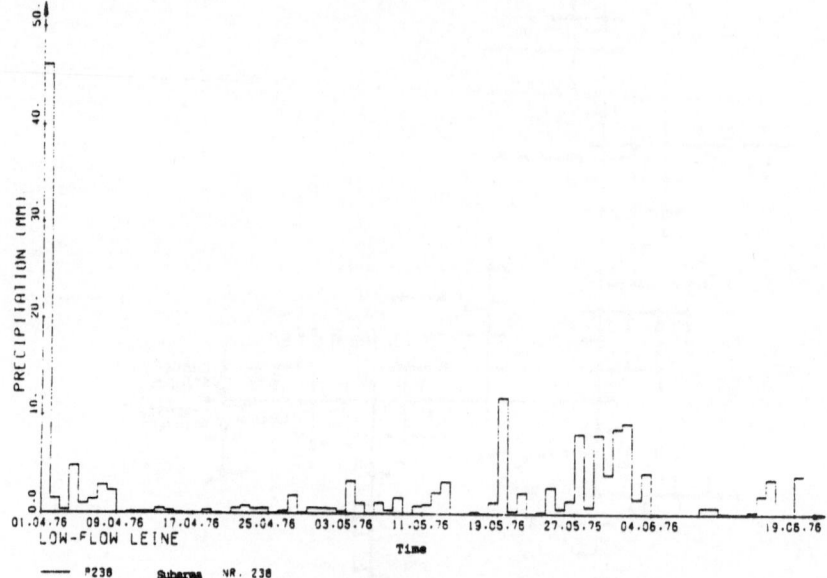

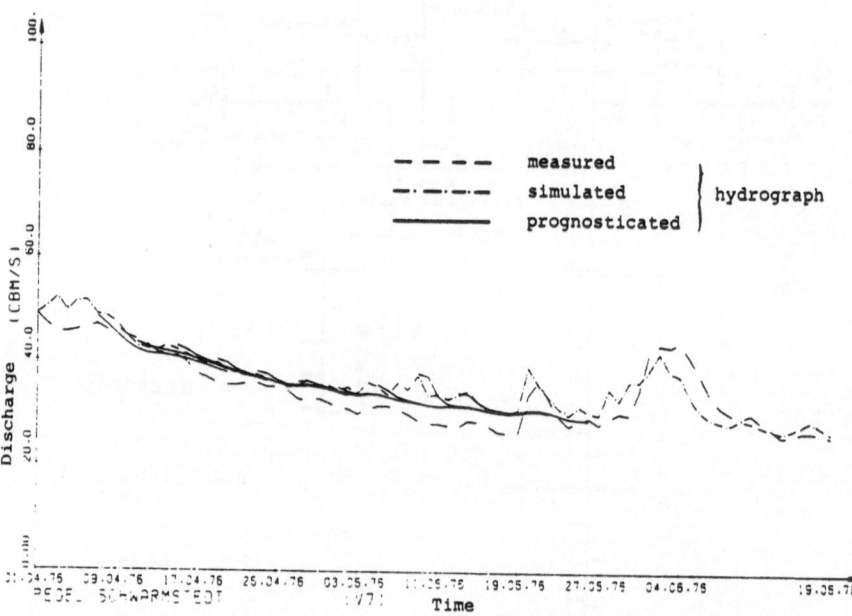

Fig. 4.3.3.   Hydrographs at a certain gauging station of the
              Leine River

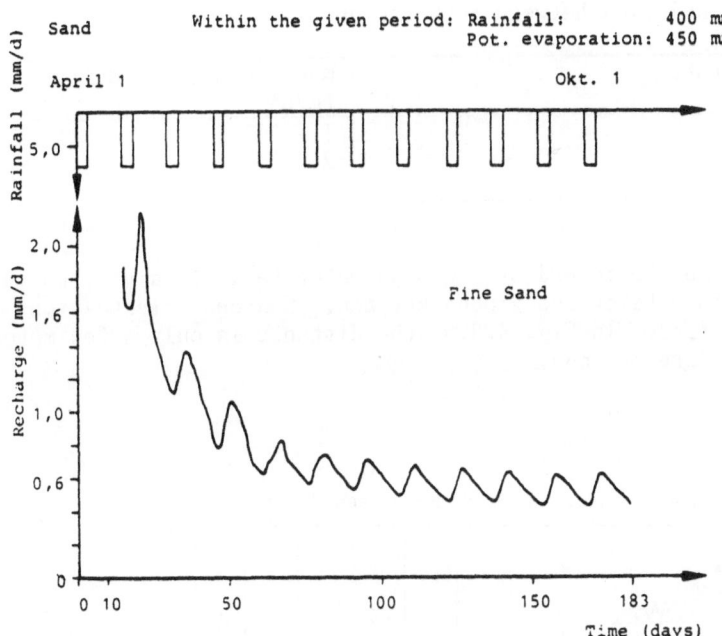

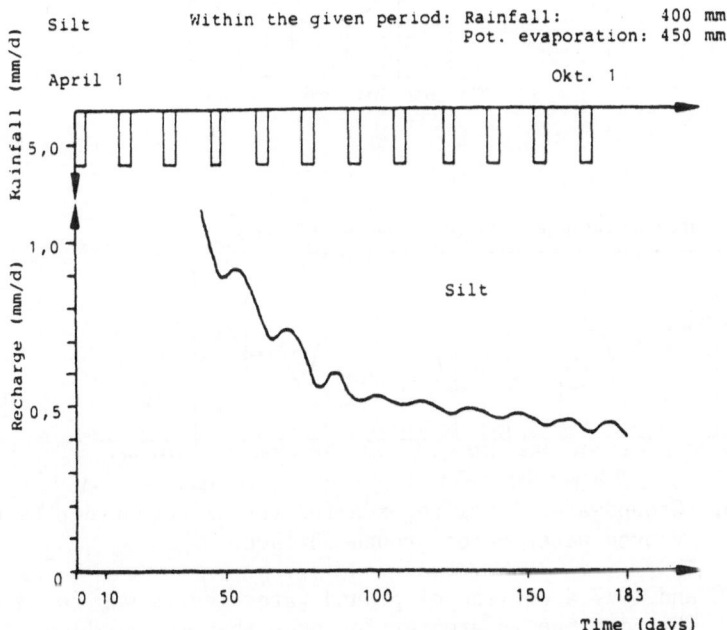

Fig. 4.3.4. Precipitation and recharge, calculations with a soil moisture model.

Table IV. Relation between recharge and seepage velocity

Recharge rate (mm/a)	Mean seepage velocity (m/a)
100	1 – 2
200	2 – 5
300	3 – 8

In Fig. 4.3.5a the record of a ground-water table is given, which is located around 50 m below the ground surface. Seasonal fluctuations cannot be identified. In Fig. 4.3.5b the distance is only a few meters. Seasonal effects are obvious.

(a)
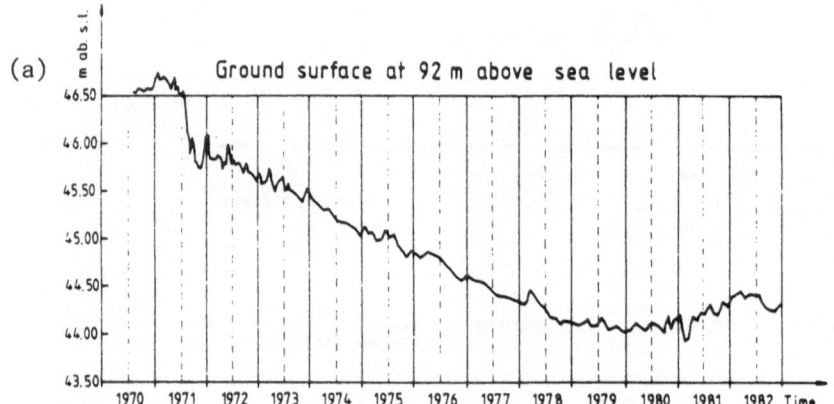

(b)
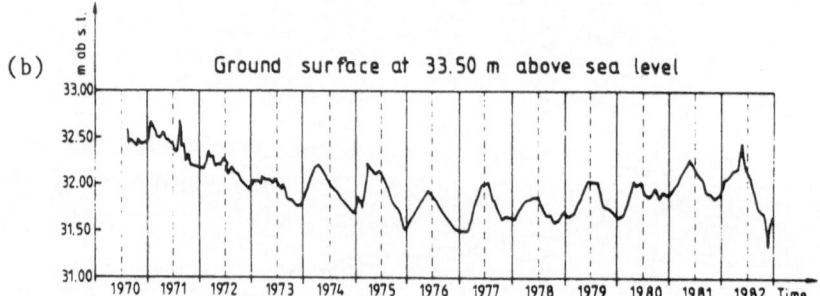

Fig. 4.3.5. Groundwater level records for two different depths of ground water below ground surface

Between 1970 and 1977 a decline of ground-water levels was found in Central Europe. This decline is greater for deep than for shallow ground water. The decline is also found for low flow as it is shown in the low flow records (lowest daily flow per month) in Fig. 4.3.6 of the rivers Ilmenau, Luhe, Seeve and Este (Table III).

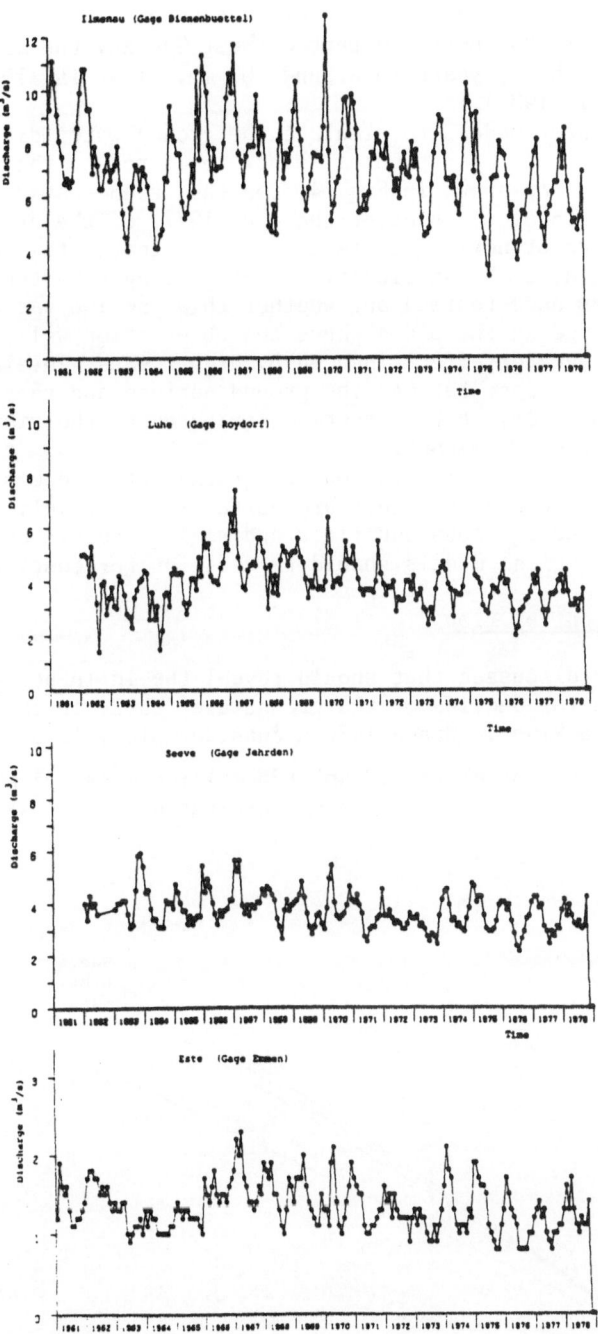

Fig. 4.3.6. Low flow in four rivers in Northern Germany (monthly values)

In the indicated period the annual rainfall was generally below the
medium except in 1974. In the northern part of West Germany the deficit
in rainfall cumulated within 7 years to around 500 mm. That is almost
the annual rainfall (Mull 1981).

When using ground-water models to simulate the ground-water discharge
to surface water systems the major problem is how to introduce the
recharge into the model. As shown in Fig. 4.3.5a the ground-water level
started rising in 1980, three years after the end (1977) of the dry
period, after three years of normal rainfall. This indicates the persistency of the system but also the difficulty of forecasting this tendency.
A ground-water model was used to find out whether this incline was due
to a higher local recharge at the place where the observation well was
located, or due to a backwater effect induced by higher water levels in
those areas, where the distance between the ground surface and the
ground-water table is smaller. Both effects contributed to the rise,
but the backwater effect was dominant.

This example illustrates the limitations of ground-water models for
low flow forecasting. Groundwater models are useful in relatively small
areas with more or less homogeneous aquifiers and small rivers. For
larger catchments hydrological models should be used for forecasting.

4.3.3 Man-made effect on low flow

Finally a study will be discussed that should reveal the influence of
ground-water pumping on the decline of the groundwater table in the
catchment of the Ilmenau River. Since 1976 a considerable volume of
water was pumped in an area of about 500 km^2 (35 million m^3/a $\simeq$ 1 m^3/s).
The influence of this ground water pumping was evaluated as follows.

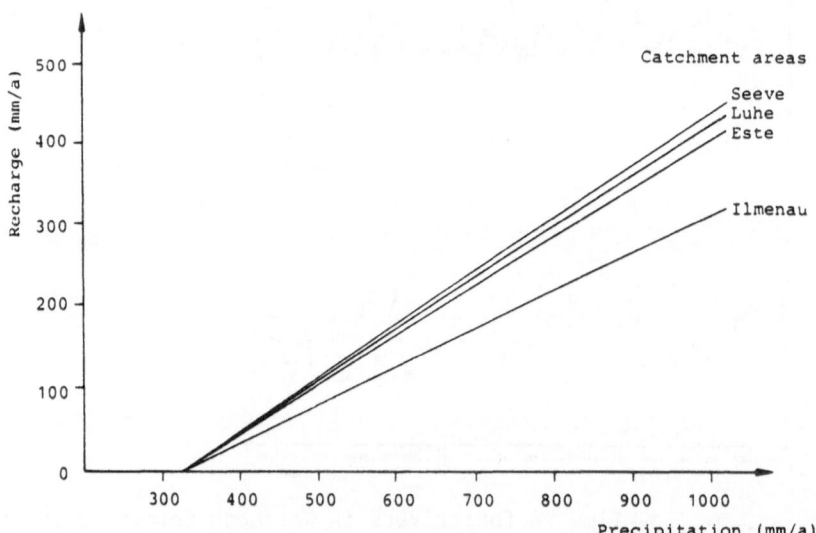

Fig. 4.3.7. Relation between recharge and precipitation for annual
            values

   The annual values of recharge and precipitation were related by linear
functions (Fig. 4.3.7). The water volume deficit given in Fig. 4.3.8 is
the difference between low flow and recharge. This is the water budget
analysis. It must be noted that the total deficit is also influenced by
ground-water extraction through pumping (Mull 1981).

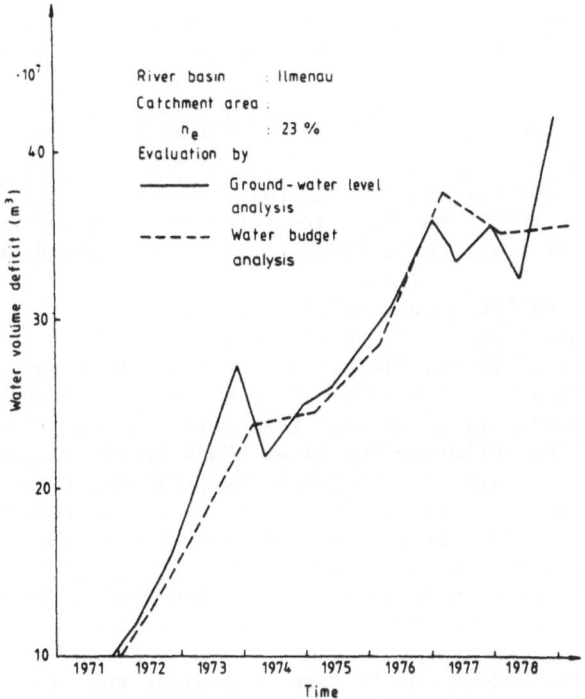

Fig. 4.3.8. Ground-water deficit, Ilmenau catchment

   The reduction of the water volume in the reservoir as a result of a
lack of rainfall during the 1971-1977 period of low rainfall was evaluated on a basis of maps of iso-lines of equal ground-water head which
were drawn every 6 months (April 1 and October 1) by interpolation of
water-levels between hundreds of observation wells. If groundwater level
records in the observation wells were influenced by nearby ground-water
extraction from pumping wells, this influence was eliminated by smoothing
the records. So the influence of local pumping was eliminated in the
maps of these lines of equal ground-water head. The volume of water
ΔV_w released from or stored in the system was found by graphical integration of Equation (4.3.1):

$$\Delta V_w = n_e \int_A h(x,y) \, dx \, dy, \qquad\qquad (4.3.1)$$

where

n_e = Effective porosity (ranging from 16-23%)

$\Delta h(xy)$ = Change of the ground-water level within 6 months L

A = Area of the catchment L^2

This is the groundwater level analysis.

The results of both evaluations were compared (Fig. 4.3.8) and it was found that about 15% of the water released from the reservoir resulted from pumping in 1976, 85% resulted from a lack of rainfall between 1971 and 1977. Low flow was affected by about 15% by this man-made input on the system.

4.3.4 Conclusions

There is a demand for low flow forecasts in certain industries, agriculture, forestry, and in authorities which are concerned with environment protection. To satisfy this demand models have to be applied which transform precipitation into flow in rivers.

For real-time forecasting the ground-water systems, from which water is released into rivers, are regarded as reservoirs. As long as the depletion curve of a reservoir is not disturbed by a rapidly changing recharge (rainfall, melting water) low flow can indeed be forecasted.

Medium and long-term forecasting can only take into consideration trends of those factors which influence low flow. Such trends originate from seasonal fluctuations of rainfall and evapotranspiration, from seasonal storage of snow, and from the storage capacity of the reservoir. Storage causes a discharge in the future which is dependent on the recharge in the past. Dependent on the storage capacity, the ground-water system has a tendency to persist in an established state. This tendency is the background for medium and long term forecasting of low flow.

Seepage and ground water models can be used to analyse the flow of ground water and the storage capacity of the aquifer. The application of these models is, however, limited to areas with more or less uniform aquifers. For larger catchments with heterogeneous aquifers in consolidated and nonconsolidated rocks this method cannot be recommended.

Man-made effects like ground water and surface water abstraction or artificial recharge can influence low flow over a wide range. An example was given but it is beyond the scope of this paper to analyse all possible influences.

Notation

S	storage	L^3
T	transmissivity	$L^2 T^{-1}$
C_s	storage coefficient	1
Q_1	inflow	$L^3 T^{-1}$
Q_2	outflow	$L^3 T^{-1}$
V_w	watervolume stored	L^3

K	reservoir constant	T
h	groundwater head	L
q	volume of water exchange via the groundwater surface per unit of area and time	LT^{-1}
Θ	moisture content	1
D	diffusion coefficient	$L^2 T^{-1}$
k_u	effective permeability	LT^{-1}
q_u	exchange rate	T^{-1}
k_f	permeability for $= 100\%$	$L^2 T^{-1}$
n_e	effective porosity	1
A	catchment area	L^2

References

Deutsches Gewässerkundliches Jahrbuch, Weser und Emsgebiet, 1980, 1981, 174, Der Niedersächsische Minister für Ernährung, Landwirtschaft und Forsten, Hannover, FRG.

Hudson, H.E. and Hazen, R. 1964 'Droughts and Streamflow', Handbook of Applied Hydrology (Ven Te Chow) McGraw-Hill, New York.

Lehmann, H. and Rubach, H. 1982, Wasser und Boden, 9, 408. Sind Gewässergütezustände in Flüssen nach Ort, Zeitpunkt, Wahrscheinlichkeit und Dauer definierbar?

Ludwig, K. 1983, Texte des Bundesumweltamtes, Systematische Berechnung von Niedrigwasserabflussvorgängen mit Flussgebietsmodellen, Institut für Wasserwirtschaft, Hydrologie und Landwirtschaftlichen Wasserbau, Universität Hannover, Hannover, FRG.

Mull, R. 1981, Wasser und Boden, 4, 176, 'Untersuchungen des Grundwasserhaushalts im nördlichen Teil der Lüneburger Heide'.

Mull, R. 1982, Schriftenreihe WAR 16, 7. Instationäres Modell zur Beschreibung der Grundwasserspeicherung, Institut für Wasserversorgung, Abwasserbeseitigung und Raumplanung, TH Darmstadt, FRG.

Philip, I.R. 1957, Austr. J. Phys. 10, 29, 'Numerical solution of equations of the diffusion type with diffusivity concentration dependent, II'.

Schulz, E.A. 1983, 'Grundwasserneubildung bei variabelem Flurabstand - Laborexperimente', Dipl.-Thesis, Institut für Wasserwirtschaft, Hydrologie und landwirtschaftlichen Wasserbau, Universität Hannover, FRG.

Wemelsfelder, P.J. 1963, Intern. Ass. Sc. Hydrol., Publ. No. 63, 90. 'The Persistency of River Discharges and Ground-Water Storage'.

5. FORECASTING MELTWATER RUNOFF FROM SNOW-COVERED AREAS AND FROM GLACIER BASINS

H. LANG

ETH-Dept. of Geography
Hydrology Section
Winterthurst. 190
CH-8057 Zürich
Switzerland

5.1 Introductory remarks

Snow and glaciers contribute greatly to the scenic variety of the Earth's surface. But they also contribute considerably to make hydrology an interesting field for research and forecasting. Snow and glaciers imply storage of water. Melting and run-off of meltwater often occur during dry weather periods. Many rivers and streams with their headwaters in mountainous regions show therefore a rather even distribution of flow from year to year and also for shorter time scales within the melt season. For example the river basins in the Swiss Alps show a minimum of the variance for annual run-off at a glacier coverage of 30-40% (Kasser 1959); for the run-off in the summer melt season the minimum variance is even observed at a much smaller percentage glacier area.

The reason for these characteristics lies in the following facts:
(1) The heat for melting is supplied mainly by net radiation and by sensible heat. In most climatic conditions, rainy weather in the melt season is usually combined with lowered net radiation and air temperature; hence run-off from rain and run-off from melting can compensate each other.
(2) In drainage basins which extend over a great range of altitudes, at the same time another compensation is effective: when the use of water by plants through evapotranspiration is high, there is a maximum of meltwater yield from snow and glacier melt at higher mountain areas, as both are mainly controlled by the available energy from net radiation.
(3) A third effect is part of the characteristics of mountain basins. It frequently happens that floods are strongly damped, because of the transition from liquid to solid precipitation at the higher altitudes, where snow covers are forming a temporary storage.

The problem of forecasting meltwater runoff from snowcovers and from glaciers requires that one or several of the following quantities be considered, depending on the type of forecast and on the catchment:
- Percentage of snow in precipitation; accumulation rates
- Areal distribution of snow-cover and glaciers, water equivalent of snowcover, thermal-physical properties of snow and ice
- Heat flux to the snow and ice surface to estimate the meltrates for

D.A. Kraijenhoff and J.R. Moll (eds.), River Flow Modelling and Forecasting, 99-127
© 1986 by D. Reidel Publishing Company.

short time forecasting; for extended lead-times meteorological forecasts may be required.
- Climatic parameters, determining the variations of snow and ice
  storage (seasonal and long-term forecasting).
The following sections give an outline of meltwater forecasting methods
and experience, with emphasis on the determination of the meltrates and
on practical applications.

5.2 The snow cover and its determination

5.2.1 Solid precipitation and snowcover

In real-time forecasting we may regard two situations in which Px/P, the
ratio of snow Px in total precipitation P (as a function of air temperature) is required.
(a) In the determination of the effective precipitation, the solid precipitation is attributed to the temporary storage of rainwater in the
basin.

In cases where snow and rain occur at the same time in a drainage
basin, the accuracy of the determination of Px/P can be decisive.
(b) In cases where no snowcover observations are available, the areal
distribution of the snowcover and its water equivalent have to be estimated on the basis of the available precipitation and temperature
measurements. During well defined cold periods or winter seasons all
precipitation is accumulated as snow, and the problem is reduced to the
problem of precipitation measurement. In all other cases, Px/P as a
function of air temperature is needed. The temporary ablation of the
snowcover during meltperiods and by evaporation processes has to be
taken into account.

In mountain regions an estimation of the total of snow water equivalent ΔS_* stored in a drainage basin can be assessed by means of the
balance equation over periods of a month or longer:

$$\Delta S^* = P - R - E. \qquad (5.2.1)$$

In any case, estimations of other storage processes (soil- and ground-
water) are also required. Often they may be negligible in comparison to
the snow storage in mountain areas. The evaporation E from the snowcover is often regarded as a negligibly small quantity during the winter
season. An overview is given in Lang (1981). See also Kuzmin (1961)
and Bengtsson (1980).

5.2.2 The ratio Px/P

The ratio of snow Px (hail and similar condensation products are not
considered here) in the total precipitation P is often taken as

$$Px/P = 1 \text{ at } T_a \leqq 0°C, \qquad (5.2.2)$$

where T_a is the air temperature. However, this threshold value tends to be 1-3°C higher on the average than the melting point.

The meteorological forecasting of the elevation range, in which for a certain precipitation event the ratio Px/P is in transition from 0 to 1, is very difficult because:
(1) the threshold temperature also depends on the intensity of the precipitation;
(2) the melting of the falling snowflakes changes the vertical temperature distribution by its cooling effect.

The problem of the Px/P ratio plays an important role in shorttime forecasting, at air temperatures around the freezing point. A change of 1-3 degrees in T_a may drastically change the whole flood situation. In steep mountain basins a change in T_a usually causes a change of Px/P in only a small part of the whole area of of the basin. In the long-term and seasonal forecasting the problem is more or less the same, if the estimation of the snow cover water equivalent is based on the continuous simulation of the snow accumulation. This requires records of P and T_a for each precipitation event and with sufficient resolution in time.

The extrapolation of T_a with altitude is assumed to be not a serious problem. However, it is recommended that this point be investigated separately in each mountain basin to discern any systematic deviation from the standard lapse rates.

The extrapolation of P with altitude is an unsolved problem and should be treated on a regional or even local scale. The reasons for this are:
(1) the systematic error in precipitation measurements is especially large in high mountain areas (wind induced error during snowfall, snowdrift).
(2) With the low density of the networks the information on P is still rather poor in high mountain areas and sometimes completely missing.

For long-term averages the relation of P to altitude in the Swiss Alps for example shows a wide range of 240 mm to 990 mm increase per 1000 m altitude for annual values (Uttinger 1951, SMA 1978). High mountain regions may show strong luff- and lee-side effects. Some areas may be situated above a maximum precipitation belt, as assumed for some of the Himalaya peak regions.

The long-term average increase of Px/P in the Alps is ca. 3% per 100 m elevation for annual values (Schuepp 1950). At an altitude of 3500-4000 m a.s.l. snowfall accounts for ca. 100% of the total precipitation.

5.2.3 Snow cover observations for operational forecasting purposes at a point

From the previous paragraph it follows that in most cases direct observations of the snowcover are necessary to achieve reliable accuracies in snowmelt run-off forecasting. Real-time operational forecasting requires the input data to be available on certain fixed dates. Without going into the details of the techniques, the following listing attempts to summarize the present situation.
(a) Manual point measurements in the field and snow courses:

- Point measurements are the classical way to obtain high precision information on snowdepth, water equivalent, snow density and snow temperatures;
- they need trained staff;
- they are very efficient and reliable especially in small basins or in simple terrain, where a few points are representative for the whole area;
- in cases where records of past years are available and provide a sufficient sample, regression relationships between meltwater flow and a few index points of snowdata may be found by means of statistical methods.

Detailed investigations of the optimum performance of snow courses and sampling have been carried out with the aim of reducing costs without sacrificing accuracy. The WMO Guide to Hydrological Practices provides complete information and recommendations on snow cover surveying and performance including snow-sampling equipment (WMO 1981).
    The disadvantages of point field measurements: they are time consuming, and often delayed because of bad weather conditions, particularly in mountain area, where safety considerations have to be taken into account. Costs may be considerable, when repeated surveys are required. In these cases recording instruments should be employed.

(b) Point measurements using recording instrumentation:
    At present there are two main principles in use to record the water equivalent of a snow cover at a point.
(1) Measuring the weight of the snow cover by snow pillows or by mechanical scales. Snow pillows and scales can be obtained at a reasonable price. Their reliability is limited as soon as stratigraphic bridges form in the snow cover.
(2) Use of radioactive gamma sources to measure the attenuation of gamma radiation between a source and a detector as a function of the water equivalent WE:

$$n = N_0 \, e^{-\mu WE}; \qquad\qquad\qquad\qquad (5.2.3)$$

$$WE = \frac{\ln N_0/N}{\mu}, \qquad\qquad\qquad\qquad (5.2.4)$$

where N_0 is the initial count,

   N = count after passing snowcover,
   μ = constant (sensitivity of instrument)
    The radioisotope snowgauges are expensive, but seem to be quite appreciated. Maintenance by trained people is essential.
    All these recording systems offer the possibility of automatic data collection and transmission from remote sites.

(c) Use of natural gamma radiation for aerial snow survey:
    Gamma radiation which emanates from natural radioactive elements in

the top layer of the soil, can also be used as a source of information
for snow surveying (WMO 1981). The method was initially developed in
the USSR (Kogan, Nazarov and Fridman 1965; Zotimov 1968; Vershinina and
Dimaksyan 1969). The ratio of the gamma radiation intensity measured
above the snow cover to that measured shortly before the beginning of
snow accumulation, provides the estimate of the water equivalent. The
accuracy depends on: the measuring equipment, fluctuations in the cosmic
radiation and the radioactivity of the near surface air layer, soil
moisture variations in the top 15 cm, uniformity of snow distribution
etc. including steady flight conditions.

The standard deviation of measurements of the water equivalent of
snow from an aircraft over a course of 10–20 km was reported to be about
8 mm and of a random nature. The method is intended for mapping the
snow in more or less flat country (Loijens and Grasty 1974). Investigations have also been undertaken in mountain areas in Norway (Andersen
and Johnsrud 1984).

5.2.4 Determination of areal values from point measurements

No general rule can be given; it is recommended that much attention be
given to the following factors which are of influence to the deposition
and depletion of the snowcover:
- topography
- altitude
- exposure
- wind conditions: influence of wind during and after snowfall on
    deposition and erosion of snowcover.

The spatial variation in the snowcover is much greater than in
precipitation and differs among various drainage basins. In large basins,
the networks usually do not provide sufficient information to determine
the total volume of water equivalent of the snowpack; economically it
would not be feasible to maintain the required density of observation
points. Here the snow surveys constitute nothing more than an index
value. These values can be used in a regression of run-off R on the
point values of the water equivalent $WE_{1...n}$ at the observation sites
1...n

$$\Sigma R = c + a_1 \, WE_1 + a_2 \, WE_2 + \, ...$$

However, if no records of sufficient sample size are available, there
has to be an estimation of the storage of snow water equivalent, using
all information available including precipitation, temperature, snow
data from other areas, and empirical relationships.

The direct determination of areal values of the snowcover volumes is
usually attempted in small research basins.

For the computation of basin values of mean snow depth and mean water
equivalent WE, the methods are similar to those for precipitation.
Depending on the spatial variation, on topography and on available information the following methods are in use:

(a) Arithmetic mean

(b) Thiessen method

(c) Application of a relation between water equivalent of snowcover and altitude z to allocate to the areas F_z of each elevation zone z the most probable water equivalent WE_z, then integrating over all elevation zones:

$$\overline{WE} = \frac{\Sigma(WE_z F_z)}{\Sigma F_z} \qquad\qquad\qquad (5.2.5)$$

(d) Subdivision of the whole basin in subareas A_i for which representative values of WE_i can be set up

$$\overline{WE} = \frac{\Sigma(WE_i A_i)}{\Sigma A_i} \qquad\qquad\qquad (5.2.6)$$

(e) Climatic regions and snow region: in larger mountain areas, such as the Alps, climatic regions and corresponding snow regions can be delineated, in addition to the altitude dependence of climate. In order to extract as much information as possible from the network observations at all the different sites, special statistical methods can be employed for the analysis of regionally different behaviour of the snowcover, providing the records at the different observation points extend over a sufficient period of time. For example in the Swiss Alps a "principal component" analysis on the snow water equivalent data and on winter precipitation was carried out (Jensen 1982).

5.2.5 Areal snow survey by photographic and photogrammetric methods

(a) Terrestrial and aircraft survey

Photographic methods can be used to yield the areal extent of the snow-cover. The use of stereophotogrammetry can provide the extent and the areal distribution of the depth of the snowcover in unforested or sparsely forested basins. Good topographic maps are important to employ these methods efficiently. Both methods can be applied as terrestrial and aerial surveys. For the stereophotogrammetric survey of snow depths, a survey should be performed before the snow season begins. A useful scale of mapping depth and extent of snow cover is 1:10000. The accuracy is within 10% (WMO 1981). Aerial photogrammetry is fairly costly, but rich information is gained that cannot be obtained in any other way. For terrestrial photographic survey, automatic cameras may be used at remote sites. The accuracy in the analysis of the snowcover extent is significantly improved if ortho-photomaps are worked out for the site where the camera is located.

The position of the temporary snowline not only defines the extent of the snowcover (or the percentage of snow coverage); particularly in the melting period, it can be used as an index for the remaining water equivalent of the snow cover in mountain regions because melt patterns

of snow areas in the zone of the snowline change little from year to
year. However, no reliable general relationship may be assumed: in an
extreme situation, the same snowline in an apline region can be generated
by (a) a single snowfall of a few centimeters of snow water equivalent
and a few days of ablation, or (b) in the same region in another year
after the accumulation of the total snow of a winter period. The limitations to hydrological interpretations of the snow coverage have been
studied in a small basin in the Swiss Alps (Martinec 1980). A summary
report on remote sensing of snow and ice has been presented by Meier
(1980).

Shafer and Leaf (in Rango and Peterson 1980) point out that the snow
cover extent is an important variable for forecast purposes after the
snowmelt season has begun, but is of limited value before that time. It
is necessary to be well informed on the melting process in each climatic
region in order to avoid misconceptions resulting from this attractive
and practical approach to snow water equivalent assessment. It can be
assumed that this method should be successful in climatic regions
(a) with well defined and regular periods of accumulation and ablation
(b) with low variability in the precipitation regime.

(b) Satellite snow survey

In the remote areas of the Earth satellite snowcover observations are
usually the only regular repeated information on the extent of the snowcover. For example, Rango et al. (1977) report on satellite observations
on the Indus River basin, Pakistan. Low resolution meteorological
satellite data and stream-gauge records for the period 1967-1973 were
analysed. The average area covered by snow at the beginning of April
was related in a simple regression analysis to run-off from 1 April to
31 July. The regression relation on the Kabul River, a tributary
stream to the Indus, showed a rather high coefficient of determination
(0.89). A set of examples of the application of satellite snowcover
observations is given in Rango and Peterson (1980). Satellite snow
surveys in the Swiss Alps are reviewed by Haefner (1980).

There is no doubt that satellite snow surveys are of the utmost
importance, as they can provide synchronous snow coverage data over
large and remote areas on a regular basis, as long as there are no
obstructions from heavy cloud cover. For regions with frequent cloud
cover, methods are being studied to derive the snow coverage of larger
regions from the snow coverage within smaller units which are detected
by the satellites as long as the cloud cover is less than 100% (Personal
communication by J. Lichtenegger 1982). Great progress is anticipated
from the application of microwave systems, which should provide information on extent and water equivalent of the snowcover under all weather
conditions; these techniques are however not yet operational and need
further investigations.

The conclusion:

In all cases of forecasting snowmelt runoff and particularly in
large areas with poor ground information, the possibilities of satellite
snowcover information should be taken into consideration.
(a) It may complement and improve the existing ground survey information,

but we cannot expect to replace it completely.

(b) In those remote areas where ground survey data are not available, satellite data are often the only source of information on the snow-cover. Simple methods as well as more sophisticated methods are available to make use of satellite information.

(c) For glacier runoff computations, the temporary snowline on the glacier surface is an important indication of the albedo conditions of the glacier surface. The albedo of snowfree ice is in the order of 0.2-0.3, and the albedo of snow ranges from 0.4 to 0.9; the albedo is the main controlling factor of the short wave net radiation and consequently of strong influence to the melt rates. Therefore, the survey of the snowlines on glaciers is essential in making accurate forecasts of the meltwater flow (see the paragraph on glacier run-off forecasting).

(d) The methods for snow cover estimation on the basin of a simulation model have to be further developed.

5.3   The determination of the meltrates

After the assessment of the snow cover extent and its water equivalent the determination of the meltrates is the second aspect of meltwater hydrology. If it could be measured in a simple way similar to that of precipitation, the meltwater runoff forecasting problem would be more or less identical to the rainfall-runoff problem. The direct measurement of the snowmelt water from snowpacks has been attempted by the use of meltwater lysimeters. However, great difficulties in the delineation of the drainage area are the main reason that these instruments are clearly not suitable for operational application. In certain conditions, with homogeneous snowcover the use of snow pillows may be appropriate. Gamma-ray snow gauges are suitable for recording accumulation and abla-tion as well; here the only limitation for operational use in networks seems to be the relatively high costs, and in some places, the security regulations on radioactive material.

Because of all the difficulties involved in the direct measurement of the meltwater yield from the snowcover, meltrates are generally determined by the energy-budget method or the degree-day method. The latter can be extended by using other available meteorological data. In principle both methods primarily provide the meltrates at the surface of the snowcover or glacier. To find the meltwater yield at the basis of the snowcover, particularly in the case of deep snowpacks, the flow of the meltwater through the snowpack forms an additional problem. This will be discussed in section 5.6.

5.3.1 The energy budget method

The energy Q_M available for melt on a horizontal snow or ice surface of a unit area and over a unit of time, can be determined by taking into account the following components:

$$Q_M = Q_{NR} + Q_S + Q_L + Q_P + Q_G. \tag{5.3.1}$$

Q_{NR} = net radiation

Q_S = sensible heat

Q_L = latent heat of condensation or evaporation

Q_P = heat provided from liquid precipitation

Q_G = heat from heat conduction in the snow pack

Q_M = heat used for melt or gained from refreezing of meltwater.

The meltrate (melted water equivalent of snow or ice per unit area and per unit time) can be calculated using

$$M = \frac{Q_M}{S},$$ (5.3.2)

this can be determined if all components of the energy budget are known. S = Latent heat of melt: 333.7 J/gr at 0°C.

(a) <u>Net radiation Q_{NR}</u>

In a basin with inhomogeneous surface-conditions, for reasons of interpolation and extrapolation, it is recommended to observe and determine each component, i.e. incoming and outgoing short wave and long wave radiation fluxes. In a homogeneous flat basin direct observations of the total net radiation with a net radiometer are feasible

$$Q_{NR} = G(1 - \alpha) + \varepsilon_a \sigma T_a^4 - \varepsilon_e \sigma T_e^4$$ (5.3.3)

G = global radiation

α = albedo

$\varepsilon_{a,e}$ = emissivity of the atmosphere a and of the surface e; for snow and ice a value of 0.98–0.99 is used

σ = Stefan-Boltzmann constant

　　 0.5673×10^{-11} (J.cm^{-2}°K^{-4}s^{-1})

$T_{a,e}$ = radiation temperature of the lower atmosphere resp. temperature of the snow or ice surface.

If no direct measurements of the radiation fluxes are available, the short wave net radiation $G(1 - \alpha)$ can be estimated from sunshine duration records, or from cloud observations, by taking into consideration latitude, time of the year, exposure and inclination of slopes. The albedo α can be based on empirical relationships between α and the age of the surface snow layer (reviews on available methods and procedures are given for example in Kuzmin (1961) and Male and Granger (1979); see also U.S. Army Corps of Engin. (1956)).

The long wave net radiation can also be estimated from empirical relations using the cloudiness. A compilation of results is given in Lang, Schaedler and Davidson (1977). The problem is discussed in detail by Kondratyev (1969).

In most cases of practical snow hydrology the snow and ice surfaces are at the melting point $T_e = 273°K$, thus the corresponding long wave emission $\varepsilon_e \sigma T_e^4$ is well defined. The problem is then reduced to the estimation of the long wave incoming radiation $\varepsilon_a \sigma T_a^4$. Since the representative emission temperature T_a of the atmosphere is not known, empirical relationships have been established using air temperature and vapour pressure observed in standard levels (1.5 to 2 m above surface). An additional difficulty is the atmospheric counterradiation under cloudy conditions. There have been many investigations about the empirical constants; a good discussion is given by Kondratyev (1969).

For reliable and accurate determinations of the radiation fluxes, measurement of all components over one meltseason is recommended, thus allowing the constants in the empirical relationships to be adapted to the local conditions.

(b) Sensible_heat Q_S_and_latent_heat_Q_L

The direct measurement of the eddy heat fluxes is based on the eddy correlation techniques. However, these techniques need sophisticated instrumentation which does not yet seem suitable for operational purposes. Therefore, the classical gradient method (or eddy diffusion method) is still employed most because it can be performed with well calibrated standard temperature and humidity sensors. The method is to a considerable extent based on the theoretical work of Prandtl (1904, 1934) Schmid (1925) and Lettau (1939). The application to ice and snow was primarily introduced by Sverdrup (1936).

According to the theoretical concept, the sensible and latent heat fluxes are proportional to the gradients of the potential air temperature θ and of the specific humidity in the atmosphere boundary layer:

$$Q_S = c_p \cdot \rho \cdot K_S \frac{d\theta}{dz} \qquad \qquad (5.3.4)$$

$$Q_L = L \cdot \rho \cdot K_L \frac{dq}{dz} . \qquad \qquad (5.3.5)$$

c_p = specific heat of dry air
ρ = density of air
K_S = eddy diffusivity for sensible heat
K_L = eddy diffusivity for latent heat
L = latent specific heat of evaporation
θ = potential air temperature; for gradients over a few meters the actual air temperature can be used
q = specific humidity (gr H_2O vapour per gr moist air)

$$q = \frac{0.622\ e}{p} \qquad \qquad (5.3.6)$$

e = vapour pressure
p = atmospheric pressure
z = sensor height above surface.

The equations are based on the assumption of constant flux with height. K_S and K_L are determined from wind profiles, which depend on the roughness of the surface (roughness parameter) and on the temperature stratification. Under neutral conditions in the constant flux layer the profiles may be expected to be logarithmic. From the analogy between eddy diffusion of momentum, heat, and water vapour, one can assume

$$K_S = K_L$$

for neutral stability.

The working formulae are derived on the basis of these concepts. The formulae most used are the following:

(1) according to the mixing length theory, used by Sverdrup (1946)

$$Q_L = L.\rho.k^2. \frac{u_2(e_2 - e_0)(0.622/p)}{\left[\ln(Z_2/Z_0)\right]^2} \qquad (5.3.7)$$

k   = von Karman constant, usually taken as 0.4
u_2, e_2 = wind velocity (cm/s) and vapour pressure at level 2
e_0 = saturation vapour pressure at surface temperature at height Z_0
Z_0 = roughness parameter (height above surface, at which u = 0).

In the same form the sensible heat flux can be determined

$$Q_S = c_p.\rho.k^2. \frac{u_2(\theta_2 - \theta_0)}{\left[\ln(Z_2/Z_0)\right]^2} \qquad (5.3.8)$$

(2) after Thornthwaite and Holzman (1939), who used the gradient method for the determination of evaporation from land and water surfaces,

$$Q_L = L.\rho.k^2. \frac{(u_2 - u_1)(e_2 - e_1)(0.622/p)}{\left[\ln(Z_2/Z_1)\right]^2} . \qquad (5.3.9)$$

In this case, the wind velocities u_2, u_1, vapour pressures e_2, e_1, and temperatures θ_2, θ_1, have to be taken from two levels Z_2, Z_1, above surface.

In both cases which assumed neutral stability, the eddy diffusivities for sensible and for latent heat K_S and K_L are equal. Deviations from the neutral conditions are taken into account through the use of the Monin-Obukhov similarity theory (Monin and Obukhov 1954) which is treated in detail among others in Male and Granger (1979). The classical way to correct for the stability is the introduction of the "Richardson" number Ri. The correction increases significantly with decreasing windspeed. L. Braun (1985) has given a nomogram showing for example the correction factors for an air temperature at 5°C of 0.9 resp. of 0.5 for windspeeds of 6 m/s resp. of 2 m/s.

(c) Precipitation as a heat source Q_P (release of heat by cooling and freezing)

Rainwater of the mass m_P (gr/cm^2) falling into a melting snow cover (T_0 = 0°C) provides heat proportional to the temperature of the drops arriving at ground level T_P:

$$Q_P = C \cdot m_P \cdot (T_P - T_0),$$ (5.3.10)

C = specific heat of water 4.1868 J/gr °C.
    A simple example shows that even a very warm and strong rain does not constitute an important heat source:

For $m_P = 4$ gr/cm^2

with a temperature of

$$T_P = 10°C$$

the maximum supply of heat can reach

$$Q_P = C \cdot m_P \cdot T_P = 167.4 \ J/cm^2$$

This heat melts only 5 mm of snow water equivalent!
    If rainfall occurs on a cold enough snowpack (temperature below 0°C) the thermal equilibrium is reached after freezing of the rainwater and releasing its latent heat of melt to the snowpack in addition to the heat released by cooling. The heat released by freezing

$$Q_M = S m_P$$ (5.3.11)

can be an efficient factor in the ripening process of a cold snowpack.
    The "cold content" C of a snowpack, i.e. the heat necessary to warm up the snowpack to 0°C, is given by

$$C = c_s \int \rho \cdot T_s(z) \ dz$$ (5.3.12)

c_s = specific heat of snow/ice: 2.10 J/gr°C (at 0°C)

ρ = density of the snowpack at depth z
T_s = temperature at depth z.

Example: Snowpack (z = 100 cm, ρ = 0.25 gr/cm^3, $\bar{T}$ = -10°C)
    Under the assumption of homogeneous density, the "cold content" is determined by:

$$C = (2.10 \text{ J/gr°C}) \cdot (0.25 \text{ gr/cm}^3) \cdot (10°C) \cdot 100 \text{ cm}$$

$$= 527.5 \text{ J/cm}^2$$

This cold content could be removed by the release of the same quantity of latent heat from freezing rain (or meltwater) having the following quantity

$$m_P = \frac{Q_M}{S} = \frac{527.5 \text{ J/cm}^2}{333.7 \text{ J/gr}} = 1.58 \text{ gr/cm}^2$$

This implies that 15.8 mm of rain at 0°C has sufficient thermal potential to bring the cold snowcover of our example to the melting temperature.
    In glacier hydrological computations, the cold content of the winter snowpack and of the surface ice layers is a considerable retention component. Ambach (1961) observed cold contents at the Hintereisferner towards the end of the winter in the order of 4100 J/cm^2, which is equivalent to the latent heat of ca. 126 mm of refreezing meltwater. About half was compensated for by refreezing meltwater, which can be observed as ice layers in the snowpack or as superimposed ice on the glacier surface. The remaining half was warmed by heat conduction.

(d) Heat_transfer_in_the_snow_cover

Conduction of heat in the snowpack occurs as long as it is below the melting point, when temperature gradients exist. In real-time forecasting of meltwater flow, this component is usually of minor importance and need not be taken into account. Mechanisms involved other than heat conduction are solar radiation penetration and latent heat turnover caused by temperature gradients and the corresponding vapour pressure gradients (condensation and evaporation processes). Details of the heat transfer equations are given in Kuzmin (1961) and in Anderson (1976).

5.3.2 The_relative_contribution_of_the_heat-budget_components_to_the
       meltrates

In selecting methods and models for practical application it is important to recognize the relative contribution of the different heat sources to the melt processes.
    The results of heat budget studies in different altitudes in the alpine region show that short wave net radiation is usually the most important source of heat causing melting. Examples of the Swiss Alps are given in Table I (Lang, Schaedler and Davidson 1977, Lang and Schoenbaechler 1967); they are typical of high alpine conditions. For the lower alpine regions combined events of melt and rain were studied in the research basin Rietholzbach. In a typical case the estimations resulted in contributions from each of the components (see Table II).

Table I. Mean daily heat fluxes and maximum, minimum values of the heat balance components in $[\text{j cm}^{-2}\text{d}^{-1}]$ and meltrates in $[\text{mm d}^{-1}]$ from field studies at the Aletschgletscher in two different altitudes: over snow at 3366 m a.s.l., and over ice at 2220 m a.s.l. (Lang and Schoenbaechler 1967; Lang, Schaedler and Davidson 1977)

	α	R_S	NR_S	NR_L	NR_T	Q_S	Q_L	Σ	mm/d
3366 m a.s.l. **3-19 August 1973**									
Mean daily value	.74	2419	636	-251	385	34	-27	392	11.8
					92%	8.0%		100%	
Max. daily value	.88	2981	887	+54	569	10	18	572	17.1
Min. daily value	.66	1431	259	-620	155	-1.7	-323	116	3.5
2220 m a.s.l. **2-27 August 1965**									
Mean daily value	.27	1863	1348	-230	1117	326	120	1563	46.9
					71%	21.0%	8.0%	100%	
Max. daily value	.42	2939	2168	+54	1704	656	741	3102	93.0
Min. daily value	.21	205	133	-578	188	42	-766	334	10.0
	Albedo	Short wave incoming radiation	Short wave net radiation	Long wave net radiation	Total radiation	Sensible heat flux	Latent heat flux	Available energy	Melt rates

Table II.   Rietholzbach: Example of a combined rain-snowmelt event
            26 January 1976.

	Net Radiation NR	Sensible Heat Q_S	Latent Heat Q_L	Rain Heat Q_P	Total Heat ΣQ_M	Melt-rate
J cm^{-2} d^{-1}	86.7	348.3	151.9	36.8	623.7	
%	14	56	24	6	100	
mm d^{-1}				21.5		18.7

In the example of a combined rain-snowmelt event it is clear that
heat provided from rain is of minor importance (6%), while the high air
humidity, resulting from rainfall and high air temperatures, produces
condensation conditions with latent heat contributing 24% to the melt
process.   Sensible heat is in this case the dominating source of heat.
Incoming solar radiation is strongly reduced by the full and high density
cloud coverage; on the other hand the long wave net radiation of a snowcover becomes slightly positive under rainy conditions; therefore total
net radiation can still contribute by 14%.

Although the energy budget method requires extensive instrumentation,
it is highly recommended to apply this method in forecasting.   In plain
areas a one point determination of the heat budget components should be
sufficient.   A reduced instrumentation should at least include a short
wave net radiometer, besides air temperature, humidity, wind velocity
and snow temperature measurements.   Empirical relationships can be derived to fit the local conditions to estimate for instance N_{RL}, if only
short wave net radiation is observed.

In real-time forecasting of meltrates in mountainous areas, however,
it is practically impossible to extrapolate the energy budget components
from one point to the whole basin.

The large spatial variation with altitude, exposure, and slope of
the different quantities, which determine the energy fluxes, cannot be
assessed with sufficient accuracy from "simple" networks of hydro-
climatological stations.   Consequently simple methods have been developed
to determine the meltrates, which confine themselves to the information
available from standard networks.

5.4   Practical methods to determine the meltrates

5.4.1   The temperature index methods

The most simple method used is to take the positive degrees of mean daily
air temperature T_+ and take one of the factors b (called degree-day-
factor DDF) from the literature and to compute the corresponding melt M
according to

$$M = b.\Sigma T_+ \qquad\qquad\qquad (5.4.1)$$

In this concept, air temperature is assumed to represent an index of all
energy available for melting from the different heat budget components.
Therefore, it is not surprising that the factor b shows large variations,
and that this method has severe limitations. However, because of the
easy availability of air temperature, this concept is used most in operational melt forecasting. The limitations are especially relevant if
this method is employed for single days. Over extended periods with a
variety of weather types the use of a mean b-value is a first approach.
Adjustments to the regional climatic conditions may be necessary.

In Table III a selection of average DDF values b (mm/°C) is given.

Table III. Temperature degree-day factors (mm/°C) from different places

Authors	Mean Value	Max. Min.	SD	Location
Zingg (1951)	4.5			snow melt Swiss Alps
U.S. Army Corps of Engin. (1956)	2.3	0.7–9.1		U.S.A.
Kuzmin (1961)	7.5	3.9–24.1		European USSR
Kuusisto (1980)	3.5 2.4	2.8–4.9 1.8–3.4		open field forest Finland
Lang et al. (1977, 1967, 1980)	5.3	5.3–47.0	12.1	Aletschglacier 3366 m a.s.l. snowmelt
T observed on glacier site	11.7	4.4–34.5	6.4	2220 m a.s.l. icemelt
T observed outside glacier	6.6	3.0–22.5	4.3	2220 m a.s.l. icemelt
T observed outside glacier	6.0	3.0–11.1	1.9	2220 m a.s.l. icemelt (only days with T > 4°C)

In the values given for the Aletsch glacier, a large part of the
variance of b occurs under conditions at temperatures between 0 and 4°C.
At higher temperatures the method seems to become more reliable, which
is to some extent self evident. The expression $M = b.\Sigma T_+$ incorporates
the assumption that at a daily mean temperature of 0°C the meltrate is
zero. Under the conditions on Alpine glaciers with its high percentage
contribution of energy to ablation by net radiation, some meltwater runoff can occur even at 0°C; on the other side the threshold air temperature T_K for snowmelt in winter and in springtime is frequently observed

in the range of 0° to 3°C; therefore, use of the following forms is
suggested:

$$M = a + b\Sigma T_+ \tag{5.4.2}$$

or

$$M = b\Sigma(T_+ - T_k) \tag{5.4.2a}$$

At the high altitude site the factor b shows a standard deviation
more than twice its mean value of 5.3. The air temperature observed at
the glacier site is strongly influenced by the transfer of sensible heat
to the ice surface; its "information content" is already reduced: therefore the mean b-value is extremely high. Furthermore it is interesting
to note that for the "normal" temperature observation site (outside
glacier), the standard deviation SD of the b-values showed a minimum at
higher temperatures. A detailed discussion of the information content
of T with respect to melt rates is given in Lang (1980).

In this context an important result was the evidence that the temperature in mountain areas observed at snowfree valley stations is
providing significantly more information regarding energy available for
melt, as compared to temperatures, observed at the glacier, snow or
mountain sites.

However, this conclusion can only be valid under conditions where
the net radiation is the most important source of melt heat. In conditions where the sensible and latent heat fluxes are predominant, one
should expect to have the optimum information from very close meteo-
stations. An interesting example of such a situation has been given by
Roald (1971) and by Ostrem (1972) for a humid and a more continental
glacier area in Norway. In any case a good knowledge of the physical
processes and relationships is helpful to make optimal use of the
observation data in empirical, statistical approaches.

5.4.2 Extended_relationships

In cases where records of the melt season are available, regression
methods should be employed to optimize the forecasting of the meltrates.
The multiple regression analysis indicated that the vapour pressure e and
global radiation G in addition to air temperature could improve the
accuracy of meltwater run-off computations significantly (Lang 1980,
Lang 1968, Jensen and Lang 1973).

As an example the relationships obtained for one melt season (July/
August 1965) at an experimental plot of 4500 m^2 at the lower snowfree
part of Aletsch glacier (2200 m a.s.l.) are given:

$$M = 17.04 + 4.18\ T_L \qquad\qquad R = 0.75 \qquad (5.4.3)$$

$$M = -2.76 + 3.06\ T_L + 0.0148\ G \qquad R = 0.89 \qquad (5.4.4)$$

$$M = -39.32 + 1.37\ T_L + 0.0193\ G + 5.36\ e_L \qquad R = 0.93 \qquad (5.4.5)$$

R   = multiple correlation coefficient
M   = daily meltrates (mm/day)
T_L, e_L = daily means of air temperature (°C) resp. vapour pressure (mb)
          observed outside the glacier
G   = global (incoming solar) radiation ($J/cm^2 day$)

Equation 5.4.5 represents the physical relations with respect to the
heat budget components well. This becomes formally more evident if
instead of e_L we consider the difference $(e_L - E_0)$, where E_0 is the
saturation vapour pressure of a melting ice or snow surface (= 6.11 mb).
Equation (5.4.5) is then modified to

$$M = -6.57 + 1.37(T_L - T_0) + 0.0193 G + 5.36(e_L - E_0) \qquad (5.4.6)$$

where $T_0 = 0°C$. This equation reproduces the meltrates, related to a
surface albedo of $\alpha' = 0.27$ (observed average albedo at the experimental
plot). Generalization of (5.4.6) consists essentially of introducing
the albedo α of an arbitrary snow or ice surface; in this case we have
to take it into account by using the ratio $(1 - \alpha)/(1 - \alpha')$ as a factor
to the radiation term:

$$M = -6.57 + 1.37(T_L - T_0) + 0.0193 G \frac{(1 - \alpha)}{(1 - \alpha')} + 5.36(e_L - E_0)$$

$$(5.4.7)$$

Following up the principle of approaching the heat budget equation and
the physical reality as closely as possible we introduce $(1 - \alpha') = 0.73$
as part of the "constant" into the radiation term and write:

$$M = -6.57 + 1.37(T_L - T_0) + 0.0264* G(1 - \alpha) + 5.36(e_L - E_0)$$

$$(5.4.8)$$

* (this coefficient for example is well approaching the physical
constant of $1/S = 0.030$ with S:latent heat of melt, 33.37 Jmm^{-1}
in the corresponding heat balance equation.)

The physical significance of the other constants in (5.4.8) is also
obvious. In its generalized "semi-empirical" form it can be applied in
practice to estimate daily meltrates (mm/day) for snow and ice surfaces
of different albedo α in places where no historical records are
available. T_L and e_L should be obtained from observation sites in a
standard position above ground (1.50-1.80 m) which provide large scale
representative values (as opposed to "oasis" information such as that
from observations taken at a glacier tongue). Depending on site conditions, adjusting the coefficients may be necessary. In cases of hilly
terrain with predominating exposures and slope angles, differences in
incoming solar radiation as compared to the horizontal surface, must also
be taken into account. Under very humid conditions with a predominance
of sensible and latent heat as the sources for melt, this regression

model will have to be modified. In particular the wind velocity will have to be included as a measure of the intensity of the turbulent exchange (expressed by the eddy diffusivities K_L and K_S in equations 5.3.4 and 5.3.5).

5.5 Operational forecasting equations for glacier basins where past records are available

In view of practical applications, a regression approach to forecasting glacier run-off was attempted for the Aletschgletscher basin (Lang 1980). The area of this basin is 194.7 km^2 and 67% is covered by glaciers; altitude range 1446 m to 4195 m a.s.l. The regression analysis was based on daily values of the melt period. Some of the equations are given together with the multiple correlation coefficient R and the ratio of the standard deviations s/s'.
s = SD of the dependent variable
s' = SD of the residuals (observed minus computed values).
Only those variables available from the standard networks were used.

Equations	R	s'/s	Eq.Nr.
$q(0) = 21.5 + 1.04 \ (T(0) + T(-1))$	0.52	0.86	(5.5.1)
$q(0) = -9.9 + 0.29 \ (T(0) + T(-1)) + 2.80 \ (e(0) + e(-1))$	0.69	0.74	(5.5.2)
$q(0) = -1.35 + 0.78 \ q(-1) + 1.28 \ T(0)$	0.80	0.61	(5.5.3)
$q(0) = -15.6 + 0.70 \ q(-1) + 3.46 \ e(0)$	0.83	0.57	(5.5.4)

$q(0)$, $q(-1)$: run-off (m^3/s) at actual day (0) and previous day (-1)
$T(0)$, $T(-1)$, $e(0)$, $e(-1)$: air temperature and vapour pressure at day
(0) and previous day (-1) in °C resp. mb.

From these equations is is clear that
(a) vapour pressure e is an important variable; this was also evident from earlier studies (Lang 1968).
(b) the discharge itself is the most important source of information in the computation of daily values for a basin of this size.

Another study (Jensen and Lang 1973) was undertaken in the area of the Z'mutt Glacier, Swiss Alps (34.5 km^2, 54% glacier area, 2213-4476 m a.s.l.). In this area the Hydropower Company Grande Dixence S.A., Sion uses daily glacier run-off forecasts for the optimum management of their reservoirs. One outstanding problem in this context is the distinct variations of the glacier basin characteristics in the course of the melting season, such as
(1) the extent and depth of the snow cover, which has a strong influence on the albedo of the glacier;
(2) the glacial drainage system.
This was taken into account by dividing the whole melting season into intervals, the processes in each interval are then assumed to be statio-

nary. The data samples of each year's melting season were divided into
three intervals according to the following criteria: extension of the
snow cover (frequently controlled by air survey) by the level of the air
temperature and of the discharge, and by the characteristics of the
daily variation (which is an indicator of the conditions in the glacial
drainage system). From preliminary equations the residuals were used as
a further control for a corrected subdivision into intervals. The
characteristics of the single intervals were described in the following
simple way:

Interval 1: <u>begin</u> (main snow melt season, June to mid July)
         2: <u>summer</u> (main ablation season with max discharges, mid July
                   to end of August)
         3: <u>end</u>   (reduced ablation, sometimes new snow on glaciers,
                   end of August to September).

In this example of the analysis the melting seasons of the 4 years
1967-1970 were considered. Some of the results are given graphically
in Figs. 5.5.1, 5.5.2 and 5.5.3.

It is interesting to note for example that the use of air temperature
as the only independent variable gives a multiple correlation coefficient
of $R = 0.84$ $(T(0).....T(-5))$ (see Fig. 5.5.3); if temperature is combined
with the discharge of the previous day $q(-1)$ the multiple correlation is
significantly improved such that $R = 0.95$.

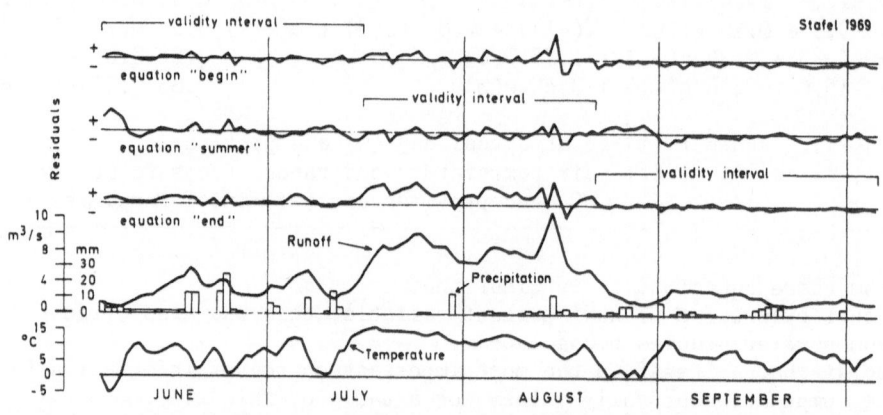

Fig. 5.5.1. Residuals (daily values of measured discharge minus
             calculated discharge), for the whole ablation season
             1969, resulting from 3 regression equations: each
             equation is valid only within the denoted interval.
             Period of reference: 4 years. The residuals and the
             discharge are in the same scale (Jensen and Lang 1973).

5.6 <u>Thermal and capillary retention capacity</u>

At the beginning of the melt season, the snowcover has not sufficiently
"ripened" to permit meltwater yield. A part of the meltwater (or rain)
is refreezing in the snowpacks when their temperatures are below the
melting point (cold content: see subsection 5.3.1 (c)). Another part

may be retained by capillary forces in a non-saturated snowpack. As
long as the free water content of the snow cover is below its maximum

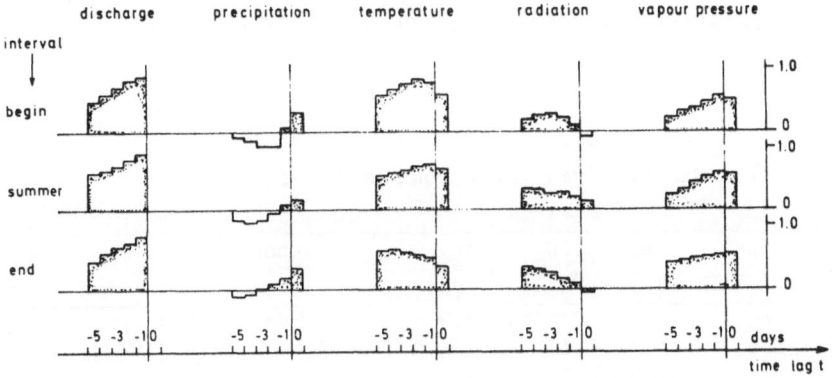

Fig. 5.5.2.   Simple correlation coefficients (bars) between dis-
              charge and several variables with various time lags
              (autocorrelation – and cross correlation function).
              Period of reference: 4 years (Jensen and Lang 1973).

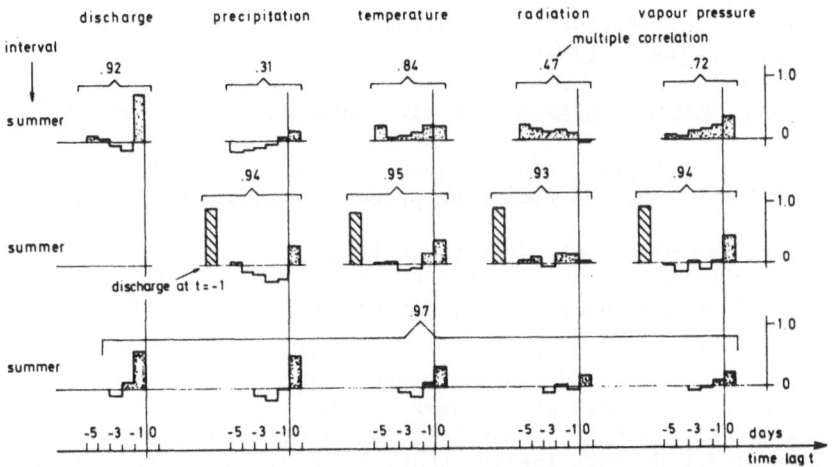

Fig. 5.5.3.   Example of partial correlation coefficients (bars)
              and multiple correlation coefficients of the regression
              equations for the summer interval (main ablation sea-
              son). First line: 5 equations, each with only one
              parameter with various time lags (for instance dis-
              charge at lag –1....–5 days) as predictor. Second line:
              4 equations, in each the discharge at lag –1 day is
              combined with only one meteorological parameter at
              lag 0....–5 days as the predictor. Third line: the
              "complete" equation with 5 parameters (19 variables)
              (Jensen and Lang 1973).

possible value R_k, the yield of meltwater from the snowpack remains
negligible. Since the snowcover is continuously under processes of
metamorphosis, R_k varies strongly. In Table IV some values of R_k as
quoted from different authors are given (for conversion into weight %:
$R_k \times \rho_* = R_k$).
(Vol.%)   (Weight %)

Table IV.   Maximum free water content R_k of a snowpack

snow type	density ρ_* (gr/cm^3)	R_k (Vol.%)	Author
old snow		5.9	Gerdel (1954)
old snow	0.4 –0.45	4–9	De Quervain (1948)
	0.45–0.5	2.3–5	
old snow		3–5.6	Ambach (1961,1963,1965)
old snow	0.15	12	Erbel (1969)

    Thus, it can be seen from Table IV that the capillary retention
capacity of a 50 cm deep snowcover with R_k = 5 Vol % is, for example,
25 mm of melt- or rainwater. For further treatment of this problem see,
for example, Wankiewicz (1979).

Combined rain-snow melt events (see also subsections 5.3.1 and 5.3.2)

Especially in flat regions floods caused by the combination of rain and
snow-melt occur often; the following processes contribute to this
phenomenon:
(a) Rainwater rapidly accelerates the "ripening" of the snowcover,
reducing the thermal and capillary retention capacity. For example (see
subsection 5.3.1(c)): Rain (at °C, intensity 1 mm/hr), will increase
the temperature of a snowpack ($\bar{T}$ = -3°C, ρ = 0.24 gr/cm^3, depth = 30 cm)
at the rate of 0.6°C/hr and can therefore have a dramatic influence on
the ripening process (Male and Granger 1979).
(b) During the ablation of a snowcover, a reduction of its retention
capacity is produced; this implies that, in addition to the accumulating
rain- and melt-water, "free water" is released, which has previously
been retained in the snowpack.
(c) In flat country the critical conditions to cause the beginning of the
melt-water yield occur simultaneously over large areas.
(d) At the end of the winter season the soil moisture in some regions
approaches the state of saturation. In other regions the soil surface
may be frozen. This indicates conditions for high run-off coefficients.
    Usually forecasting of meltwater yield from the snowcover is not very
efficient because of a lack of data from observations of the physical
conditions of the snowcover (depth, temperature, density, free water
content). In some models the whole process of accumulation and ripening
of the snowcover is simulated (for example Bergström 1976).

Some remarks on the problem of the "best model"

The competition and the efforts of the model builders have been great
through the last two decades. In the case of meltwater runoff models
the particular problem is the determination of the meltrates (in addition
to the modelling of the runoff process, which is not the subject of this
chapter). Since the meltrates are determined by the components of the
energy balance equation, it seems obvious that the model component to
determine the meltrates should be an energy balance approach. However,
it needs observation data such as net radiation etc. which are generally
not available. In mountain river basins the interpolation of the observation data under consideration of all topographic and climatological
variations is an additional difficulty. If the energy balance approach
is used with inaccurate input data, optimization of all the free parameters has become the universal method for overcoming the data problems;
but this is no longer a real energy balance approach.

The other way is to use a simple model, which only uses operationally
available data. All experiences so far have confirmed that the more
sophisticated models do not provide better results than network-adapted
simple models, if the data basis is the same. Charbonneau et al. (1981),
for example, have shown that some variables such as temperature gradients
or spatial distribution of precipitation are much more crucial than
possible choices between different approaches for modelling snowmelt.
A snowmelt-runoff study in the Swiss Alpine region clearly showed the
insignificant influence of the model structure to the overall performance
of runoff simulation; however, an improvement is possible by moving from
a simple temperature index method model to an extended model, which uses
wind speed and vapour pressure as additional input variables (Braun,
1985). In the use of simple models, however, it is important to recognize that a temperature index model may provide good results in a time
scale of weeks or months, while it cannot account for example for day to
day variations in evaporation which can affect the meltrates significantly because of the high specific energy of latent heat fluxes (Lang
1968, 1980). Any information about the real physical processes and
about the climatological conditions of a basin can be helpful in the
appropriate use of simple and of sophisticated models. The model problem
is further discussed in chapter 8, where O'Connell et al. state "not to
accept a priori that the more complex model is best"; see also chapter
12 (J. Nemec).

5.7 Long range, seasonal forecasting

One of the few successful possibilities of long range forecasting is
provided by the seasonal snowcovers in mountain areas. The complete
melting of a seasonal snowcover is generally within a more or less
forseeable length of time, which is the end of the summer ablation
season. There is some uncertainty with respect to the upper position of
the snowline in those regions, where perennial snow or glaciers exist.
Forecasts of variations of the snow line and of glaciers will not be
possible until reliable climatic long range forecasts are available.

Therefore seasonal forecasting is not really successful in strongly

glacierized areas. Depending on the relative contribution of summer
seasonal precipitation to run-off, the success of seasonal forecasting
depends in part on the accuracy of the assessment of the water equivalent
of the snow cover at the end of the accumulation period. The ideal
conditions exist in regions with a dry warm summer climate and a winter
precipitation season with a well defined accumulation period.

5.7.1 Seasonal forecasting in the Swiss Alps

The water equivalent WE of the winter snow cover in a river basin at
the beginning of the ablation period is illustrated by the water balance
equation:

$$WE = P_W - R_W - E_W \pm \Delta S_W \qquad (5.7.1)$$

P_W winter precipitation

R_W winter runoff

E_W winter evaporation

ΔS_W winter storage change of soil moisture, of groundwater, and of the
storage in natural and artifical lakes and reservoirs.
If no direct observations are available, this equation can be used
to estimate WE; E_W and ΔS_W must be based on certain assumptions.

Example of seasonal forecasting for the River Rhine/Rheinfelden

The drainage area of this basin is 34550 km^2, 1.8% glacier area, 3.6%
lakes, mean altitude 1085 m a.s.l. (258–4277 m a.s.l.). There is a network of ca. 50 stations for the measurements of the water equivalent WE.
An absolute determination of the total WE-volume in such a big river
basin is difficult. Since 28 years of records are available, a regression scheme could conveniently be used as a basis for the forecasting
(Vischer and Jensen 1978, Jensen 1974). The following regression
equation is an example of the runoff estimation for the month June R_6:

$$R_6 = a_0 + a_1 V_{31.5} + a_2 WE_{(1)31.3} + a_3 WE_{(2)31.3} + a_4 WE_{(3)31.3} +$$

$$+ a_5 P_{1.4-31.5} + a_6 R_{1.4-31.5}. \qquad (5.7.2)$$

$V_{31.5}$ volume of storage of 10 lakes in the river basin at May 31

$WE_{(1)31.3}$ water equivalent of the snow cover of the highest station at
March 31.

$WE_{(2)31.3}$ water equivalent average of 4 selected stations

$WE_{(3)31.3}$ water equivalent of 7 other stations

$P_{1.4-31.5}$ precipitation of 32 stations April–May

$R_{1.4-31.5}$ run-off of the Rhine-Rheinfelden, April–May

$a_0, \ldots, a_6$ regression coefficients.

Storage changes in the soil moisture and in the groundwater or of variations in the evaporation are not explicitly considered. The natural uncontrolled lakes can, to a certain degree, provide indirect information on groundwater and soil moisture conditions. Evaporation is a rather small quantity and is regarded as a quasi constant varying only slightly from year to year.

Experience shows that no general rule can be given concerning the choice of the best combination of variables. An optimum representative assessment of the snow cover undoubtedly is the most important key to the problem.

As a criterion of the forecasts we may take the ratio $(s - s')/s$, which is a measure for the reduction of the uncertainty.

s = SD of the runoff (1949-1977)
s' = SD of the error of the forecasts

This ratio decreases from 55% (April) to 30% (April to September) for the forecasts of the cumulative run-off, indicating the declining influence of melt water run-off to the river towards the end of the season. If the observed precipitation of the summer season is included as an additional variate (for the purpose of testing the model), the ratio $(s - s')/s$ shows values of ca. 70%. The remaining 30% uncertainty represents all the deficiences of the model, the data and networks.

SYMBOLS

Px/P	ratio of snow in total precipitation	1
ΔS^*	total of snow water equivalent	L
R	run-off	L
T_a	air temperature	Θ
WE	water equivalent	L
N	count after passing snowcover	1
N_0	initial count	1
μ	sensitivity of instrument	L^{-1}
F_z	area at elevation z	L^2
z	elevation	L
A_i	subarea	L^2
Q_M	energy available for melt per unit area and time	MT^{-3}
Q_{NR}	net radiation	"
Q_S	sensible heat	"
Q_L	latent heat of condensation or evaporation	"
Q_P	heat provided from precipitation	"
Q_G	heat from conduction in the snow peak	"
M	meltrate	$ML^{-2}T^{-1}$
S	latent heat of melt	L^2T^{-2}

G	global radiation	MT^{-3}
α	albedo	1
σ	Stefan–Boltzmann constant	$MT^{-3}\Theta^{-4}$
ε_a	emissivity of the atmosphere a	1
ε_e	emissivity of the surface e	1
T_e	temperature of snow surface	Θ
Θ	potential air temperature	Θ
c_p	specific heat of dry air	$L^2T^{-2}\Theta^{-1}$
ρ	density of air	ML^{-3}
K_s	eddy diffusivity for sensible heat	L^2T^{-1}
K_L	eddy diffusivity for latent heat	L^2T^{-2}
L	latent specific heat of evaporation	L^2T^2
q	specific humidity	1
e	vapour pressure	$ML^{-1}T^{-2}$
p	atmospheric pressure	"
k	von Karman constant	1
u	wind velocity	LT^{-1}
Z_o	roughness parameter	L
m_P	mass of rainwater	ML^{-2}
C	specific heat of water	$L^2T^{-2}\Theta^{-1}$
T_P	temperature of precipitation at ground level	Θ
T_0	temperature of melting snow cover	Θ
C	cold content of a snowpack	MT^{-2}
c_s	specific heat of snow/ice	$L^2T^{-2}\Theta^{-1}$
T_s	temperature at depth z	Θ
T_+	mean daily air temperature (positive or zero)	ΘT^{-1}
b	degree day factor	Θ^{-1}
R	multiple correlation coefficient	1
M	daily meltrates	LT^{-1}
E_0	saturation vapour pressure of a melting ice or snow surface	$ML^{-1}T^{-2}$
T_L	daily mean of air temperature	Θ
e_L	daily mean of vapour pressure	$ML^{-1}T^{-2}$
α'	observed average albedo	1
q	run-off	L^3T^{-1}
s	standard deviation of dependent variable	L^3T^{-1}
s^1	standard deviation of residual variable	L^3T^{-1}

R_k	max. value of free water content	1
P_w	winter precipitation	L
R_w	winter run-off	L
E_w	winter evaporation	L
ΔS_w	winter storage change	L
$V_{31.5}$	volume of storage in 10 lakes at 31st May	L^3

REFERENCES

Ambach, W. 1961 Z. f. Gletscherkunde 4, 169-189. 'Die Bedeutung des aufgefrorenen Eises für den Massen- und Energiehaushalt eines Gletschers' ('The role of frozen ice in the mass and energy balance of a glacier').

Ambach, W. 1963 Med. om Grönland, 174, 4. 'Untersuchungen zum Energie-umsatz in der Ablationszone des Grönländischen Islandeises' ('Investigations in the energy transformations in the meltzones of the Greeland ice caps').

Ambach, W. 1965 Archiv f. Meteorologie, Geophysik u. Bioklimat. B, 14, 148. 'Untersuchungen des Energiehaushaltes und des freien Wasser-gehaltes beim Abbau der winterlichen Schneedecke' ('Investigations in the energy balance and the free water content during the reduction of the snowpack').

Andersen, T. and Johnsrud, M. 1984: Proc. 5th Northern Research basins Symposium. Vierumäki, Finland, 2.1-2.13. Experiences of the gamma-ray snow survey method after ten years of operational use.

Anderson, E.A. 1976: 'A point energy and mass balance model of a snow cover', NOAA Techn. Rep. NWS 19, U.S. Dep. of Commerce.

Bengtsson, L. 1980: Nordic Hydrology 11, 221-234. 'Evaporation from a snow cover'.

Bergström, S. 1976: 'Development and application of a conceptual run-off model for Scandinavian catchments', Dep. of Water Resources Engin., Lund Inst. of Technology, Univ. of Lund Bull. Ser. A, No. 52, 134 p.

Braun, L. 1985: 'Simulation of snowmelt-runoff in lowland and lower alpine regions of Switzerland' Züricher Geographische Schriften, No. 21 ETH Zürich, 157 p.

Charbonneau, R., Lardeau, J.P. and Obled, C. 1981: 'Problems of model-ling a high mountainous drainage basin with predominant snow yields'. Hydrol. Sciences Bull. 26, 4, 345-361.

De Quervain, M. 1948: IUGG/IAIIS General Ass. Oslo, ICSI, Vol. II, 55-68. 'Ueber den Abbau der alpinen Schneedecke' ('On the structure of the alpine snowpack').

Erbel, K. 1969: Mitt. aus dem Institut für Wasserwirtschaft, Univ. Stuttgart, 12, 251 p. 'Ein Beitrag zur Untersuchung der Metamorphose von Mittelgebirgsschneedecken unter besonderer Berücksichtigung eines Verfahrens zur Bestimmung der thermischen Schneequalität'.

Gerdel, R.W. 1954: Trans. Am. Geoph. Union, 35, No. 3. 'The trans-mission of water through snow'.

Haefner, H. 1980: 'Snow surveys from earth resources satellites in the

Swiss Alps'. Remote Sensing Series Vol. 1, 65 pp., Dep. of Geography, Univ. of Zürich.

Jensen, H. and Lang, H. 1973: Proc. Banff Symposia 1972, 'On the role of snow and ice in hydrology', UNESCO-WMO-IAHS Vol. 2, 1047-1054, 'Forecasting discharge from a glaciated basin in the Swiss Alps'.

Jensen, H. 1974: VAW Mitt. Nr. 12, 'Hydrologische Prognosen für die Wasserwirtschaft', 137-164. 'Anwendung der Regressionsanalyse' ('An application of regression analysis').

Jensen, H. 1982: Jahresbericht der VAW 1981, 48-50. Versuchsanstalt für Wasserbau, Hydrologie und Glaziologie, ETH Zürich. 'Räumliche Analyse der Winterniederschläge in den Alpen' ('Spatial analysis of winter precipitation in the Alps').

Kasser, P. 1959: Wasser und Energiewirtschaft 6. 'Der Einfluss von Gletscherrückgang und Gletschervorstoss auf den Wasserhaushalt' ('The influence of glacier decline and growth on the hydrological regime').

Kogan, R., Nazarov, M. and Fridman, Sh.D. 1965: Soviet Hydrology: Selected Papers, No. 2, AGU, 183-187. 'Determination of water equivalent of snow cover by method of aerial gamma-ray survey'.

Kondratyev, K. 1969: Radiation in the Atmosphere Acad. Press.

Kuzmin, P.P. 1961: Melting of snow cover, Gidrometeorologicheskoe Izdatel'stvo, Leningrad 1961: translated JPST Jerusalem, 1972.

Lang, H. and Schoenbaechler, M. 1967: 'Heat balance studies and runoff at the Aletschgletscher 1965'. VAW, ETH Zürich, Internal Report.

Lang, H. 1968: IUGG General Assembly 1967, Berne. IAHS Publ. No. 79, 429-439. 'Relations between glacier run-off and meteorological factors observed on and outside the glacier'.

Lang, H., Schaedler, B. and Davidson, G. 1977: Zeitschr. f. Gletscherkunde u. Glazialgelogie XII, 2,109-124. 'Hydroglaciological investigations on the Ewigschneefeld - Gr. Aletschgletscher'.

Lang, H. 1980: Proc. Sympos. Tbilisi 1978. Acad. of Sciences of the USSR, Geophysical Committee, Data of Glaciol. Studies, Publ. No. 38, 187-194. Theoretical and practical aspects in the computation of run-off from glacier areas.

Lang, H. 1981: Nordic Hydrology 12, 217-224. 'Is Evaporation an Important Component in High Alpine Hydrology?'

Lettau, H. 1939: Atmosphärische Turbulenz (Atmospheric Turbulence). Akad. Verlagsges. Leipzig.

Loijens, H.S. and Grasty, R.L., 1974: 'Airborne measurement of snow-water equivalent using natural gamma radiation over Southern Ontario, 1972-1973'. Environemnt Canada, Scient. Series, No. 34, Inland Waters Directorate, Ottawa.

Male, D.H. and Granger, R.J. 1979: Proceed. Modeling Snow Cover Run-off. CRREL Hannover, N.H. Energy mass fluxes at the snow surface in a prairie environment. (Ed.: Colbeck and Ray).

Martinec, J., 1980: Nordic Hydrology 11, 209-220. 'Limitations in hydrological interpretations of the snow coverage'.

Meier, M.F. 1980: Hydrological Sciences Bull. 25, 3, 9, 307-330. 'Remote sensing of snow and ice'.

Monin, A.S. and Obukhov, A.M. 1954: Geophys. Inst. Acad. Sci. USSR, 24, 151, 163-187 (English Translat. American Met. Soc., T-R-174, 1959). 'Basic laws of turbulent mixing in the ground layer of the atmosphere'.

Ostrem, G. 1972: 'Runoff forecasts for highly glacierized basins'.
IAHS Publ. No. 107, Vol. 2, 1111-1129 (Proc. Banff Symposia).
Prandtl, L. 1904: Proc. Mathem. Congress, Heidelberg 1904, Leipzig
1934: 'On fluid motions with very small friction'.
Prandtl, L. 1934: The Mechanics of Viscous Fluids (Aerodynamic Theory)
Durand (ed.) Vol. III, G. Berlin.
Rango, A., Salomonson, V.V. and Foster, J.L., 1977: Water Resources
Research, 13, 109-112. Summary report by A.T.C. Chang and A. Rango
during 50th Annual Meeting, Western Snow Conference, 1982, 204-207).
'Seasonal streamflow estimation in the Himalayan region employing
meteorological satellite snow cover observations'.
Ibid., 1980: Operational Applications of Satellite Snow Cover Observa-
tions, (Eds.: A. Rango, R. Peterson), NASA Conference Publ. 2116.
Roald, L. 1971: 'Glacier discharge as a function of meteorological
parameters'. In: Glasiologiske undersokelser; Norge, 1970, Rep. No.
2/71, N.V.E., Hydrology Div. Oslo.
Schmid, W. 1925: Der Massenaustausch in freier Luft und verwandte
Erscheinungen (Mass Exchange in Open Air and Related Phenomena),
Henri Grand, Hamburg.
Schüepp, M. 1950: Wetter, Wind und Wolken (Weather, Wind and Clouds)
Zürich, 1950.
SMA 1978: Klimatologie der Schweiz. Bd. 2 Regionale Klimabeschreibg.
1. Teil. Beiheft zu den Annalen der Schweizer. Meteorol. Anstalt.
Sverdrup, H.U. 1936: Geophys. Publ. 11, 7, 1-69. 'The eddy conductivity
of the air over a smooth snow field'.
Sverdrup, H.V. 1946: J. of Meteorol. 3, 1-8. The humidity gradient
over the sea surface.
Thornthwaite, C.W. and Holzman, B. 1939: Monthly Weather Review, 67,
4-11. The determination of evaporation from land and water surfaces.
Ibid., 1956: 'Snow Hydrology', U.S. Army Corps of Engineers, Summary
Report of the Snow investigations, North Pacific Corps Eng. Portland,
Oreg.
Uttinger, H. 1951: Archiv. für Meteorol., Geoph. und Bioklim. II, 4,
360-382. 'Zur Höhenabhängigkeit der Niederschlagsmenge in den Alpen'
('The dependence of precipitation on elevation in the Alps').
Vershinina, L.K. and Dimaksyan, A.M. (ed.), 1969: Determination of the
water equivalent of snow cover, JPST, No. 5779, Jerusalem.
Vischer, D. and Jensen, H. 1978: Wasserwirtschaft 68 Jg. 9, 259-264.
'Langfristprognosen für den Rheinabfluss in Rheinfelden' ('Long range
discharge forecasts for the Rhine at Rheinfelden').
Wankiewicz, A. 1979: 'A review of water movement in snow'. In: Proc.
Modelling of snow cover runoff. Ed.: Colbeck and Ray. CRREL, Hannover
N.H., 222-252.
WMO 1981: 'Guide to Hydrological Practices', WMO Publication No. 168,
Geneva.
Zingg, T. 1951: IUGG/IAHS Gen. Ass. Bruxelles, Publ. No. 32, Vol. 1,
266-269. 'Beziehungen zwischen Temperatur und Schmelzwasser' ('Rela-
tions between temperature and meltwater').
Zotimov, N.V. 1968: Soviet Hydrology: Selected Papers No. 3, AGU,
254-266. 'Investigation of a method of measuring snow storage by
using the gamma radiation of the earth'.

6. TIME-SERIES METHODS AND RECURSIVE ESTIMATION IN HYDROLOGICAL SYSTEMS ANALYSIS

Peter C. Young

University of Lancaster
Dept. of Environmental Sciences
Lancaster LA14 YQ
UK

6.1 Introduction

Previous chapters have shown how models of hydrological systems can be formulated in many different ways and with various levels of complexity. In this chapter, we will see how these kinds of models can be considered within a unified stochastic setting and how it is then possible to treat model calibration as a problem of time-series analysis. In this manner, powerful time-series techniques, such as recursive estimation (Young 1984) can be used in the identification, estimation and validation of the models. And, because ot their inherently stochastic nature, such models can subsequently provide a natural vehicle for real-time flow forecasting. Moreover, the recursive approach to estimation allows for continuous updating of the model parameter estimates and the possibility of more advanced "self-adaptive" forecasting and control procedures.

6.2 The simplest first order, linear hydrological model

In Chapters 2 and 3, both linear and nonlinear models have been intro-duced for flow and rainfall-flow processes. Let us first consider linear formulations of such models and investigate how they can be considered in time-series terms.

The basic lumped parameter representation for the changes in storage of a single reach in a river system is given by Equation (2.4.2) of Chapter 2 or (3.12.2) of Chapter 3. Considering Equation (3.12.2), we see that the rate of change of storage S in the reach is defined by the following equation

$$\frac{dS(t)}{dt} = Q_1(t) - Q_2(t) + R(t) \qquad (6.2.1)$$

where $Q_1(t)$ is the upstream (input) flow into the reach; $Q_2(t)$ the downstream (output) flow out of the reach; and $R(t)$ the total inflow (or outflow if negative) that occurs within the reach.

There are various possible forms for the relationship between the

D.A. Kraijenhoff and J.R. Moll (eds.), River Flow Modelling and Forecas-ting, 129–180.
© 1986 by D. Reidel Publishing Company.

storage $S(t)$ and the variables on the right hand side of (6.2.1). Limiting discussion initially to linear representations, the simplest relationship is given by Equation (3.13.1) of Chapter 3, i.e.

$$S(t) = K.Q_2(t)$$

where $S(t)$ is considered a linear function of the output flow $Q_2(t)$ and K is the "storage delay time". If $R(t) = 0$, (i.e. no inflow within the reach), then this assumption yields Equation (3.13.2) which can be written

$$\frac{dQ_2(t)}{dt} = -\frac{1}{K} Q_2(t) + \frac{1}{K} Q_1(t) \qquad (6.2.2)$$

Time-series analysis can be carried out in continuous-time and applied directly to Equation (6.2.2) as discussed by Young (1981) and Young and Jakeman (1980); or in discrete-time, when it is applied to discrete-time (sampled data) equivalents of (6.2.2). Here we will consider only discrete-time representations, since the analysis is much more straightforward in discrete-time terms. Moreover, in this age of microprocessor controlled data acquisition systems and the digital computer, it is much more likely that the hydrologist will be confronted with sampled data.

If we assume, for analytical convenience, that $Q_1(t)$ remains constant between sample measurements[*] then the <u>exact</u> discrete-time equivalent of (6.2.2) takes the form

$$Q_{2,k} = -a_1 Q_{2,k-1} + b_0 Q_{1,k-1} \qquad (6.2.3)$$

where the subscripts k, $k-1$ etc. denote that the variables (Q_2 or Q_1) are sampled measurements with the value of k indicating the sample number, where $k = 1,2,\ldots,N$ and N is the total sample size. In (6.2.3) a_1 and b_0 are defined as follows:

$$a_1 = -\exp(-\Delta t/K); \qquad b_0 = 1 + a_1 \qquad (6.2.4)$$

where Δt is the sampling interval in appropriate time units. A simple way of obtaining an equation of similar form to Equation (6.2.3) is to directly discretize Equation (6.2.2) by finite difference approximation, in the following manner,

----

[*] This is the simplest assumption possible: since we measure Q_1 only every Δt time unit, it is equivalent to assuming that no change in Q_1 occurs over this sampling interval. More complicated assumptions lead to more complicated analysis and results (see Section 3.15 of Chapter 3).

$$\frac{Q_{2,k} - Q_{2,k-1}}{\Delta t} = - \frac{1}{K} Q_{2,k-1} + \frac{1}{K} Q_{1,k-1}$$

$$Q_{2,k} = \left[1 - \frac{\Delta t}{K}\right] Q_{2,k-1} + \frac{\Delta t}{K} Q_{1,k-1}$$

This is an approximate representation of (6.2.2) and the degree of approximation is a function of the sampling interval Δt. It will be noted that, not surprisingly, it is also an approximation to Equation (6.2.3) as we can see if we introduce an infinite dimensional expansion for the exponential in the definition of a_1 and replace the exponential by the first two terms of this expansion.

It is clear that a number of alternative discrete-time approximations to the differential Equation (6.2.2) can be derived even for a fixed sampling interval Δt. For example, if $Q_1(t)$ and $Q_2(t)$ are approximated by their mean value between samples, i.e. in the case of $Q_1(t)$,

$$Q_1(t) \approx \frac{Q_{1,k} + Q_{1,k-1}}{2}$$

then the more complicated discrete-time model takes the form

$$Q_{2,k} = - a_1 Q_{2,k-1} + b_0 Q_{1,k} + b_1 Q_{1,k-1} \qquad (6.2.3a)$$

where now

$$a_1 = - \frac{2k - \Delta t}{2k + \Delta t}; \qquad b_0 = b_1 = \frac{\Delta t}{2k + \Delta t} \qquad (6.2.4)$$

This is a more accurate representation of the differential Equation (6.2.2) and has been used by hydrologists in the computer modelling of such equations. But this does not mean it is a more accurate representation of physical reality since Equation (6.2.2) is itself only an approximation of the real world.

A minor extension of the model (6.2.3) is obtained by reference to Equation (3.13.9) of Chapter 3, in which the flow behaviour is represented by a "single reservoir" model such as (6.2.2), together with a pure transportation time-delay (or lag) of T time units (in Chapter 3, this is referred to as the "delay time of the linear channel" or the "pure lag"), to allow for pure transitional effects (the "lag and route method" of Meijer, 1941). We can write this model in the equivalent form

$$\frac{dQ_2(t)}{dt} = - \frac{1}{K} Q_2(t) = \frac{1}{K} Q_1(t - T) \qquad (6.2.5)$$

The discrete-time representation then takes the form (cf. Equation 6.2.3),

$$Q_{2,k} = - a_1 Q_{2,k-1} + b_0 Q_{1,k-\tau} \qquad (6.2.6)$$

where $\tau = T/\Delta t$ is the pure time delay in sampling intervals (usually τ defined in this manner is rounded to the nearest pure integer for computational purposes).

Of course, the need to choose an integer value for T can lead to an approximation which may be important in certain circumstances. However, in such situations, it is possible to compensate for the approximation, to some extent, by introducing an additional delayed term in $Q_{1,k-\tau-1}$, where the two discrete values for Q_1 (i.e. at $k - \tau$ and $k - \tau - 1$) then span the pure time delay. In other words, Equation (6.2.6) is modified to the form

$$Q_{2,k} = - a_1 Q_{2,k-1} + b_0 Q_{1,k-\tau} + b_1 Q_{1,k-\tau-1}$$

where the true time delay T lies between $(\tau - 1)$ t and $\tau \Delta t$ time units. This equation is similar to a time delayed version of (6.2.3a) but here the differential weighting on the two terms in Q_1, provided by the coefficients b_0 and b_1, provides the required compensation for the non-integral time delay.

As an example of this kind of discrete-time or sampled data representation, let us consider a flow model which is discussed in more detail later in Section 6.7. This is a model for flow behaviour between two measurement locations on the River Wyre near Lancaster. As we shall see, time-series analysis reveals that the "best identified model" (a term we will discuss later) is of the form

$$Q_{2,k} = 0.75 A_{2,k-1} = 0.89 Q_{1,k-2} \tag{6.2.7}$$

where the sampling interval $\Delta t = 1$ hour and the pure time delay $\tau = 2$ sampling intervals (i.e. 2 hours). From Equations (6.2.4) we can note that this implies that the storage delay time (or "time-constant" in systems terminology) is given by

$$K = - \Delta t/(\ln 0.75) = 3.48 \text{ hours}$$

However, we see that b_0 is 0.89 in this case and that this is not equal to $1 + a_1$, which is only 0.25 from equation (6.2.7).

The reason for this discrepancy lies in the inflow term R(t), which we omitted from Equations (6.2.2) and (6.2.5). The relationships (6.2.4) only apply for the case R(t) = 0 and we need to develop alternative relationships if there is inflow into the reach between the upstream and downstream measurement locations. However, since, in general, R(t) will not be measured, we need to examine its effect on the relationships (6.2.2) and (6.2.5) rather than explicitly including it as an additional term in the model.

The presence of R(t) will mean that the downstream flow $Q_2(t)$ will be either greater (for positive R(t)) or less (for negative R(t)) than would be expected by considering the upstream inflow effects $Q_1(t)$ alone. Thus, if we assume that $R(t)$ is some linear function of inflow $Q_1(t)$, i.e.

$$R(t) = \alpha \, Q_1(t)$$

then Equation (6.2.5), for example, can be written as,

$$\frac{dQ_2(t)}{dt} = -\frac{1}{K} Q_2(t) + \frac{1}{K} Q_1(t - T) + \frac{\alpha}{K} Q_1(t - T)$$

or,

$$\frac{dQ_2(t)}{dt} = -\frac{1}{K} Q_2(t) + \frac{1}{K} (1 + \alpha)Q_1(t - T) \qquad (6.2.8)$$

The discrete-time equivalent of Equation (6.2.8) is of the same form as Equation (6.2.6) but now b_0 is defined by the following relationship

$$b_0 = G(1 + a_1) \qquad (6.2.9)$$

where the "steady state gain" G is defined as

$$G = b_0/(1 + a_1) = 1 + \alpha \qquad (6.2.10)$$

This model can be considered as a "lag and route model with gain", to differentiate it from the simple lag and route model which has inherent unity gain.

If we now return to the River Wyre example, it is easy to see that

$$G = 0.89/(1 - 0.75) = 3.56$$

so that

$$\alpha = 2.56$$

In other words, there is strong evidence for lateral inflow (α is positive) of the order of 2.56 times the upstream measured inflow into the reach. This is, of course, quite obvious from the upstream and downstream hydrographs in this case, where the upstream measured flow is clearly much smaller than that measured downstream (see Fig. 6.2.1).

It should be noted that three quite important properties of linear first order, time-delay systems are revealed by this simple analysis. These are best visualised in relation to the <u>unit step response</u> of the system, i.e. the response of the outflow $Q_2(t)$ to a sustained unit step increase of the inflow $Q_1(t)$, as shown in Fig. 6.2.2. Although a little artificial in hydrological terms, this response is very important in

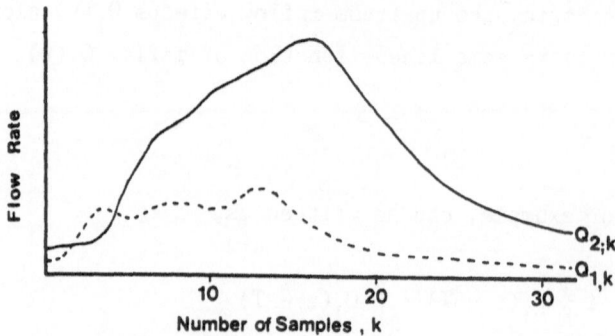

Fig. 6.2.1. Wyre data

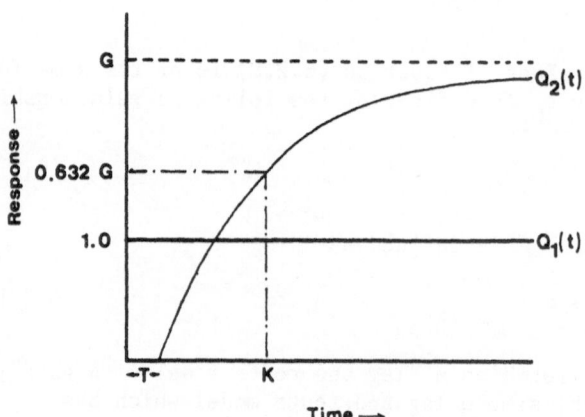

Fig. 6.2.2. Unit step response

applied systems analysis. The step response $\left[Q_2(t)\right]$ of the model
(6.2.8) with T = 0 is given by

$$\left[Q_2(t)\right]_s = G\left[1 - \exp(-t/K)\right] \tag{6.2.11}$$

which can be compared directly with the unit impulse response given by
Equation (3.13.4) of Chapter 3.

The step response for T > 0 is simply this same response shifted
(i.e. delayed) by T time units. The time delay T is then seen to be the
time taken for the effects of the upstream variation in flow to first
appear at the downstream location. The steady state gain G is the final
level reached by the downstream flow in response to the upstream sus-
tained unit step change (i.e. $\left[Q_2(t)\right] \rightarrow G$ as $t \rightarrow \infty$).

Alternatively, for a general step input of arbitrary magnitude, the
steady state gain is the ratio of the final steady downstream flow magnitude to the steady upstream magnitude, as $t \rightarrow \infty$. It is also interes-

ting to note that, in hydrological terms, the steady state gain is the ratio of the area underneath the downstream flow hydrograph to that underneath the upstream hydrograph. As a result, if a model such as (6.2.6) is obtained directly from flow data, in the manner discussed in subsequent sections, then the associated value of G provides a quick method for determining this area ratio and the 'flow gain' of the reach, provided the data are explained well by the model. However, as pointed out later in Section 6.5, this estimate of G can be sensitive to small errors in the parameter estimates $\hat{a}_1$ and $\hat{b}_0$ of the parameters a_1 and b_0, respectively.

Finally the <u>time constant</u> K is the time taken for the downstream flow to reach $\{1 - \exp(-1)\} = 0.632$ of its final steady state level G; it is, in other words, a primary measure of the <u>transient response</u> characteristics of the system. On the other hand, the inverse of K, which is usually denoted by w_n and termed the <u>break frequency</u>, is indicative of the <u>frequency response</u> characteristics of the system; i.e. the manner in which the system output responds to input signals in the form of sinusoidal perturbations at all frequencies. For input signal components with frequencies less than w_n, there is little attenuation or phase lag associated with the output signal; while, for components with frequencies greater than w_n, there is progressively greater attenuation and increasing phase lag, up to a maximum of 90° (see e.g. Truxal 1955). The importance of such frequency response characteristics is quite obvious: by Fourier analyses, we can consider a quite general class of signals to be composed of sinusoidal components and so the dynamic behaviour of the system (reach) in response to fairly arbitrary input flow changes is well defined by the frequency response characteristics of the model.

6.3 More complicated linear hydrological models

Various alternative linear hydrologic models are discussed in Chapter 3. Perhaps the most important of these, in a hydrological sense, is the Muskingum model of Equation (3.13.6) where the storage S(t) is defined as

$$S(t) = K\left[x\, Q_1(t) + (1 - x)Q_2(t)\right], \qquad (6.3.1)$$

so that the differential equation representation of the reach behaviour takes the form

$$\frac{dQ_2(t)}{dt} = - \frac{1}{K(1 - x)}\, Q_2(t) + \frac{1}{K(1 - x)}\, \{Q_1(t) - Kx\, \frac{dQ_1(t)}{dt}\} \qquad (6.3.2)$$

The discrete-time equivalent of this model, as obtained by the simplest finite difference approximation, can be written*

---

* Note that the sign convertion used here for the a_1 coefficient is different from that used in Chapter 3 (e.g. Equation (3.15.6)): the present sign convertion is appropriate to the control, systems and time-series estimation literature.

$$Q_{2,k} = - a_1 Q_{2,k-1} + b_0 Q_{1,k} + b_1 Q_{1,k-1}, \qquad (6.3.3)$$

where now

$$- a_1 = 1 - [\Delta t / K(1 - x)]$$

$$b_0 = - [x/1 - x]$$

$$b_1 = [1 + \Delta t / Kx] x / (1 - x).$$

As in the case of the simpler model (6.2.2), it is possible to use more complicated finite-difference approximations. For example, if similar approximations to those used to develop (6.2.3a) are used in the present situation, then the modified expressions for a_1, b_0 and b_1 are as follows (cf. equations 3.15.7–3.15.10)

$$a_1 = - \frac{K(1 - x) - \frac{\Delta t}{2}}{K(1 - x) + \frac{\Delta t}{2}}; \quad b_0 = - \frac{Kx - \frac{\Delta t}{2}}{K(1 - x) + \frac{\Delta t}{2}}; \quad b_1 = \frac{Kx + \frac{\Delta t}{2}}{K(1 - x) + \frac{\Delta t}{2}}$$

However, the same remarks still apply: while the resulting model is a better approximation to Equation (6.3.2), it is not necessarily a better representation of reality.

If x in the definition of the parameters a_1, b_0 and b_1 is assumed to be positive and less than unity, then we see that the b_0 coefficient in (6.3.3) will be negative. As a result, for positive increments in $Q_{1,k}$ the initial effect on $Q_{2,k}$ will be negative, as pointed out in Chapter 3. In systems terms, such a model is said to possess "non-minimum phase (NMP) characteristics" (because of peculiarities in its frequency response; see e.g. Truxal (1955). In hydrological terms, such NMP behaviour is difficult to justify from a physical standpoint (see discussion in Chapter 3, following Equation (3.13.8)) and this tends to throw some doubt on the validity of the assumption (6.3.1) and the Muskingum model (6.3.2). But we shall have more to say on this aspect of the Muskingum model when we consider the River Wyre results in Section 6.7.

Second and higher order dynamic models, such as that defined by Equation (3.13.9) of Chapter 3, can be interpreted as logical extensions of the first order model (6.2.2). In their most general form, these models can be formulated in the following manner

$$Q_2(t) + \alpha_1 \frac{dQ_2(t)}{dt} + \ldots + \alpha_r \frac{d^r Q_2(t)}{dt^r}$$

$$= \beta_0 Q_1(t-T) + \beta_1 \frac{dQ_1(t - T)}{dt} + \ldots + \beta_s \frac{d^s Q_1(t - T)}{dt^s} \qquad (6.3.4)$$

where a pure time delay of T time units has been introduced for complete generality. In the case of the simple linear model (6.2.2), $r = 1$, $s = 0$ and $T = 0$; for the delay-differential equation (6.2.5), $r = 1$, $s = 0$ and T is non zero; for the Muskingum model (6.3.2) $r = 1$, $s = 1$ and $T = 0$; and finally, for the Kalinin-Milyukov model (3.13.9) of Chapter 3, $s = 0$, $T = 0$, and r is non zero.

The most useful discrete-time representation of a linear delay-differential equation such as (6.3.4) has the following form

$$Q_{2,k} + a_1 Q_{2,k-1} + \cdots + a_n Q_{2,k-n}$$

$$= b_0 Q_{1,k-T} + b_1 Q_{1,k-T-1} + \cdots + b_m Q_{1,k-T-m} \qquad (6.3.5)$$

where we see that the output (downstream) flow Q_2 at the kth sampling instant is a linear sum of past output, and present and past input (upstream) flows. The relationship between the coefficients $(a_1,\ldots,a_n, b_0,\ldots,b_m)$ of Equation (6.3.5) and the coefficients $(\alpha_1,\ldots,\alpha_r, \beta_0,\ldots,\beta_s)$ of Equation (6.3.4) for this general case need not concern us here: it is sufficient to note that continuous-time linear systems described by differential equations can be represented in discrete-time terms by an equation of the form (6.3.5) and the coefficients $a_1,\ldots,a_n, b_0,\ldots,b_m$ will need to be estimated by reference to upstream and downstream hydrograph records.

A convenient form for Equation (6.3.5) is obtained by introducing the backward shift operator z^{-1}, where $z^{-1} Q_{2,k} = Q_{2,k-1}$. With this operator notation, Equation (6.3.5) can be written alternatively as

$$\left[1 + a_1 z^{-1} + \cdots + a_n z^{-n}\right] Q_{2,k} = \left[b_0 + b_1 z^{-1} + \cdots + b_m z^{-m}\right] Q_{1,k-T}$$

so that $Q_{2,k}$ is related to $Q_{1,k}$ by the expression,

$$Q_{2,k} = \frac{b_0 + b_1 z^{-1} + \cdots + b_m z^{-m}}{1 + a_1 z^{-1} + \cdots + a_n z^{-n}} Q_{1,k-T}$$

or

$$Q_{2,k} = \frac{B(z^{-1})}{A(z^{-1})} Q_{1,k-T} = \frac{z^{-T} B(z^{-1})}{A(z^{-1})} \qquad (6.3.6)$$

Here, the ratio of the two polynomials in the backward shift operator z^{-1} is termed the "transfer function" of the system. In the case of

the Muskingum model, for example, we see from Equation (6.3.3) that the
transfer function takes the form,

$$\frac{B(z^{-1})}{A(z^{-1})} = \frac{b_0 + b_1 z^{-1}}{1 + a_1 z^{-1}}$$

6.4 Recursive estimation of a simple time-series model

As its name implies, time-series analysis is concerned with the statistical investigation and modelling of stochastic time-series data; in
other words, data that are uncertain and are naturally arranged in some
temporal order. In the present context, the analysis is usually involved
with the investigation of the assumed causal relationships between two
or more time-series: for example, the upstream and downstream hydrographs of a river system; or the rainfall and runoff time-series for a
catchment. The investigation of a single time-series (univariate analysis) is also important in hydrological modelling but will not be dealt
with in any detail in this chapter, although some reference to univariate analysis will appear later in Section 6.7.

In the previous section it has been shown that all of the common
linear representations used in hydrological modelling can be considered
as special cases of the general differential-delay model (6.3.4). And
we have also seen that such a model has a useful discrete-time, sampled
data representation in the form of Equation (6.3.5) or its operational
equivalent (6.3.6). Equation (6.3.6) provides a basis for time-series
analysis, providing as it does a parametrically efficient representation
of the linear dynamic relationship between two time-series. But it is
a purely deterministic relationship and it does not allow for any
uncertainty associated with either the measured time-series or the
nature of the relationship.

In order to introduce some measure of uncertainty into the problem
formulation, it is most straightforward and analytically convenient if
we assume that the all stochastic effects can be lumped into a single
stochastic variable ξ_k that appears additively in the model equations.

Thus we modify Equation (6.3.6) to the following form

system equation: $x_k = \dfrac{B(z^{-1})}{A(z^{-1})} u_{k-T}$ (i)

output or meas-
urement equation: $y_k = x_k + \xi_k$ (ii)

(6.4.1)

where, for generality, $Q_{2,k}$ and $Q_{1,k}$ have been replaced by the variables
x_k and u_k, respectively, which represent the noise free sampled input
and output of a general linear dynamic system. The measured output y_k
is seen to be the sum of the noise free output x_k and the 'noise'

variable ξ_k, which is assumed to represent that part of the output measurement y_k that is not causally related and cannot, therefore, be explained by the input u_k. This noise term can be due to random effects such as measurement errors as well as the effects of other unmeasured flow and rainfall inputs to the river system. The transfer function $B(z^{-1})/A(z^{-1})$, on the other hand, describes the assumed causal relationship between u_k and the noise free but unmeasurable output x_k.

It should be noted that, in certain circumstances, baseflow effects can be considered as additive noise. In general, the baseflow changes are of a long period, low frequency type and it may not always be possible to relate the input and output baseflow components by the same model that relates the short term changes in discharge. In such circumstances, the statistical methods discussed in this chapter will still work since the resulting baseflow discrepancy can, if necessary, be considered as a noise component. This means that pre-processing procedures, such as baseflow removal, are not essential to such modelling exercises (although sometimes they can be an advantage in statistical terms (Young (1985)). In the examples discussed in this chapter, for instance, the data were used directly in their 'raw' form and, in line with the convention in flood routing analysis, no baseflow removal was attempted.

Clearly Equations (6.4.1) (i) and (ii) can be combined to provide a single stochastic representation of the form

$$y_k = \frac{B(z^{-1})}{A(z^{-1})} u_{k-T} + \xi_k \qquad (6.4.2)$$

or, on expansion back into discrete-time equation terms,

$$y_k + a_1 y_{k-1} + \cdots + a_n y_{k-n} = b_0 u_{k-T} + \cdots + b_m u_{k-T-m} + \xi_k +$$
$$+ a_1 \xi_{k-1} + \cdots + a_n \xi_{k-n} \qquad (6.4.3)$$

This equation can then be written in the following more compact vector notational terms,

$$y_k = \underline{z}_k^T \underline{a} + \eta_k \qquad (6.4.4)$$

where

$$\underline{z}_k^T = \left[-y_{k-1}, \ -y_{k-2} \ \cdots \ -y_{k-n}, u_{k-T}, \dots, u_{k-T-m} \right];$$

$$\underline{a} = \left[a_1, a_2, \dots, a_n, b_0, \dots b_m \right]$$

and

$$\eta_k = \xi_k + a_1\xi_{k-1} + \ldots + a_n\xi_{k-n}. \qquad (6.4.5)$$

In its simplest form, time-series analysis is concerned with answering the following questions:
(1) given data in the form of the two time series u_k, y_k for k = 1,2, ...,N, how do we identify the most appropriate structure (i.e. the values of n, m and T) for the model (6.4.4),
(2) given information on the model structure, how do we then estimate the values of the coefficients $a_1, a_2, \ldots, a_n, b_0, b_1, \ldots, b_m$ that characterise this structure?

The first question is rather difficult to answer and can depend, in part, on the answer to the second question. It is simpler, therefore, to consider this second question initially and deal with the problem of model structure identification later in Section 6.6. In order to understand the estimation problem, it is convenient at first to refer to a very simple example of the model (6.4.4) when n = 0, m = 0, and only T is non zero; in other words,

$$y_k = b_0 u_{k-T} + \xi_k \qquad (6.4.6)$$

Furthermore, we will assume initially that $\xi_k = e_k$, where e_k has a very simple stochastic description and is in the form of a zero mean, serially uncorrelated sequence of random variables with variance σ^2, i.e.

$$E\{e_k\} = 0; \qquad E\{e_k e_j\} = \sigma^2 \delta_{kj}$$

Here E{ } is the expected value operator (Young 1984) while δ_{kj} is the so-called <u>Kronecker delta</u> function and is defined as unity if k = j, and zero if k ≠ j. A sequence such as e_k is usually termed <u>discrete white noise</u>.

The physical interpretation of this simple model (6.4.6) is quite straightforward; the system is such that the deterministic variable x_k is related to the delayed input u_{k-T} by a gain coefficient b_0; and the measured output is simply the "noise free" output plus the white noise e_k. By "deterministic" here, we mean that part of the measured output which is assumed to be <u>causally related</u> to the delayed input u_{k-T}. In a typical hydrological system, (6.4.6) could represent the relationship between coarsely sampled measurements of upstream or downstream flow in a river. The coarse sampling (i.e. Δt is large in comparison with the dominant time-constants associated with the system transient response) means that only long term dynamic behaviour is discernible and, therefore, identifiable from the measured data. In other words, if the sampling frequency is much lower than the break frequency w_n then the resulting time-series will only reveal the low frequency behaviour of the system.

The white noise e_k is the lumped effect of all stochastic disturbances operating on the system and will include both actual measurement errors and modelling inaccuracies (i.e. the fact that the model (6.4.6) is unlikely to be a perfect representation of the system). Clearly, in practice, it is unlikely that the very simple assumption of white noise for ξ_k will be justified and we will need to investigate the consequences of any contravention of this assumption later.

The best known statistical method for estimating b_0 in (6.4.6) is the method of <u>least squares regression analysis</u> (Kendall and Stuart 1951). This can be considered as an optimisation problem in which the "best estimate" $\hat{b}_0$ of b_0 is chosen as the value of b_0 which minimises the following "least squares cost function" J,

$$J = \sum_{i=T+1}^{N} \left[y_i - \hat{b}_0 u_{i-T} \right]^2 \qquad\qquad (6.4.7)$$

This optimization problem is solved by setting the partial derivative of J with respect to $\hat{b}_0$ to zero and solving for $\hat{b}_0$ in the normal manner, i.e.

$$\frac{\partial J}{\partial \hat{b}_0} = -2 \sum_{i=T+1}^{N} \left[y_i - \hat{b}_0 u_{i-T} \right] u_{i-T} = 0$$

so that

$$\hat{b}_0 = \sum_{i=T+1}^{N} y_i u_{i-T} \bigg/ \sum_{i=T+1}^{N} u_{i-T}^2 \qquad\qquad (6.4.8)$$

This type of solution is fairly well-known. But what if we wish to obtain a <u>recursive</u> or <u>sequentially updated</u> solution? In other words, rather than obtain an estimate $\hat{b}_0$ based on N time-series samples, we wish to continuously update the estimate as fresh input and output data samples are received, as might be the case in flow forecasting applications.

To develop the recursive algorithm, note that we can write (6.4.8) after k samples as

$$\hat{b}_{0,k} = p_k q_k \qquad\qquad (6.4.9)$$

where,

$$p_k^{-1} = \sum_{i=T+1}^{k} u_{i-T}^2 = p_{k-1}^{-1} + u_{k-T}^2 \tag{6.4.10}$$

and,

$$q_k = \sum_{i=T+1}^{k} y_i u_{i-T} = q_{k-1} + y_k u_{k-T} \tag{6.4.11}$$

while $\hat{b}_{0,k}$ denotes the least squares estimate of b_0 obtained on the basis of k samples.

Rearranging (6.4.10), we obtain

$$p_{k-1} = p_k + p_k u_{k-T}^2 p_{k-1} \tag{6.4.12}$$

$$p_{k-1} u_{k-T} = p_k u_{k-T} + p_k u_{k-T}^3 p_{k-1}$$

$$= p_k u_{k-T}\left[1 + p_{k-1} u_{k-T}^2\right]$$

so that,

$$p_{k-1} u_{k-T}\left[1 + p_{k-1} u_{k-T}^2\right]^{-1} = p_k u_{k-T}.$$

Now, multiplying by $p_{k-1} u_{k-T}$ and using (6.4.12) we find that

$$p_{k-1}^2 u_{k-T}^2\left[1 + p_{k-1} u_{k-T}^2\right]^{-1} = p_k u_{k-T}^2 p_{k-1} = p_{k-1} - p_k$$

Consequently,

$$p_k = p_{k-1} - p_{k-1}^2 u_{k-T}^2\left[1 + p_{k-1} u_{k-T}^2\right]^{-1} \tag{I(2)}$$

Substituting this result in (6.4.9) and using (6.4.11)

$$\hat{b}_{0,k} = \left\{p_{k-1} - p_{k-1}^2 u_{k-T}^2\left[1 + p_{k-1} u_{k-T}^2\right]^{-1}\right\}\left\{q_{k-1} + y_k u_{k-T}\right\}$$

Noting that $\hat{b}_{0,k-1} = p_{k-1} q_{k-1}$, this expression can be expanded to yield

$$\hat{b}_{0,k} = \hat{b}_{0,k-1} + K_k\{y_k - \hat{b}_{0,k-1} u_{k-T}\} \tag{I(1)}$$

where,

$$K_k = P_{k-1}u_{k-T}\left[1 + P_{k-1}u_{k-T}^2\right]^{-1} \qquad I(3)$$

Now, I(3) can be written as

$$K_k = \left[P_k P_k^{-1}\right]P_{k-1}u_{k-T}\left[1 + P_{k-1}u_{k-T}^2\right]^{-1}$$

$$= P_k\left[P_{k-1}^{-1} + u_{k-T}^2\right]P_{k-1}u_{k-T}\left[1 + P_{k-1}u_{k-T}^2\right]^{-1}$$

$$= P_k\left[u_{k-T} + u_{k-T}^2 P_{k-1}u_{k-T}\right]\left[1 + P_{k-1}u_{k-T}^2\right]^{-1}$$

Therefore,

$$K_k = P_k u_{k-T} \qquad I(4)$$

so that an alternative to equation I(1) is the following,

$$\hat{b}_{0,k} = \hat{b}_{0,k-1} + P_k\{u_{k-T}y_k - \hat{b}_{0,k-1}u_{k-T}^2\} \qquad I(5)$$

where it will be noted that the term in parentheses {.} is the proportional negative gradient of the instantaneous form J_I of the cost function J, i.e. if

$$J_I = \left[y_k - b_{0,k-1}u_{k-T}\right]^2$$

then

$$\frac{\partial J_I}{\partial \hat{b}_{0,k-1}} = -2\left[u_{k-T}y_k - \hat{b}_{0,k-1}u_{k-T}^2\right]$$

Consequently I(4) can be written as

$$\hat{b}_{0,k} = \hat{b}_{0,k-1} + P_k g_k \qquad I(6)$$

where g_k is the negative gradient of J_I. Thus both I(5) and its equivalent I(1) can be recognised as discrete stochastic gradient (or stochastic approximation) algorithms as discussed, for example, by Wilde (1964) and Young (1984).

The major computational equations in the recursive algorithms (I) are I(1), I(2) and I(3); and I(5) and its equivalent I(6) are useful mainly for indicating the nature of the algorithmic solution.

Equation I(1) also has a nice physical interpretation, for we see that the term

$$\hat{y}_{k/k-1} = \hat{b}_{0,k-1} u_{k-T}$$

can be considered as an _a priori_ prediction of the observation y_k at the kth sampling instant, given the estimate $\hat{b}_{0,k-1}$ at the previous, (k-1)th, instant and the new input measure u_{k-T}. Thus the term in the parenthesis {.} in I(1) can be interpreted as the prediction error, $y_k - \hat{y}_{k/k-1}$, and we see that the algorithm adjusts the estimate on the basis of this latest prediction error.

The physical nature and performance of algorithm I is demonstrated by the properties of the gain term K_k. It can be seen from I(4) and the definition of p_k that K_k is a strictly decreasing function of the sample size: as a result, if p_0 (and, therefore, K_0) are initially selected to be large, then the algorithm takes considerable notice of the prediction error, which is most likely due to error in the initial estimate of b_0, and adjusts this initial estimate accordingly. On the other hand, as more samples are processed and the estimate improves, so the algorithmic gain K_k reduces and less and less notice is taken of the prediction error, which is more likely to be due to the stochastic influence, e_k.

6.5 Recursive estimation of general linear time-series models

The simple recursive least squares algorithm I can be practically useful, but (6.4.6) is obviously a rather simple model with rather restricted application potential. The algorithm is more useful in demonstrating the nature of recursive estimation and for indicating how more complex and real problems can be solved in a recursive manner.

Consider, for example, the model (6.4.4) and, initially, assume that η_k is white noise, i.e.

$$y_k = \underline{z}_k^T \underline{a} + e_k$$

A recursive least squares estimate $\hat{\underline{a}}_k$ of the parameter vector $\underline{a}$ can then be obtained by considering the least squares cost function.*

$$J = \sum_{i=T+m+1}^{N} \left[y_i - \underline{z}_i^T \hat{\underline{a}}_i \right]^2 \qquad (6.5.1)$$

The recursive algorithm is developed by following a similar procedure to that used in the previous section, but being careful to remember that vectors and matrices are involved and that these require close

---

* Strictly this applies for $T + m > n$; summation lower limit would be $i = n + 1$ for $T + m < n$.

adherence to the laws of matrix algebra. The resultant algorithm takes the form (Young 1984, Young 1974)

$$\hat{\underline{a}}_k = \hat{\underline{a}}_{k-1} + P_{k-1}\underline{z}_k \left[1 + \underline{z}_k^T P_{k-1}\underline{z}_k\right]^{-1} \{y_k - \underline{z}_k^T \hat{\underline{a}}_{k-1}\} \qquad \text{II(1)}$$

$$P_k = P_{k-1} - P_{k-1}\underline{z}_k \left[1 + \underline{z}_k^T P_{k-1}\underline{z}_k\right]^{-1} \underline{z}_k^T P_{k-1} \qquad \text{II(2)}$$

where now P_k is an $(n + m + 1) \times (n + m + 1)$ square, symmetric matrix which is defined as the inverse of the matrix C_k where,

$$C_k = \sum_{i=T+m+1}^{k} \underline{z}_i \underline{z}_i^T \qquad (6.5.2)$$

C_k is a "cross product" matrix since its elements represent the sums of squares (along the diagonal) and cross-products (off diagonal) of the elements of $\underline{z}_k$. This is clear if we consider the case where $n = 1$, $m = 1$, $T = 0$; then

$$C_k = \sum_{i=2}^{k} \begin{bmatrix} -y_{i-1} \\ u_{i-1} \end{bmatrix} \begin{bmatrix} -y_{i-1} & u_{i-1} \end{bmatrix}$$

so that

$$C_k = \begin{bmatrix} \Sigma y_{i-1}^2 & -\Sigma y_{i-1} u_{i-1} \\ -\Sigma y_{i-1} u_{i-1} & \Sigma u_{i-1}^2 \end{bmatrix}$$

Because of this interpretation of P_k, II(2) is often termed the "matrix inversion lemma".

The similarity between the vector-matrix algorithm II and its scalar equivalent algorithm I should help the reader to understand both the development and functioning of the more complex algorithm. A detailed discussion of the algorithm is available, however, in reference (Young 1984) and the interested reader should consult this or reference (Young 1974) for further information. Here, it will suffice to say that the algorithm II can be interpreted in exactly the same way as I. Also it is worth noting that the assumption about the statistical nature of e_k can be used to extend the algorithm into a more obvious statistical form. In particular, it can be shown (Young 1984) that the covariance matrix P_k^* associated with estimate $\hat{\underline{a}}_k$ is defined by

$$P_k^* = \sigma^2 P_k \qquad\qquad (6.5.3)$$

Here the covariance matrix is defined statistically as

$$P_k^* = E\{\tilde{\underline{a}}_k \tilde{\underline{a}}_k^T\}$$

where $\tilde{\underline{a}}_k = \hat{\underline{a}}_k - \underline{a}$ is the estimation error after k samples.

The covariance matrix P_k^* for $k = 1,2,\dots,N$ provides a continuously updated measure of the confidence associated with the estimate $\hat{\underline{a}}_k$: the smaller the elements of P_k^*, the better the definition of the estimate and the greater the confidence the analyst can have in its value. Indeed, the square root of the diagonal elements provides a direct measure of the $\pm$ standard error associated with the estimates ($\hat{\underline{a}}_k$), i.e. the standard error of the ith element of $\underline{a}_k$ is defined as $\sqrt{(\sigma^2 (p_{ii})_k)}$, where $(p_{ii})_k$ is the ith diagonal element P_k.

This interpretation of P_k can be exploited algorithmically by substituting for P_k from equation (6.5.3) to yield the following least squares regression algorithm

$$\underline{a}_k = \underline{a}_{k-1} + P_{k-1}^* \underline{z}_k [\hat{\sigma}^2 + \underline{z}_k^T P_{k-1}^* \underline{z}_k]^{-1} \{y_k - \underline{z}_k^T \hat{\underline{a}}_{k-1}\} \qquad III(1)$$

$$P_k^* = P_{k-1}^* - P_{k-1}^* \underline{z}_k [\hat{\sigma}^2 + \underline{z}_k^T P_{k-1}^* \underline{z}_k]^{-1} \underline{z}_k^T P_{k-1}^* \qquad III(2)$$

This algorithm provides a continuously updated estimate $\hat{\underline{a}}_k$, together with an estimate P_k^* of its associated covariance matrix. In practice, however, it does require a (possibly updated) estimate $\hat{\sigma}^2$ of the white noise variance σ^2.

Unfortunately, despite the elegance and computational attractiveness of the recursive least squares algorithms II and III, they are not generally reliable as tools for the analysis of hydrological time-series (although in many practical circumstances, they can provide quite acceptable results, as we shall see later in the River Wyre example). The main limitation of the algorithms in practical terms arises because of the required assumption that the stochastic disturbance η_k can be considered as white noise, e_k.

Clearly η_k will not often be in the form of white noise: even if ξ_k is white noise, we see from (6.4.5) that η_k will be "coloured" because it is formed as a linear combination of past values of ξ_k. And if it is not white noise, then it can be shown that the estimate $\hat{\underline{a}}_k$ obtained from algorithm II or III will be <u>asymptotically biased</u> away from the true value of the parameter vector $\underline{a}$ (Young 1966), so that no matter how much data the analyst has available for estimation, he will never get an unbiased estimate of $\underline{a}$. The asymptotic bias is, however, a direct function of the noise level (i.e. the variance of ξ_k) and so acceptable estimates with only minor bias errors can be obtained by least squares analysis, if the noise level is low. This requires a well formulated model, which is able to closely explain the data.

In practical terms, it is more satisfactory to utilise recursive algorithms which do not have the limitations of the least squares equations. Probably the simplest and most robust algorithm of this type is based on the concept of "instrumental variables". The instrumental variable (IV) approach originated in the statistical literature (see e.g. Kendall and Stuart 1961) and it has since been developed into a recursive time-series form by the present author and others (Young 1984, Young 1966, Wong and Polak 1967, Young 1970, Young 1976, Young and Jakeman 1979, Jakeman and Young 1979, Soderstrom and Stoica 1983, Young, Jakeman and McMurtrie 1980).

The recursive IV algorithm is quite similar in form to the recursive least squares algorithm II. Its two constituent equations take the following form

$$\underline{\hat{a}}_k = \underline{\hat{a}}_{k-1} + \hat{P}_{k-1}\underline{\hat{x}}_k \left[1 + \underline{z}_k^T\hat{P}_{k-1}\underline{\hat{x}}_k\right]^{-1}\{y_k - \underline{z}_k^T\underline{\hat{a}}_{k-1}\} \qquad \text{IV(1)}$$

$$\hat{P}_k = \hat{P}_{k-1} - \hat{P}_{k-1}\underline{\hat{x}}_k \left[1 + \underline{z}_k^T\hat{P}_{k-1}\underline{\hat{x}}_k\right]^{-1} \underline{z}_k^T \hat{P}_{k-1} \qquad \text{IV(2)}$$

where $\hat{P}_k$ is the inverse of the instrumental cross-product matrix (IPM) $\hat{C}_k$ and

$$\hat{C}_k = \sum_{i=T+m+1}^{k} \underline{\hat{x}}_i\underline{z}_i^T \qquad (6.5.4)$$

$\hat{C}_k$ here can be compared with C_k introduced earlier in (6.5.2). Clearly the major difference between algorithms IV and II is the introduction of the vector $\underline{\hat{x}}_k$ in IV. This is termed the instrumental variable (IV) vector, and it is defined as

$$\underline{\hat{x}}_k = \left[-\hat{x}_{k-1}, -\hat{x}_{k-2}, - \cdots -\hat{x}_{k-n}, u_{k-T}, \ldots, u_{k-T-m}\right]^T$$

where $\hat{x}_k$ is a variable generated within the algorithm so that it is as highly correlated as possible with the noise free output x_k but completely statistically independent of the noise ξ_k.

The mechanism used to generate $\hat{x}_k$ in the algorithm is quite simple:

the input u_k is passed through a model of the system based on the latest parameter estimates and the output of this model, which is an estimate of x_k, is the source of the instrumental variables, i.e.

$$\hat{x}_k = \frac{\hat{B}(z^{-1})}{\hat{A}(z^{-1})} u_{k-T} \tag{6.5.5}$$

where $\hat{B}(z^{-1})$ and $\hat{A}(z^{-1})$ are estimates of $B(z^{-1})$ and $A(z^{-1})$. The procedure for updating these estimates is outside the scope of this chapter but it is described fully in Young (1984) and associated references.

The effect of the IV modification is to remove the asymptotic bias from the estimates and to make the algorithm inherently robust, in the sense that it retains this good performance for a very wide variety of stochastic disturbances. Moreover, with further straightforward modifications, it is possible to obtain a "refined" IV algorithm that is statistically optimal in the sense that the parameter estimates are both consistent and asymptotically efficient (minimum variance). Once again, such optimum algorithms are discussed in Young (1984) and associated references. In this regard, it is interesting to note that the refined IV estimates of the Wyre model parameters were a little different from the ordinary IV estimates given in equation (6.2.7). As a result, the steady state gain for the model is modified from G = 3.56 to G = 3.36. The reader will notice the sensitivity of the G estimate to small changes in the parameter estimates, as noted previously. In this case, we might conclude that the refined IV estimates are more reliable than the ordinary IV estimates and accept G = 3.36 as a better estimate of G in this case.

6.6 Model structure (order) identification

Before we can make efficient use of the recursive estimation algorithms, it is necessary to establish the structural form of the model (6.4.2) i.e. the values of n, m and T. This problem is discussed by Young et al. (1980) who develop a method based on two statistics: a coefficient of determination R_T^2; and an error variance norm EVN, which are defined as follows:

$$R_T^2 = 1 - \frac{\sum\limits_{i}^{N} \hat{\xi}_i^2}{\sum\limits_{i}^{N} (y_i - \bar{y})^2} \tag{6.6.1}$$

$$EVN = \frac{1}{n + m + 1} \sum\limits^{n+m+1} \hat{P}_{kk}^* \tag{6.6.2}$$

Here $\hat{\xi}_i$ is the estimate of ξ_i obtained from the algorithm, i.e.
$\hat{\xi}_i = y_i - \hat{x}_i$; $\bar{y}$ is the mean value of y_i; and $\hat{p}^*_{kk}$ is the kth diagonal
element of the $\hat{P}^*_N$ matrix, where $\hat{P}^*_N = \hat{\sigma}^2 \hat{P}_N$ and $\hat{\sigma}^2$ is the estimated noise
variance. In the structure identification analysis, these two statistics
are computed for a whole range of plausible model structures (i.e.
various combinations of values for n, m and T).

In the normal manner, R_T^2 is a measure of model fit: if the model
explains the data well then $\hat{\xi}_i$ is small in relation to $(y_i - \bar{y})$ and R_T^2
approaches unity; if there is poor explanation of the data $\hat{\xi}_i$ tends
towards $(y_i - \bar{y})$ and R_T^2 tends to zero. The EVN, on the other hand, is
an indication of the overall (or average) variance of the parameter
estimates in the (m + n + 1)th order model. It is a sensitive indicator
of over-parameterisation, particularly in relation to the $A(z^{-1})$ polynomial: if the model has too many parameters, then it can be shown that
the instrumental product matrix $\hat{C}_N$ defined in (6.5.4) tends to singular-
ity and its inverse $\hat{P}_N$, as computed by the IV algorithm, tends to
increase sharply in value, with a consequent increase in the EVN. This
increase can often be several orders of magnitude and so it is usual to
quote the natural logarithm of EVN, i.e. ℓn EVN.

In practice, the analyst monitors both R_T^2 and ℓn EVN and chooses the
model which has the best combination of the two statistics: usually R_T^2
will have reached a 'plateau' level, with little further increase for
any increase in model order; while ℓn EVN will have a very low value in
relation to·that obtained for higher order models. A good example of
the efficacy of such a procedure is given in the next section where
different model structures are evaluated in relation to the River Wyre
example. Other examples are given in Young et al. (1980).

6.7   Flow modelling for the river Wyre

The time-series approach to river flow modelling discussed in previous
sections of this chapter has been applied successfully to numerous sets
of hydrological data from the United Kingdom, Australia and Europe. The
results for the River Wyre, South of Lancaster, are typical of those
obtained in most cases. The data for this example are in the form of 32
flow measurements made at two gauging stations over a period of 32 hours,
with a sampling period of one hour: the data are shown in Fig. 6.2.2.
The two gauging stations in question are located 20 km apart at Abbeystead
(upstream) and St. Michaels (downstream). It is clear from Fig. 6.2.2
that there is considerable inflow to the river between the gauging
stations, with the upstream hydrograph much smaller than that measured
downstream. We have already seen the effect of this on the steady-state
gain of the model in Section 6.2. Let us now look at the complete
estimation results for this example.

Table I. MICROCAPTAIN Results for River Wyre Flow Data

Model	no. of A parameters	no. of B parameters	time delay	R_T^2	ℓn EVN
1	1	1	2	0.997	−7.138
2	1	2	2	0.996	−4.817
3	2	1	2	0.995	−4.227
4	2	2	2	0.995	−2.696
5(Muskingum)	1	2	0	0.985	−3.706

Table I provides a summary of the MICROCAPTAIN (Appendix 1) results for a selection of different models, including the Muskingum model, which is designated model 5 in the Table. Model 1 with one 'A' parameter (a_1), one 'B' parameter (b_0) and a pure time delay $T = 2$ time periods (2 hours) is clearly the best identified model, with an EVN much smaller than for any of the other models. The coefficient of determination R_T^2 is about the same as that for models 2 to 4 (0.997) and somewhat better than the Muskingum model (0.985). In other words, all the models 1 to 4 fit the data well but model 1, with its superior parametric efficiency (3 parameters including T compared with 4 for the other models) and better defined parameter estimates, has clear advantages in statistical terms.

The superiority of model 1 in relation to the Muskingum model is emphasized in Fig. 6.7.1(a) to (e). Fig. 6.7.1(a) and (b) show part of the MICROCAPTAIN output for the best identified model 1 and compares the graphical results with those obtained for the Muskingum model 5. Model 1 has a noticeably better fit, although the Muskingum model still provides a reasonable explanation of the data. Indeed, it is quite likely that the model would have been considered acceptable if it had not been critically compared, in statistical terms, with the other candidates. Note that Fig. 6.7.1(b) shows the estimate $\hat{\xi}_k$ of the 'noise' ξ_k, i.e. that part of the data not explained by the model. We shall discuss the possibility of modelling this remaining "lack of fit" later in this section.

Fig. 6.7.1(c) to (e) show the recursive estimates for model 1 and the impulse responses for both model 1 and the Muskingum alternative. The recursive estimates converge rapidly and are very stable, indicating that they are well identified. In contrast, the b_1 estimate for model 2, as shown in 6.7.1(c), is slow to converge and tends to 'wander' about. This is typical of the recursive estimation results for an over-parameterised model.

Although the impulse responses ("one interval unit hydrographs") shown in Fig. 6.7.1(e) are not too meaningful in physical terms, since it is not easy to visualise an impulsive flow input or its effect on downstream flow, they do provide a good means of evaluating the estimated transient response characteristics of the models. The NMP behaviour is quite clear from the impulse response for the Muskingum model, with a well defined flow decrease following the positive impulse excitation and preceding the subsequent flow increase to the peak flow value after 2 hours. It is possible to interpret this aspect of the Muskingum model behaviour as providing a "Padé" approximation (Truxal (1955)) to the time-delay, but it seems unlikely that this was the primary motivation for its inclusion in the model. Rather the NMP characteristics are

```
          FINAL ITERATION RESULTS
       COEFF. OF DET.=.997109584
          LN EVN=-7.13874793
       H-R CRITERION=1.1899883
       OUTPUT MEAN=49.702436

    OUTPUT VARIANCE=911.230173

    NOISE VARIANCE=2.63383442

       PARAMETER ESTIMATES

    A(1)=-.748898965  (.0116385552)
    A(2)=.886216357  (.0381055721)
```

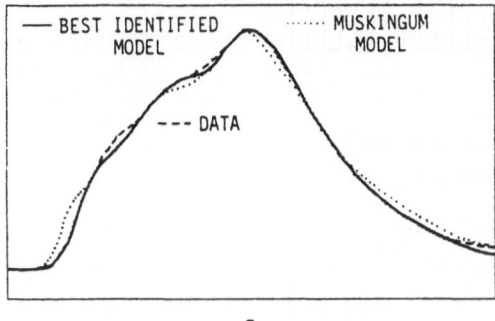

a

```
PLOT NOISE ESTIMATE ? :-YES
COPY REQUIRED ? YES
```

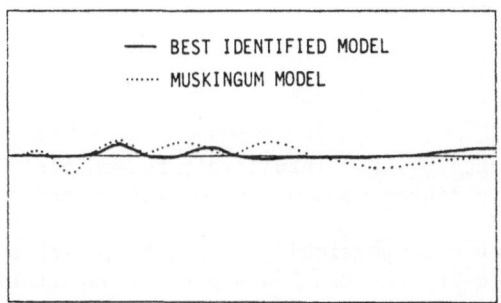

b

Fig. 6.7.1.

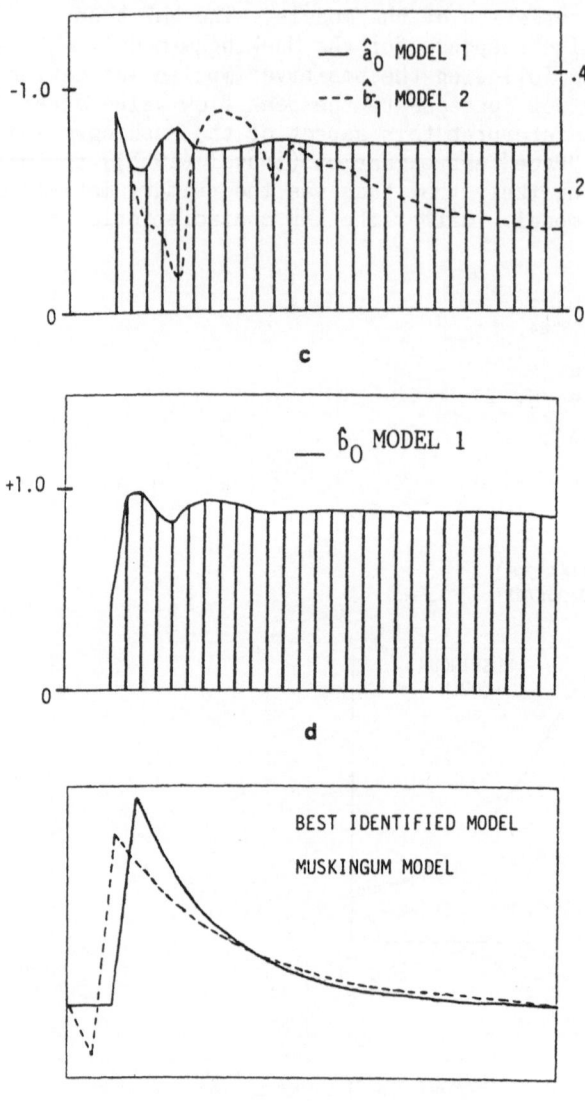

Fig. 6.7.1. contd.

probably an accidental consequence of the 'intuitively reasonable'
assumption that the storage is a linear combination of inflow and outflow.

The model 1 response is much more physically reasonable, with an
immediate increase in flow following the two hours pure transportation
time delay. Note that the areas under these unit hydrographs are quite
similar, as indicated by the comparable steady state gains for the two
models (3.53 and 3.63). But we would, of course, have more confidence

in the steady state gain and the time constant of the better identified model 1.

From a statistical standpoint, the conclusions from this example are quite unambiguous. The Muskingum model would be rejected as a theory of flow behaviour in favour of the alternative model 1. Also, since similar results have been obtained in many other applications, the objective analyst is forced to seriously question the Muskingum model, in statistical terms, despite its accepted success and popularity in flood routing terms; see NERC (1975).

But, to most hydrologists, model validity and credibility (Young 1982) are not just matters of statistics; they depend upon the practical experience of the user, which is rather more difficult to quantify. Perhaps the best conclusion, at this time, is that the Muskingum model should not be accepted without question. Rather its performance should always be compared with other competing models which, while not so popular heretofore, do seem to have superior dynamic characteristics and can be evaluated objectively using statistical methods. In this manner, practical experience of the time-series model alternatives will be gained and their relative merits, in relation to the Muskingum and other hydrologic models, will become more apparent.

This re-appraisal of the Muskingum model suggests that modifications to the routing procedure would be appropriate. For example, the unity gain form of the basic model (6.3.2) is overly restrictive and forces the analyst to introduce compensatory flows at node points (i.e. between the Muskingum type elemental reach models). Also the selection of the overall flood routing model structure (i.e. the number of Muskingum model elements used in series to represent a given length of river) could clearly be guided by the statistical structure identification results.

A practical alternative to the Muskingum flood routing approach would be to utilise the statistical based models discussed in this chapter in a digital simulation program, with the pure transportation time delays intentionally modelled by Padé approximations. In this way, the simulation problems introduced by variations in the transportation delay could be obviated and, in contrast to the Muskingum model, higher order Padé approximations could be utilised to minimise the 'negative' flow excursions associated with the first order approximation. This approach would combine the advantages of the stochastic modelling and Muskingum methods in a straightforward manner, whilst allowing the flood forecasting problem to be posed directly in statistical terms.

This statistical identification procedure is one of the advantages of the time-series approach. In the Wyre case, for example, there is no doubt that the best model, whether the time-series model 1 or the Muskingum-type model 5, has first order dynamics, i.e. the output flow at the kth sampling instant is a function of the output flow only at the previous (k - 1)th sampling instant. But this will not always be the case: if the sampling locations are far enough apart or, alternatively, if the flow is at a very low level, then the best identified model may be of higher order. Experience with these kinds of models suggests that, in such cases, the model $A(z^{-1})$ polynomial normally factorises into n real roots, where n is the order of $A(z^{-1})$. As a result, the model represents n "reaches" in series, each with first order dynamic characteristics.

The primary advantage of the recursive time-series approach is that the model order identification criteria will 'inform' the analyst of the higher order possibility by indicating that the higher order representation is the best identified model form. This kind of result has been particularly useful in the development of models for the dispersion of pollutants in river systems. Beer and Young (1983), for example, have shown that an aggregated dead-zone (ADZ) model, similar to the flow models discussed here, appears to have some advantages over the classical Fickian diffusion model.

We see from the above that statistical model identification provides an objective method of defining the number of reaches in a river system and, consequently, the number of first order model elements that are required in the complete flood routing model between any two spatial locations on the river. As such, it has certain potential attractions for the practitioner: it is not easy to decide on the number of reaches simply by observing the physical nature of the river and it is possible, therefore, to over-specify the model in these terms and finish up with a computer implementation which is more complex than it need be in order to adequately represent the flow dynamics. Objective statistical methods, such as those discussed in this chapter, can help the analyst to avoid these possibilities.

Of course, like any other method of statistical inference, the objective statistical procedure for defining model structure and order has its disadvantages. In particular, it is likely that different model structures (i.e. different numbers of elemental reach models in series) will be required under different flow conditions. The analyst will probably be advised, therefore, to choose an appropriate fixed model structure which makes reasonable sense under all flow conditions. Although this structure will undoubtedly be overly complex for some high flow conditions, it will provide a more practical routing model.

So far we have discussed only the relationship between u_k and x_k, where x_k is the part of the downstream flow which appears directly related (in a dynamic manner) to the upstream flow. If we want a complete stochastic-dynamic model we must also model the 'noise' ξ_k via its estimate $\hat{\xi}_k$. It is not possible to go into the details of such analysis here but methods for performing univariate time-series analysis of this type are discussed elsewhere by the present author (Young 1984, Young 1985). It will suffice to point out that this type of analysis indicates a model for ξ_k of the form,

$$\xi_k = \frac{1 + 0.28z^{-1}}{1 - 0.73z^{-1} + 0.26z^{-2}} e_k$$

where e_k is discrete white noise with estimated variance $\hat{\sigma}^2 = 1.06$. We discuss this kind of stochastic model later in Section 6.10.

Finally, it should be noted that, over a long flow record, it is likely that all the parameters in the hydrological model, whatever its form, will vary as functions of the flow magnitude. For example, it is to be expected that the time constant, steady state gain and the pure time delay (or equivalently the Muskingum model x parameter) will all be functionally related to the discharge Q. If the magnitude of the flow variations are small, then it could well be that a set of constant parameters could provide a reasonable forecasting model. If flow changes are large, however, this is unlikely and the model will need to be updated if acceptable flow predictions are to be obtained. In the next section, we discuss how the recursive estimation methods can be modified to allow for the estimation of parametric change and how, in this manner, it may be possible to evaluate the functional form of the relationships between the variations in the parameters and the changes in flow.

6.8 Time-variable parameter estimation

Implicit in the time-series model (6.4.4) is the assumption that the parameters in the vector $\underline{a}$ are time-invariant, i.e.

$$\underline{a}_k = \underline{a}_{k-1} \qquad\qquad (6.8.1)$$

If the assumption is not valid and the model parameters can vary over the observation interval spanned by the time-series data set, then it is necessary to modify the estimation algorithms in some manner. There are basically two methods of doing this: first, a "forgetting factor" can be built into the recursive algorithms, so that the most recent data are weighted more heavily and past data are gradually forgotten in some manner; second, the assumption (6.8.1) can be modified to allow for the possibility of parameter variations.

There are various procedures for introducing a forgetting factor or "fading memory" into the recursive algorithms (Young 1984) but the most popular involves weighting the data exponentially into the past. This approach can be considered in optimisation terms as the selection of those estimates which minimise a cost function of the form

$$J = \sum_{i=1}^{k} \left[y_i - \underline{z}_i^T \underline{\hat{a}}_i \right]^2 \gamma^{k-i} \; : \; \text{for } k = n \text{ to } N \qquad (6.8.2)$$

where $0 < \gamma < 1.0$ controls the nature of the exponential memory. The resulting recursive algorithm is similar in form to the normal recursive algorithms but now includes the factor γ, which can be constant or variable, i.e. γ_k (Young 1984). In the former case, the algorithm IV takes the following form

$$\underline{\hat{a}}_k = \underline{\hat{a}}_{k-1} + \hat{P}_{k-1}\underline{\hat{x}}_k \left[\gamma + \underline{z}_k^T \hat{P}_{k-1}\underline{\hat{x}}_k \right]^{-1} \left\{ y_k - \underline{z}_k^T \underline{\hat{a}}_{k-1} \right\} \qquad V(1)$$

$$\hat{P}_k = \frac{1}{\gamma}\{\hat{P}_{k-1} - \hat{P}_{k-1}\hat{x}_k[\gamma + z_k^T\hat{P}_{k-1}\hat{x}_k]^{-1}z_k^T\hat{P}_{k-1}\} \qquad \text{V(2)}$$

The effect of γ (or γ_k) is simply to prevent $\hat{P}_k$ from decreasing to zero, so ensuring that the algorithm always continues to modify $\hat{\underline{a}}_{k-1}$ in proportion to the error $\{y_k - z_k^T\hat{\underline{a}}_{k-1}\}$. In this manner, the effect of any changes in $\underline{a}$ will be detected and used to improve the estimate $\hat{\underline{a}}_k$ on a continuing basis. Other details of the algorithm and its use can be found in Young (1984).

A more flexible approach to the time-variable parameter estimation problem can be obtained by assuming that the condition (6.8.1) is modified to allow for temporal variations in the parameters. The simplest stochastic representation which allows for this is the following "random walk" (RW) model

$$\underline{a}_k = \underline{a}_{k-1} + \underline{\mu}_k, \qquad (6.8.3)$$

where $\underline{\mu}_k$ is vector of serially uncorrelated random variables with zero mean and covariance matrix Q, i.e.

$$E\{\underline{\mu}_k\} = 0; \qquad E\{\underline{\mu}_k\underline{\mu}_k^T\} = Q\delta_{kj}.$$

This model simply assumes that there will be random variations in all the parameters between samples and that the magnitude of and inter-relationship between the random variations is defined by Q. Typically the variations are assumed to be uncorrelated and so Q has a diagonal form, i.e.

$$Q = \begin{bmatrix} q_{11} & & & \\ & q_{22} & & \\ & & \ddots & \\ & & & q_{n+m+1} \end{bmatrix} \qquad (6.8.4)$$

The magnitude of the elements q_{ii} in (6.8.4) represents the expected variance of the perturbations in the ith parameter of $\underline{a}_k$. If they are all chosen to be equal to a constant scalar δ, i.e.

$$Q = \delta I,$$

where I is the unit diagonal matrix (i.e. with all diagonal elements unity) then the algorithm performs similarly to the exponential forgetting algorithm V. However, if the q_{ii} are chosen differently, it allows the analyst to inform the algorithm of different expected rates of variation in different parameters; for example, he can indicate constant parameters by setting the appropriate q_{ii} element to zero.

The recursive algorithm for the parametric variation model (6.8.3) is once again basically similar to the normal recursive algorithms but it is usually written a little differently. In the case of algorithm IV, for example, it takes the form

Prediction
between
samples

$$\hat{a}_{k/k-1} = \hat{a}_{k-1} \qquad\qquad\qquad VI(1)$$

$$\hat{P}_{k/k-1} = P_{k-1} + Q; \qquad\qquad\qquad VI(2)$$

correction
at kth
sample

$$\hat{a}_k = \hat{a}_{k/k-1} + \hat{P}_{k/k-1}\underline{\hat{x}}_k \left[\hat{\sigma}^2 + \underline{z}_k^T \hat{P}_{k/k-1}\underline{\hat{x}}_k\right]^{-1} \times$$

$$\times \left\{ y_k - \underline{z}_k^T \hat{a}_{k/k-1} \right\} \qquad\qquad VI(3)$$

$$\hat{P}_k = \hat{P}_{k/k-1} - \hat{P}_{k/k-1}\underline{\hat{x}}_k \left[\hat{\sigma}^2 + \underline{z}_k^T \hat{P}_{k/k-1}\underline{\hat{x}}_k\right]^{-1} \underline{z}_k^T \hat{P}_{k/k-1}, \quad VI(4)$$

where $\hat{a}_{k/k-1}$ and $\hat{P}_{k/k-1}$ are the _a priori predictions_ of $\hat{a}_k$ and $\hat{P}_k$ at the kth sampling instant, based on the assumed model form (6.8.3). This exposes nicely the Bayesian character of the algorithms (Bryson and Ho 1969) with the prior assumed knowledge of the parameter variation being used to make prior predictions of the variations; and then the new information, in the form of the latest data, being used to provide _a posteriori corrections_ to these predictions. Note that the algorithm is, however, still quite simple: the only modification in real terms to algorithm IV is the addition of Q to the $\hat{P}_{k-1}$ matrix at each recursion. As in the case of the forgetting factor γ, this simply prevents $\hat{P}_k$ from getting too small and allows for continual monitoring of the error term $\left\{ y_k - \underline{z}_k^T \hat{a}_{k/k-1} \right\}$.

More complex and sophisticated time-variable parameter estimation algorithms than VI can be developed (Young 1984) but the algorithm, as it stands, is attractive because of its simplicity and limited require-ments as regards prior knowledge about the nature of the parameter variations. In a practical sense, the algorithm can be used in two major ways. First, it can provide the basis for self adaptive forecas-ting, where the model parameters are continually updated so as to ensure that flow forecasting is always based on the most up-to-date model infor-mation. Second, the algorithm provides a means of identifying and estimating nonlinear characteristics in the time-series data. We shall consider the problem of forecasting subsequently in Sections 6.10 and 6.11. The next section, however, is concerned with a hydrological example where non-linear behaviour is extremely important: namely, char-acterising certain aspects of the hydrodynamic behaviour of an estuarine system.

6.9 <u>Salinity variations in the Peel Inlet-Harvey Estuary Western
     Australia</u>*

There are a number of practical examples where time-variable parameter
estimation has proven useful for the investigation of non-linear phenomena. In the present context, for example, the modelling of rainfall-
flow behaviour in the Bedford-Ouse river system of the U.K. is particularly pertinent and is described by Whitehead <u>et al</u>. (1979). Here,
however, we will consider another example which is interesting in hydrodynamic terms; namely, the identification of a suitable model for
characterising the weekly changes of salinity in the Peel Inlet-Harvey
Estuary system of Western Australia. The research outlined here was
carried out as part of a larger multidisciplinary study of the system
commissioned by Estuarine and Marine Advisory Committee (EMAC) of the
Department of Conservation and Environment, Western Australia, and it
has been described in detail by Humphries <u>et al</u>. (1981).

There have been many attempts at modelling estuarine dynamics
ranging from the very simple (e.g. Ellis <u>et al</u>. 1977) to the highly
esoteric (e.g. Smith 1980). In the Peel-Harvey Study, an intermediate
route was taken and the model was chosen so that it was capable of
describing the behaviour on a weekly time-scale, which was appropriate
to the Study objectives, but without the fine detail normally demanded
by classical hydrodynamic analysis.

The system was decomposed into seven zones or compartments, associated with the seven sampling sites, as shown in Fig. 6.9.1 and each
compartment was assumed to be well mixed in the sense that the sampled
salinity was considered representative of the salinity in the whole of
the compartment. Using a lumped parameter, ordinary differential
equation modelling approach (rather than the more usual distributed
parameter, partial differential equation representation) it is easy to
formulate the following mass conservation equation,

$$\frac{d(VS)}{dt} = Q_i S_i - QS \tag{6.9.1}$$

| rate of change | mass | mass |
| of mass | in | out |

where V is the reach volume, S the salinity in the well mixed compartment, Q the flow out of the compartment, Q_i the flow into the compartment, and S_i the salinity of the inflowing water. Clearly, in an estuary
"inflow" and "outflow" will be determined by factors such as riverine
inputs and tidal exchange characteristics. Note also that, because of
the simplicity of the representation and the great degree of aggregation
inherent in this particular mathematical description, we cannot assume
that Q, Q_i and V are directly related to equivalent <u>real</u> measurable
variables in the system. Rather they represent the <u>effective</u> values of
these variables appropriate to the simplicity of the formulation and the
degree of aggregation.

---

* This section has been reprinted with permission by Springer Verlag from
  <u>Uncertainty and Forecasting of Water Quality</u>, edited by M.B. Beck and
  G. van Straten.

This concept of effective rather than real variables is extremely important in systems analysis. The intent to represent the system mathematically in the simplest manner consistent with the objectives of the analysis means that model variables are often defined at the aggregate or macro-level. At this level, it is important not to necessarily consider the internal model variables in relation to equivalent variables defined at the micro-level; they must be considered within the context of the model formulation and at the level of aggregation appropriate to that formulation. But we will have more to say on this topic as the analysis proceeds.

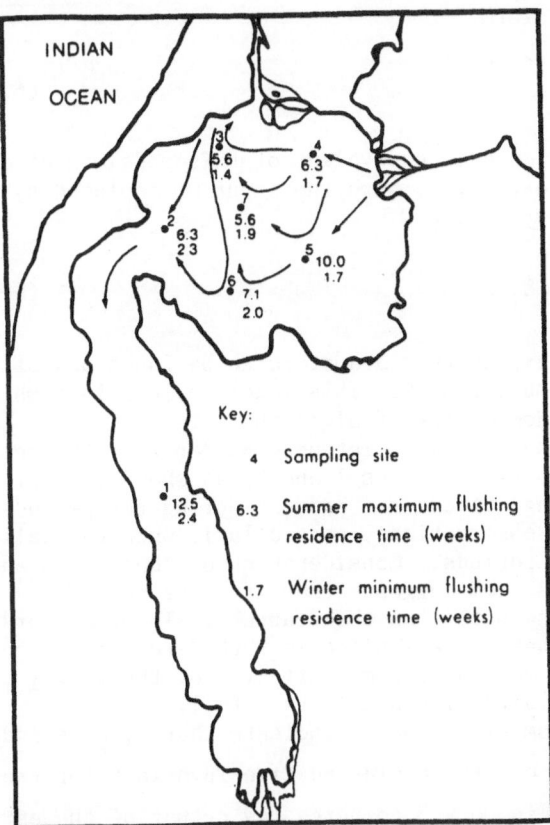

Fig. 6.9.1.   Circulation directions and flushing residence times in Peel Inlet and Harvey Estuary

Since both the salinity S and the volume V are time variable quantities, it is necessary to expand the left hand side of (6.9.1) to yield

$$V\frac{dS}{dt} + S\frac{dV}{dt} = Q_i S_i - QS$$

or,

$$V\frac{dS}{dt} = -\left(Q + \frac{dV}{dt}\right)S + Q_sS_i$$

so that the rate of change of salinity, under the assumption of complete mixing, is given by

$$\frac{dS}{dt} = -\left\{\frac{Q}{V} + \frac{1}{V}\frac{dV}{dt}\right\} S + \frac{Q_i}{V} S_i \qquad (6.9.2)$$

This is a first order ordinary differential equation with <u>time variable coefficients</u> of the general form

$$\frac{dS}{dt} = -\alpha(t)S + \beta(t) S_i \qquad (6.9.3)$$

The discrete-time, sampled data equivalent of this model is of a first order form, similar to (6.2.3), with the input Q_1 replaced by S_i and the output Q_2 replaced by S, i.e.

$$S_k = -a_{1,k}S_{k-1} + b_{0,k}S_{i,k-1} \qquad (6.9.4)$$

Note, however, that now we expect the parameters to be functions of time, so that they are given the subscript k. This model is in a form which can be investigated by the use of the IV algorithm VI.

In order to exemplify the analysis, let us consider only the relationship between salinities measured at sites 1 and 2, as shown in Fig. 6.9.2. Obviously dynamic lag effects are present during the period covered by the 104 weeks of observations, particularly when the salinity is rising during the Summer periods. Consideration of these lags and the physical nature of the system suggests that it makes sense to consider site 2 as the "input" and site 1 as the "output". In other words, although water movement probably takes place in both directions, the <u>aggregate</u> effect of all the various dynamic effects, <u>on the weekly time-scale</u>, is dominantly in the direction site 2 to site 1.

The results obtained from Algorithm VI indicate that $a_{1,k}$ is indeed time variable but that $b_{0,k}$ can be considered time-invariant for the purpose of this analysis. Fig. 6.9.3 compares the output of the estimated model with the observed salinity at site 1, and we see that the data are explained rather well, with the residual series having zero mean value and showing only minor correlation. Fig. 6.9.4 shows the recursive estimate $\hat{\alpha}(t)$ of $\alpha(t)$, as obtained by reference to the recursive estimate $a_{1,k}$ of $a_{1,k}$. The dotted sinusoidal curve indicates that the estimated variation is dominated by a periodic component with a period of one year. Considerable fluctuations about this sinusoid occur, however, particularly during weeks 1-12, 48-72 and 96-104: these periods correspond to Winter periods in Western Australia when fluvial inputs to the system are dominant, as shown in Fig. 6.9.5.

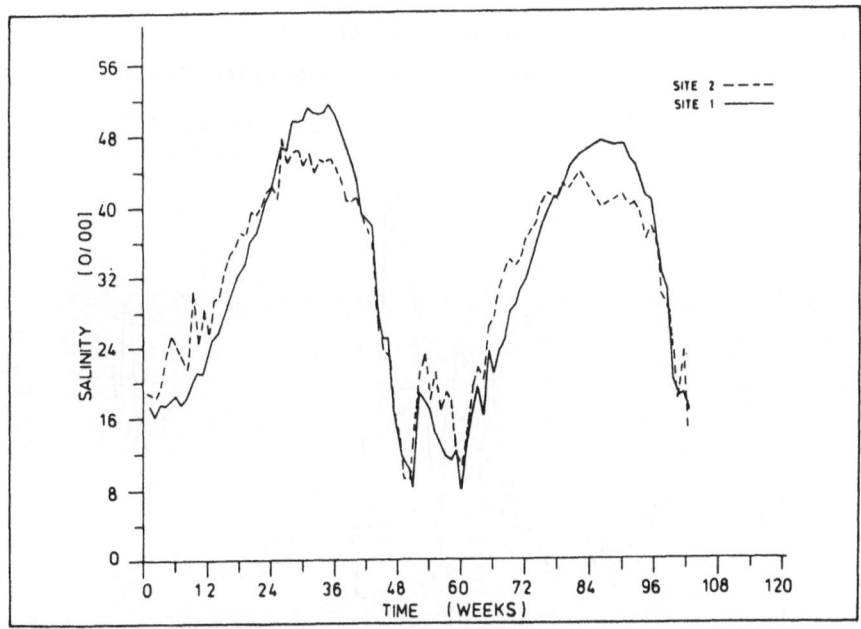

Fig. 6.9.2.   Comparison of salinity concentration at two sites

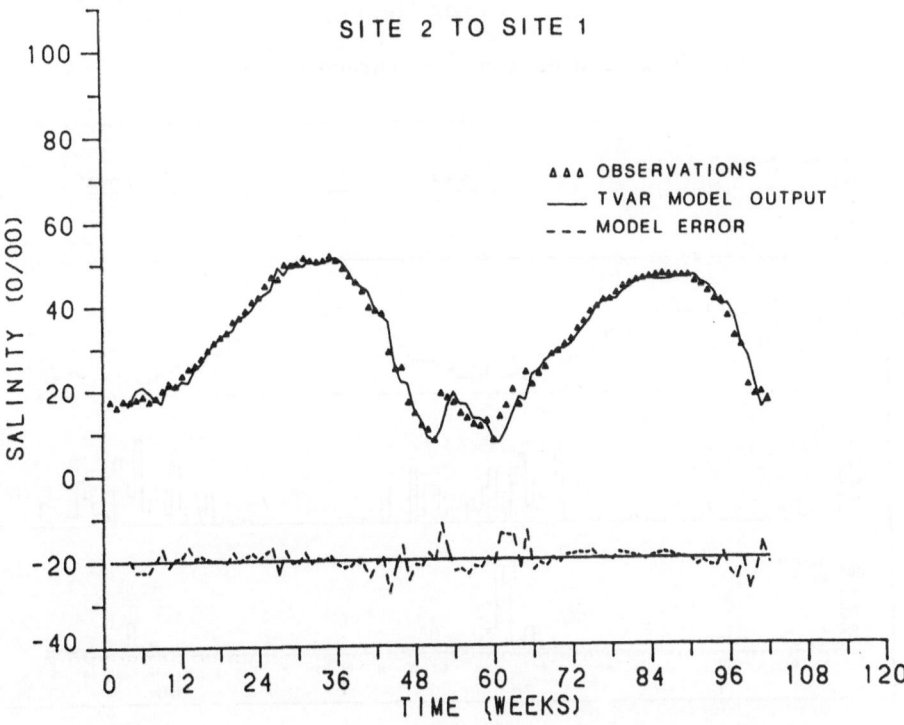

Fig. 6.9.3.   Comparison of model estimations and observations

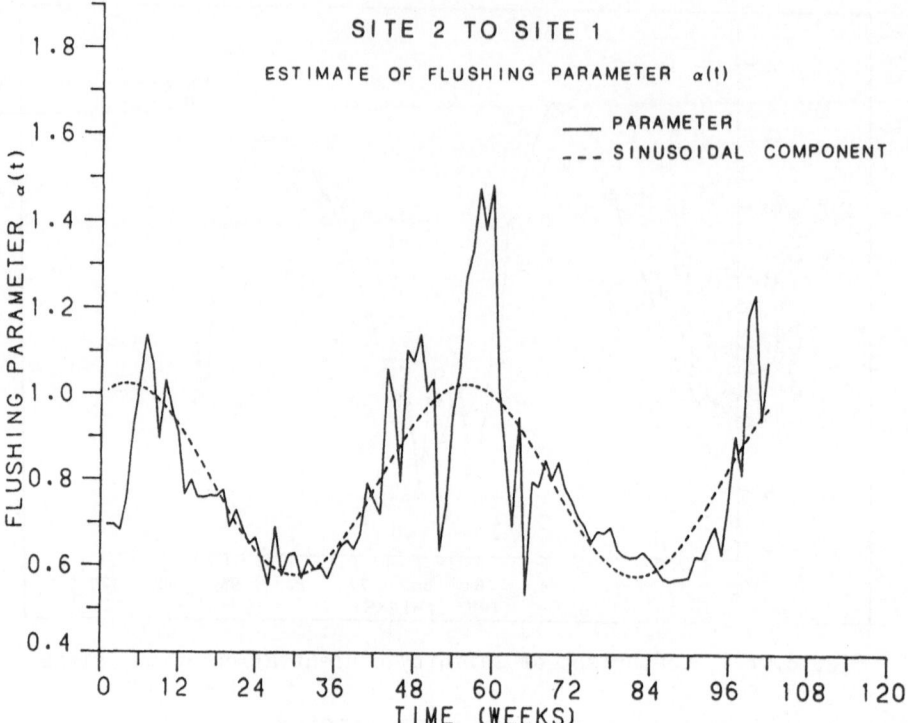

Fig. 6.9.4.   Recursive estimate of parameter α(t).

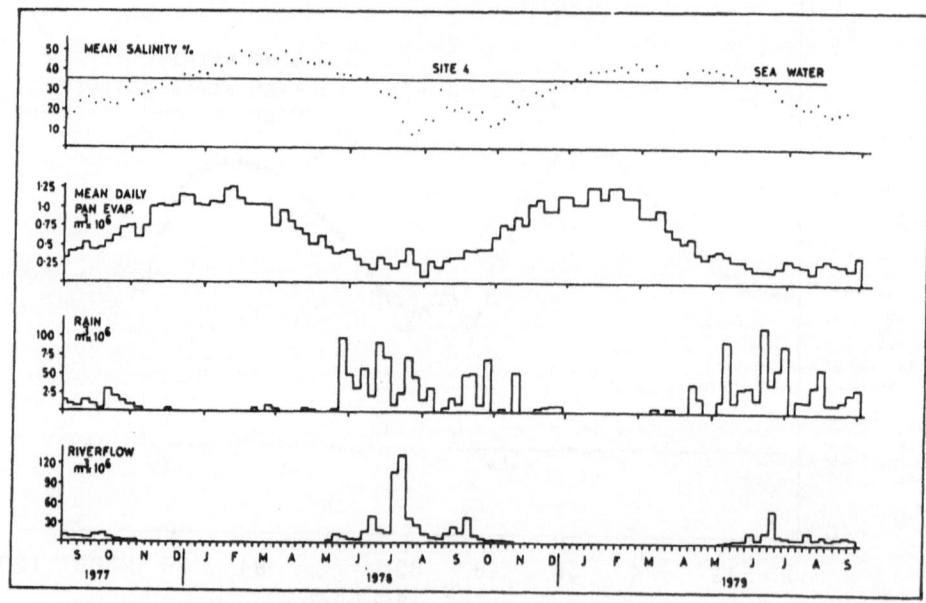

Fig. 6.9.5.   Data for the Peel Harvey study

Bearing in mind Equation (6.9.2), these results make good sense: changes in volume will occur because of periodic evaporation changes, seasonal rainfall effects and the differences in tidal height between the compartments. Put mathematically, the small perturbation equation can be written

$$\frac{dV}{dt} = \frac{d(Ah)}{dt} = \frac{h_i - h}{R} + Q_R - Ae$$

or

$$\frac{dh}{dt} = - \left[\frac{1}{AR} + \frac{1}{A}\frac{dA}{dt}\right]h + \frac{1}{AR}h_i + \frac{1}{A}Q_R - e \qquad (6.9.5)$$

where h and A are, respectively, the depth of water in and the surface area of the compartment appropriate to the definition of V; h_i is the depth of water associated with the input location and measured with respect to the same height datum as h; Q_R is the river flow; e the effective evaporation (evaporation minus rainfall), and R a "reservoir coefficient".

The identification analysis suggests that (6.9.2) and (6.9.5) provide a reasonable a priori model structure in this case and it would be interesting to pursue the analysis further on this basis. However, in relation to the Study objectives (and given the usual time restrictions on any practical Study), this did not prove necessary. The estimated variation of $\alpha(t)$, in itself, provides sufficient information both to assess the overall nature of the flushing dynamics and to help in the evaluation of nutrient budgets, as required by the Study objectives. Fig. 6.9.1, for example, shows the estimated maximum and minimum flushing times (obtained by performing the above analysis at each site in turn) together with inferred circulation patterns: the details of this analysis are given in Humphries, Young and Beer (1981). Fig. 6.9.6 is a plot of the innovations series (i.e. observed - predicted nitrogen (N) load) obtained in subsequent nutrient budget analysis which made use of the flushing information to estimate Ocean exchange of nutrients. The fact that this series has zero mean, serially uncorrelated characteristics is a further, independent check on the efficacy of the analysis. The two large transient deviations in the innovations series in July (negative) and November (positive) can be accounted for by biological activity in the Estuary; the negative deviation is probably due to apparent gross sedimentation of inorganic N from the water column by a phytoplankton bloom during the Winter riverine enrichment of the estuarine water column; the positive deviation occurred during a massive nodularia bloom which fixed about 270 tonnes of N in the Estuary.

Finally in this example, it should be stressed that the analysis does not represent a complete modelling exercise; rather it is the "identification" stage of a statistical evaluation of the data. The analysis has, effectively, generated a plausible hypothesis about the nature of the dynamic relationship between salinities in the estuary (and, therefore,

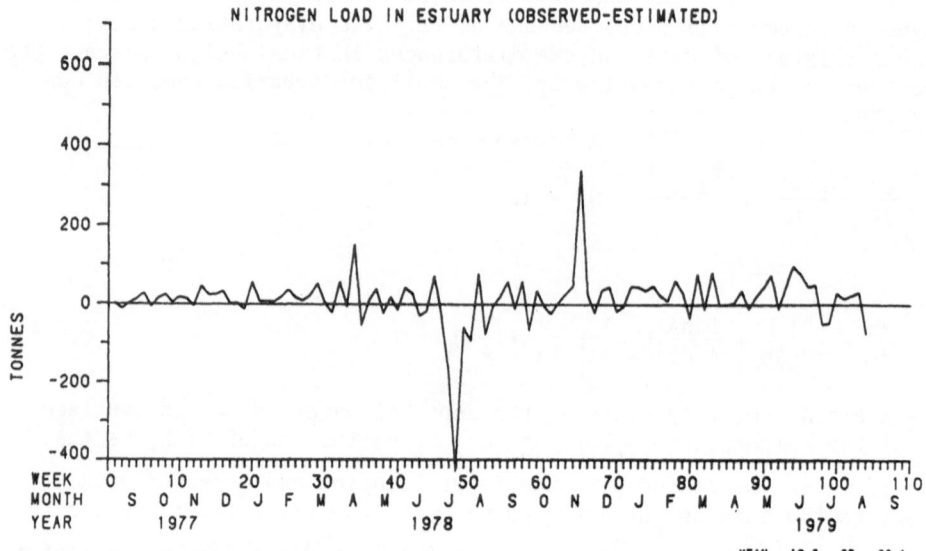

Fig. 6.9.6. Model residuals.

the underlying hydrodynamic behaviour) in the form of a simple differential equation with parameters that are functions of other variables
(evaporation and input river flows). Further research is now required
to completely test out this hypothesis, and so try to obtain a fully
estimated and validated model which would be useful for forecasting
purposes, on the weekly time scale appropriate to the data base.

6.10 Time-series analysis and flow forecasting

In their book "Time-Series Analysis Forecasting and Control", Box and
Jenkins (1970) show how stochastic models of the type discussed in this
chapter can be used for forecasting time-series. In order to utilise
their approach to forecasting, however, it is necessary to extend the
model (6.4.2) by making certain assumptions about the statistical nature
of the stochastic disturbance ξ_k. In particular, ξ_k is assumed to be a
stochastic sequence with <u>rational spectral density</u>. In the present
context, this simply means that it is generated from a discrete white
noise sequence e_k by a transfer function composed of the ratio of two
rational polynomials in z^{-1}. A simple example of this kind of model
was introduced in Section 6.7. In its most general form, such an auto-
regressive moving average (ARMA) model takes the form,

$$\xi_k = \frac{D(z^{-1})}{C(z^{-1})} e_k \qquad\qquad (6.10.1)$$

where it is usual to assume that,

$$C(z^{-1}) = 1 + c_1 z^{-1} + \ldots + c_p z^{-P}$$

and   (6.10.2)

$$D(z^{-1}) = 1 + d_1 z^{-1} + \ldots + d_q z^{-q}$$

while e_k is a zero mean, serially uncorrelated sequence with variance σ^2. (Clearly if $p = q = 0$, then $\xi_k = e_k$ and we have the situation considered earlier in Section 6.4, with the "coloured noise" ξ_k now simplified to uncorrelated white noise, e_k).

With these assumptions, it is possible to develop optimum recursive methods of identifying and estimating the complete time series model. Here all of the parameters characterising the polynomials $A(z^{-1})$, $B(z^{-1})$, $C(z^{-1})$ and $D(z^{-1})$ in the full model, i.e.

$$y = \frac{B(z^{-1})}{A(z^{-1})} u_{k-T} + \frac{D(z^{-1})}{C(z^{-1})} e_k \tag{6.10.3}$$

are estimated simultaneously and the resulting estimates are both statistically consistent and asymptotically efficient (minimum variance). Such methods are described fully elsewhere (Young 1976, Young and Jakeman 1979, Jakeman and Young 1979, Young 1984).

Sub-optimum versions of these optimum methods are also available for identifying and estimating the model (6.10.3) and these are often quite good enough for general day-to-day use. They were, in fact, used to analyse the Wyre data and the complete model in the form of equation (6.10.3) is as follows,

$$y_k = \frac{0.89z^{-2}}{1 - 0.75z^{-1}} u_{k-2} + \frac{1 + 0.28z^{-1}}{a - 0.73z^{-1} + 0.26z^{-2}} e_k; \qquad \hat{\sigma}^2 = 1.06$$

This is simply the combination of the 'deterministic' model (6.2.7) (with $Q_{2,k}$ and $Q_{1,k}$ represented by y_k and u_k respectively) and the noise model for ξ_k presented in Section 6.7.

It should be noted that the above stochastic model was estimated by the ordinary IV method and somewhat different results are obtained from the refined IV procedure (see earlier remarks). However, in the present "low noise" situation (i.e. practically the whole of the downstream flow explained by the upstream inputs) the results are probably 'good enough' for most practical purposes. Moreover, in this case, it may not be possible to justify the assumptions on the noise process ξ_k required by the optimal method (i.e. that ξ_k has rational spectral density).

The use of a model such as (6.10.3) for forecasting is quite straight-forward but outside the scope of this chapter. The interested reader is encouraged to consult either the Box-Jenkins book (1970) or other, simpler, texts, such as Chatfield (1975) to find the details of the forecasting procedures. Put simply, the model (6.10.3) provides infor-mation on two aspects of the time-series data: first, the relationship between the inputs u_k (upstream flows or rainfall inputs in the present context) and the measured output y_k (downstream flow); and, second, the statistical patterns characterising the stochastic disturbances, ξ_k. Thus, at sampling instant k, knowledge of upstream flows (u_k) can be used, via the model, to predict the noise-free or deterministic output of the system x_k (see Equation 6.4.1) a certain number of sampling inter-vals, say L, into the future. The most likely future effects of the stochastic disturbances ξ_k can also be predicted from a knowledge of the model (6.10.1) and these can then be used to modify the deterministic prediction and produce the final forecast for y_L. It is usual to denote this forecast by $\hat{y}_{L/k}$, where L is termed the <u>lead time</u>: in other words, $\hat{y}_{L/k}$ is the forecast of y_L based on the time-series data up to sampling instant k and knowledge of the model (6.10.3).

Since the model (6.10.3) is statistical in form, it is possible to do more than simply forecast future values of flow. In addition, a <u>confidence interval</u> can be computed and used to indicate the level of uncertainty associated with the forecast for any lead time L. If the model is well defined and the signal-noise ratio (i.e. the ratio of the variance of x_k to the variance of ξ_k) is large, then this confidence band will be quite small for reasonable lead times. However, if the noise level is high or the lead time is large, then the uncertainty can be very high and the analyst will need to use the forecast with care.

The magnitude of the pure time delay T is quite important in flow forecasting terms. If it is large, then advanced warning of flow changes is provided by the upstream measurements. Thus very accurate forecasts are possible for lead times less than or equal to T (i.e. $L \leq T$).

It we consider the River Wyre model, for example, then forecasting with a 2 hour lead time is straightforward: not only is the transporta-tion time delay 2 hrs but also the noise ξ_k is quite small. Thus a good (but sub-optimum) 2 hr ahead predictor $\hat{Q}_{2,k+2/k}$ can be obtained in the form (cf. Equation (6.2.7))

$$\hat{Q}_{2,k+2/k} = 0.75 \ Q_{2,k+1/k} + 0.89 \ Q_{1,k}$$

where

$$\hat{Q}_{2,k+1/k} = 0.75 \ Q_{2,k} + 0.89 \ Q_{1,k-1}$$

so that on substitution

$$\hat{Q}_{2,k+2/k} = 0.563\ Q_{2,k} + 0.89\ Q_{1,k} + 0.668\ Q_{1,k-1} \qquad (6.10.4)$$

Table II compares the predictions obtained with this equation with the measured flows. It should be noted that forecasting is not nearly so easy with the Muskingum model because of the absence of the pure time delay.

Table II. 2 hr. ahead forecasts of flow in River Wyre

Measured flow (m^3/sec)	2 hr. ahead forecast (m^3/sec)	forecast error (m^3/sec)
19	19	0
37	36	1
50	48	2
59	55	4
63	63	0
69	73	-4
77	78	-1
82	81	1
85	83	2
89	86	3
94	95	-1
99	102	-3
98	100	-2
92	92	0
82	82	0
72	72	0
62	62	0
53	53	0
45	45	0
39	39	0
33	33	0
29	29	0
26	25	1
23	22	1
21	20	1
19	18	1
18	16	2
16	15	1

For lead times longer than the pure time delay, however, it is necessary to forecast the upstream flow into the future in order to forecast subsequent downstream flow changes. This is clearly not an

easy task: sometimes rainfall measurements can be used to forecast the
upstream flows via rainfall-flow models (see Young and Whitehead (1975)
and Whitehead et al. (1979)) for the upper catchment; or, alternatively,
a time-series model for the upstream flow (e.g. a rational spectral
density description such as (6.10.1) can be developed by statistical
analysis of the upstream flow alone). These latter models are usually
termed "univariate" time-series representations. It is difficult to
generalise on this point, however, since each flow forecasting problem
will have its own special characteristics and these will need to be
taken into account in deciding upon the most appropriate statistical
forecasting procedure.

But whatever forecasting method is employed, it is possible to
improve its performance by using recursive estimation. It is unlikely,
in practice, if a single, time-invariant parameter such as (6.10.3) or
its equivalent will remain applicable over long periods of time. If the
incoming time-series data (rainfall, upstream and downstream flows etc.)
are continuously processed in a recursive manner, however, it is possible
to continually update the model parameters using a recursive algorithm
such as VI (or one of the many alternative algorithms that have been
suggested over the past few years, e.g. (Young 1981, Soderstrom 1973,
Ljung 1979)). In this manner, the forecast can always be made on the
basis of the most recently estimated model parameters and it should
benefit accordingly. Such self-adaptive or self-tuning flow forecasting
systems have not been used much in practice up to the present time, but
they hold the promise of improved statistical flow forecasting in the
future (e.g. Ambrus and Szollosi-Nagy in Unny and McBean (1982)).

Perhaps the major disadvantage of such systems is their complexity:
if all the parameters in the model (6.10.3) have to be recursively estimated on a continuous basis, then the resultant algorithms can be quite
complicated. And they can be even more complicated if, as might be
expected in practice, it proves necessary to use more complex "multivariable" models with several inputs and outputs (Jakeman and Young 1979).

In the next section, we consider other recursive approaches to the
problem which, while potentially less sophisticated than the self-
adaptive procedures discussed above, do have the advantage of relative
simplicity. These approaches are based on the most famous recursive
algorithm, the Kalman Filter.

6.11   Flow forecasting and the Kalman Filter

Recursive estimation, as discussed in this chapter, was first suggested
by Gauss at the beginning of the nineteenth century (see Appendix 2 of
Young 1984) and was rediscovered by Plackett in 1950. But recursive
estimation was first brought into real prominence by the systems theorist
R.E. Kalman in 1960. Kalman proposed a time-domain solution to the
optimal filtering problem posed some years earlier by Wiener (1949).
Since 1960, this optimal estimation algorithm has, almost universally,
been known as the "Kalman filter".

The Kalman filter is an algorithm for estimating the state variables
that characterise a linear, stochastic dynamic system on the basis of
measurements of the input and noisy output of the system. In fact we

have seen an algorithm which is very similar to the Kalman filter in algorithm VI. Here the model of the system is the simple random walk model (6.8.3) and the observation of the system is via the measurements y_k and u_k and the model of the system (6.4.4). In the Kalman filter, the system model is nominally more complex and usually takes the form of the following Gauss-Markov process

$$\underline{X}_k = \phi\underline{X}_{k-1} + \Gamma\underline{v}_{k-1} + \underline{\mu}_k \tag{6.11.1}$$

where $\underline{X}_k$ is the $(n \times 1)$ vector of state variables, $\underline{v}_k$ is an $(m \times 1)$ vector of known inputs, $\underline{\mu}_k$ is again a white noise vector with covariance matrix Q; and ϕ, Γ are two known (and possibly time-variable) matrices that define the dynamic characteristics of the system.

The observation equation, in contrast, is simpler than in the case of algorithm VI: it is assumed that the system is observed via a vector $\underline{y}_k$ which is related linearly to $\underline{X}_k$ by a known (but again possibly time-variable) matrix H, i.e.

$$\underline{y}_k = H\underline{X}_k + \underline{e}_k \tag{6.11.2}$$

where H is of dimension $p \times n$ and $\underline{e}_k$ is a $(p \times 1)$ white noise vector which is assumed statistically independent of $\underline{\mu}_k$ and has covariance matrix R. Usually p will be of lower dimension than n. Although (6.11.2) related the vectors $\underline{y}_k$ and $\underline{X}_k$ it is simpler than (6.4.4) in the sense that H is a matrix with exactly known elements, while the vector $\underline{z}_k$ in (6.4.4) is composed, in part, of noisy measurements. Thus while it is quite common to call algorithms such as VI "Kalman filter" algorithms, it is somewhat misleading and is not to be recommended.

The Kalman filter does, however, closely resemble algorithm VI. It is in the form of the following prediction-correction equations

prediction between samples
$$\hat{\underline{X}}_{k/k-1} = \phi\hat{\underline{X}}_{k-1} + \Gamma\underline{v}_{k-1} \qquad \text{VII(1)}$$

$$P_{k/k-1} = \phi P_{k-1}\phi^T + Q \qquad \text{VII(2)}$$

correction at kth sample
$$\hat{\underline{X}}_k = \hat{\underline{X}}_{k/k-1} + P_{k/k-1} H^T[R + HP_{k/k-1}H^T]^{-1}\{\underline{y}_k - H\hat{\underline{X}}_{k/k-1}\} \qquad \text{VII(3)}$$

$$P_k = P_{k/k-1} - P_{k/k-1} H^T[R + HP_{k/k-1}H^T]^{-1} HP_{k/k-1} \qquad \text{VII(4)}$$

where it can be shown that P_k is the covariance matrix associated with the estimation error $\underline{\tilde{X}} = \underline{\hat{X}} - \underline{X}$, i.e.

$$P_k = \{\underline{\tilde{x}}_k \underline{\tilde{x}}_k\} \tag{6.11.3}$$

Thus, at each recursion, both the estimate $\underline{\hat{x}}_k$ and the covariance matrix P_k are updated and the user is given not only an improved estimate $\underline{\hat{x}}_k$ but also an indication of confidence he can associate with this estimate.

In algorithm VII it is interesting to note that the term $H\underline{\hat{x}}_{k/k-1}$ in VII(3) can be considered an estimate $\underline{\hat{y}}_{k/k-1}$ of $\underline{y}_k$ i.e.

$$\underline{\hat{y}}_{k/k-1} = H\underline{\hat{x}}_{k/k-1} \tag{6.11.4}$$

The algorithm is, therefore, a forecasting "filter" in the sense that the observation $\underline{y}_k$ is filtered by the algorithm to produce the one step ahead forecast $\underline{\hat{y}}_{k/k-1}$ from which much of the observation noise e_k has been removed. In this sense, it is similar in function to the earlier Wiener filter, although it is algorithmically much different. But the algorithm also provides estimates of certain non-measured or hidden signals; namely the "state variables" of the system.

At first sight, algorithm VII seems to demand so much prior knowledge about the system (in the form of the model (6.11.1)) that it is difficult to see how it can be practically useful. This certainly can be a deterrent to its practical application but there are special situations where such difficulties can be circumvented, and practically useful algorithms can result.

To demonstrate how the Kalman filter can be used in flow forecasting, let us consider a typical hydrograph over an extended period of time. In general, we can denote such a measurement mathematically as a signal y_k of the form

$$y_k = T_k + S_k + e_k \tag{6.11.5}$$

where T_k are low frequency components (in other contexts, these are sometimes referred to as "trends"); S_k represents some periodic behaviour ("seasonal" components); and e_k are the remaining stochastic (or random) parts of the signal.

In Kalman filter terms, if we assume that e_k is white noise then (6.11.5) can be considered as equivalent to the observation equation (6.11.2) with p = 1. In order to use the Kalman filter, however, we need a model of the signal generation process; in other words a model for the variables T_k and S_k in the form of Equation (6.11.1). There are, of course, many ways of modelling these components in this form, but whatever we choose must be quite simple, in the sense that the matrices Φ and Γ must be <u>exactly known</u> in order to implement the filter VII.

The clue to tackling this problem lies in the simple random walk model (6.8.3) we have considered previously to model parameter variations.

It is clearly a simple model with $\phi = I$ and $\Gamma = 0$; and it also allows for quite wide variations in the variables ($\underline{a}_k$ in (6.8.3); $\underline{X}_k$ in the present K.F. context). T_k is easily modelled in this form: we can, for instance, assume simply that

$$T_k = T_{k-1} + \mu_k \qquad (6.11.6)$$

Or, alternatively, we might assume that it is a slightly more complex integrated random walk, i.e.

$$
\begin{aligned}
T_k &= T_{k-1} + U_{k-1} & \text{(i)} \\[2mm]
U_k &= U_{k-1} + \mu_{1,k} & \text{(ii)}
\end{aligned}
\qquad\qquad (6.11.7)
$$

where T_k in (i) is now the sum of the random walk variations in (ii).

This model should allow for even smoother and more wide-ranging changes in T_k, while still not requiring too much increased complexity. Equation (6.11.7) is easily put into the form of (6.11.1) by defining $\Gamma = 0$ and

$$
\phi = \begin{bmatrix} 1 & 1 \\ 0 & 1 \end{bmatrix}; \quad \mu_k = \begin{bmatrix} 0 \\ \mu_{1,k} \end{bmatrix}
$$

in other words,

$$
\begin{bmatrix} T_k \\ U_k \end{bmatrix} = \begin{bmatrix} 1 & 1 \\ 0 & 1 \end{bmatrix} \begin{bmatrix} T_{k-1} \\ U_{k-1} \end{bmatrix} + \begin{bmatrix} 0 \\ \mu_{1,k} \end{bmatrix} \qquad (6.11.8)
$$

The seasonal component S_k is not so straightforward to model in this form, although it is still possible in various ways (Young 1984). However, full discussion of this is outside the scope of the present chapter and it will suffice to mention that, in the most straightforward case, the model is chosen so that it is still quite simple but has inherent dynamic properties which allow for periodic phenomena. The simplest such model is the following

$$S_k = S_{k-\tau} + \mu_{2,k} \qquad (6.11.9)$$

which exhibits periodic behaviour with period τ sampling intervals. In the case $\tau = 2$, this model can be put in the form (6.11.1) by defining $\Gamma = 0$ and

$$\phi = \begin{bmatrix} 0 & 1 \\ 1 & 0 \end{bmatrix} \quad ; \quad \mu_k = \begin{bmatrix} \mu_{2,k} \\ 0 \end{bmatrix}$$

in other words,

$$\begin{bmatrix} S_k \\ S_{k-1} \end{bmatrix} = \begin{bmatrix} 0 & 1 \\ 1 & 0 \end{bmatrix} \begin{bmatrix} S_{k-1} \\ S_{k-2} \end{bmatrix} + \begin{bmatrix} \mu_{2,k} \\ 0 \end{bmatrix} \qquad (6.11.10)$$

It will be noticed that, once again, ϕ and Γ are very simple known matrices, as required.

It is now possible to consider the complete model of the signal generation process by combining the constituent Equations (6.11.8), (6.11.10) and (6.11.5) in the following manner

$$\underset{(\underline{X}_k)}{\begin{bmatrix} T_k \\ U_k \\ S_k \\ S_{k-1} \end{bmatrix}} = \underset{(\phi)}{\begin{bmatrix} 1 & 1 & 0 & 0 \\ 0 & 1 & 0 & 0 \\ 0 & 0 & 0 & 1 \\ 0 & 0 & 1 & 0 \end{bmatrix}} \underset{(\underline{X}_{k-1})}{\begin{bmatrix} T_{k-1} \\ U_{k-1} \\ S_{k-1} \\ S_{k-2} \end{bmatrix}} + \underset{\underline{\mu}_k}{\begin{bmatrix} 0 \\ \mu_{1,k} \\ \mu_{2,k} \\ 0 \end{bmatrix}} \qquad (6.11.11)$$

$$y_k = \underset{(H)}{\begin{bmatrix} 1 & 0 & 1 & 0 \end{bmatrix}} \underset{(\underline{X}_k)}{\begin{bmatrix} T_k \\ U_k \\ S_k \\ S_{k-1} \end{bmatrix}} + e_k$$

This is of the desired form for the application of the Kalman filter and it is straightforward to program the algorithm VII with $\underline{\hat{X}}_k = \begin{bmatrix} \hat{T}_k \hat{U}_k \hat{S}_k \hat{S}_{k-1} \end{bmatrix}^T$, $\Gamma = 0$ and the above definitions for ϕ and H. In this implementation, the matrix Q is the covariance of $\underline{\mu}_k$, i.e.

$$Q = E\{\underline{u}_k\underline{u}_k^T\} = \begin{bmatrix} 0 & 0 & 0 & 0 \\ 0 & q_{22} & q_{23} & 0 \\ 0 & q_{23} & q_{23} & 0 \\ 0 & 0 & 0 & 0 \end{bmatrix}$$

while R is a scalar σ^2 denoting the variance of e_k.

The Kalman filter programmed in this manner will generate continuously updated estimates of T_k, U_k, S_k and S_{k-1}. More importantly, however, it will provide forecasts of y_k into the future. For example, the one step ahead forecast at sampling instant $k + 1$ from the previous kth instant is given as $\hat{y}_{k+1/k}$ where (cf. (6.11.4) and VII(1) with $\Gamma = 0$)

$$\hat{y}_{k+1/k} = H\hat{x}_{k+1/k} \qquad\qquad (6.11.12)$$

with,

$$\hat{\underline{x}}_{k+1/k} = \phi\hat{\underline{x}}_k$$

If we wish to forecast more steps ahead, then it is necessary to apply the prediction equation VII(1) repeatedly <u>without intermediate updating</u> (since no future observations are available for updating at the kth sampling instant). In the present case with $\Gamma = 0$, for example,

$$\hat{\underline{x}}_{k+1/k} = \phi\hat{\underline{x}}_k$$

$$\hat{\underline{x}}_{k+2/k} = \phi\hat{\underline{x}}_{k+1/k} = \phi^2\hat{\underline{x}}_k \qquad\qquad (6.11.13)$$
$$\vdots$$
$$\hat{\underline{x}}_{k+L/k} = \phi^L\hat{\underline{x}}_k$$

where ϕ^L is easily computable because of the simple form for ϕ in (6.11.11).

The efficacy of this simple 'univariate' approach to flow forecasting will tend to depend upon the nature of the flow data. It is not likely to be sufficient on its own for the forecasting of hourly data, where it is likely that transfer function modelling, as discussed in previous sections of the paper, would be required. However, it might prove useful, in modified form, for forecasting upstream flows for a few hours ahead, if rainfall information was not available and rainfall-flow modelling could not be used. This statistical approach to forecasting can be compared with the deterministic method considered in Section 3.15

of Chapter 3. The simple approach considered there to predict the upstream input flow (i.e. linear extrapolation) is equivalent, in statistical terms, to the assumption that the upstream flow can be modelled
by the integrated random walk model given in equation (6.11.7). This is
discussed fully by Young (1984 p. 74-75). In such an application, it
would probably be necessary to change the model (6.11.11) by removing
the seasonal component model (which would not be relevant) and enhancing
the low frequency component (trend) model using, for example, a smoothed
random walk, a multiple integrated random walk, or some more complex
alternative (Young 1984).

However, the model (6.11.11) and its associated Kalman filter-based
forecasting scheme should be directly useful for application to long
term flow data, particularly where periodic behaviour predominates (e.g.
in large tropical rivers). Indeed, in such applications, it would
probably be worth enhancing the model (6.11.11) by including more sophisticated seasonal component models.

In general terms, the exact form of the model (6.11.11) is likely to
be determined by a mixture of experience and analysis. Any time-series
that we wish to model in this manner can be analysed using univariate
time-series modelling methods, such as those mentioned earlier in
Section 6.7 and described elsewhere by the present author and others
(Young and Jakeman 1979, Young 1985, Box and Jenkins 1970). Such analysis will often reveal patterns of behaviour that will assist in initially
prescribing the model (6.11.1). But final specification, as so often in
practical applications, is likely to depend upon the results of exhaus-
tive tests, where the filter parameters are 'tuned' in some manner in
order to obtain 'practically optimum' performance for the widest range
of flow conditions.

The simple KF forecasting procedure described above can, of course,
be made more complex and capable of handling more diverse multivariable
situations. For example, the simple model (6.11.11) can easily be
replaced by a more general stochastic model, as discussed by Young and
Wallis (1985). And, if necessary, the effects of other inputs (upstream
flows and rainfall) can be introduced by defining additional components
in (6.11.5) modelled by the kind of general linear model discussed in
previous sections (or multi-input, multi-output, versions of these).
Such KF-based forecasting can be very effective, as we see in Fig.
6.11.1 (Ng and Young, 1985), where the airline passenger data of Box
and Jenkins (1970) is forecasted 3.5 years ahead with good accuracy.

The Kalman filter is, of course, now very well known and it has
been applied successfully to many different problem areas from aerospace
to economics. Indeed, one of its most attractive features is its flexibility. But it has one major weakness: as we have seen it requires
complete information on the nature of the stochastic signal generation
process (i.e. the stochastic hydrologic model in the present situation).
Kalman himself recognised this limitation and indicated (Kalman 1960)
that "the two problems (state and parameter estimation) should be solved
jointly if possible". One way of obviating this difficulty, the so-
called Extended Kalman Filter (EKF), is discussed in the next section.
However, by using similar procedures of recursive parameter estimation
to those discussed in previous sections of this chapter, the Kalman

filter can be made self-adaptive by employing the recursive methods to continuously update the constituent models in the filter. Such an adaptive river forecasting procedure is adumbrated in Young (1979) and is currently under investigation at Lancaster (Ng and Young, 1985).

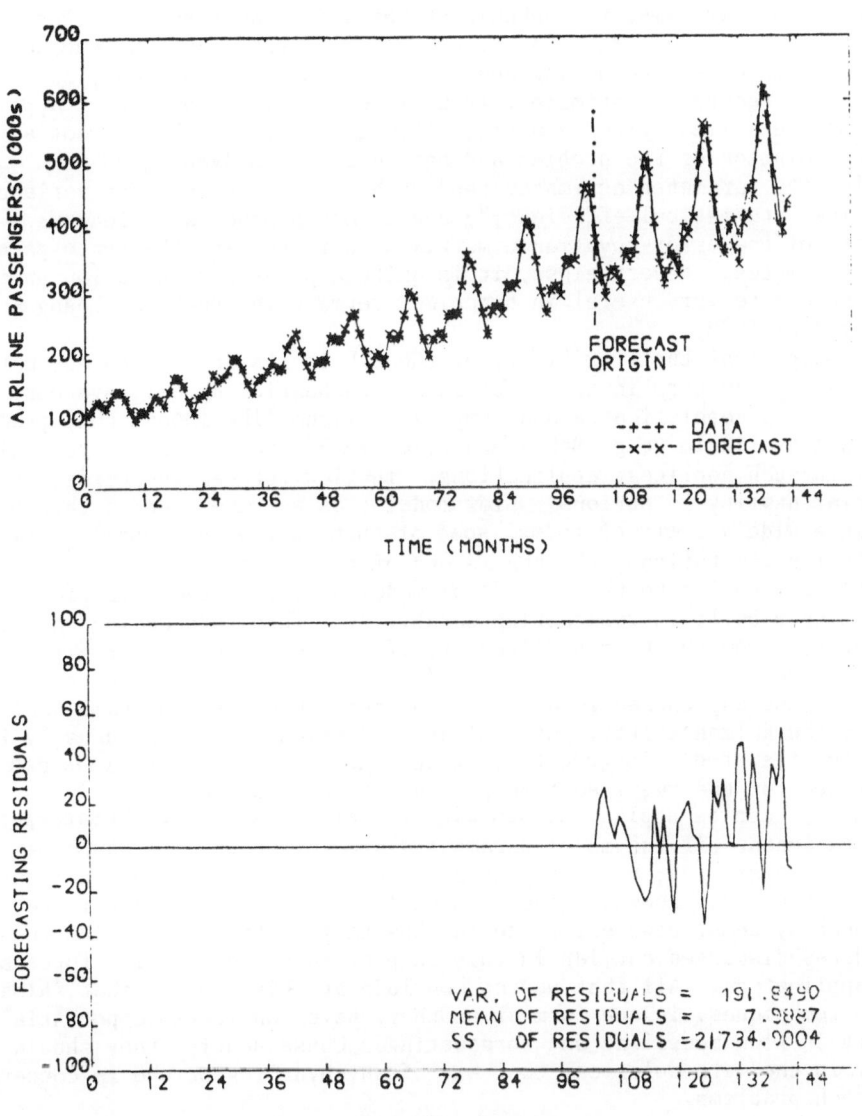

Fig. 6.11.1.   Recursive forecasting of a nonstationary time-series

6.12   The Extended Kalman Filter

Since the publication of Kalman's famous paper, various methods for

simultaneously estimating states and parameters in linear, stochastic
dynamic systems have been proposed, of which the most popular and flexible
is the "extended" or "re-linearised" Kalman filter (EKF) (e.g. Young
1974, Beck 1978). The EKF recognises the inherent nonlinearity of the
joint state-parameter estimation problem which arises because, even in
linear dynamic systems, the unknown states and parameters appear as
product terms in the system equations. Thus a process of successive re-
linearisation of the resulting non-linear estimation equations, about
the current recursive estimates, is utilised to allow for direct application of the Kalman filter equations (Young 1984). The EKF is not a
perfect solution of the problem and has several disadvantages; for
example, its parameter estimates tend to have quite high error variance
(i.e. low "statistical efficiency") and it may produce a misleading
estimate of the error-covariance matrix associated with the state-parameter estimates. Nevertheless, it is quite a powerful estimation procedure and can be very useful in practical terms (e.g. Beck and Young
1976, Todini 1978).

One aspect of the flexibility of the EKF is its ability to estimate
states and parameters in truly nonlinear stochastic, dynamic systems.
Although more sophisticated and complex "maximum likelihood" (ML) procedures are available (e.g. Mehra and Tyler 1973) the EKF has proven quite
useful in such nonlinear applications. In flow forecasting terms, its
potential utility is obvious: flow models, as we have seen in this book,
come in a wide variety of forms, some of them complex and non-linear.
Despite its limitations, the EKF is one of the few estimation procedures
that can be applied to these kinds of model while, at the same time,
possessing a built-in forecasting capability. This forecasting capability derives from the Kalman filter itself: as we have seen, the filter
involves prediction equations, which can be used directly to generate
forecasts for any chosen lead time. The EKF, which has the same basic
form as the Kalman filter, has similar prediction equations, usually in
the form of a prediction update in which the non-linear equations are
integrated for the required time period into the future.

The EKF can be applied to non-linear systems either as a state-parameter estimator (if the model parameters are not known or need to be
updated recursively), or simply as a state estimator (if the model parameters are assumed known). There is still not sufficient experience
with such systems, however, to decide how they might compare with the
procedures discussed earlier in this chapter in any given flow forecasting application. All that we can conclude at this time is that Kalman
filter techniques, in one form or another, have considerable potential
utility in flow modelling and forecasting. Consequently, they should
appear in the methodological tool kit of any hydrologist who is concerned
with such problems.

6.13 Conclusions

This tutorial chapter has introduced the concept of recursive estimation
and shown how recursive methods of time-series analysis can be applied
to both flow modelling and forecasting. It has not been possible to
deal with all aspects of the subject and the reader is urged to study

the many papers on recursive estimation that have appeared in the hydrology literature, and elsewhere, during the past ten years. Hopefully, the introduction to the subject provided here will aid in the assimilation of this extensive literature and will help to guide the reader through its many superficial complexities. In turn, this may encourage greater practical evaluation of these techniques, which appear to have considerable potential in hydrological system analysis.

ACKNOWLEDGEMENTS

The author is grateful to Professor Terence O'Donnell and the North West Water Authority for providing the flow data from the River Wyre.
    The author is also grateful to Dr. Boyd, Professor D.A. Kraijenhoff and Professor G. Hornberger, for their helpful comments on the original version of this chapter.

SYMBOLS

S	storage	L^3
Q	flow	$L^3 T^{-1}$
R	total lateral inflow	$L^3 T^{-1}$
K	reservoir constant	T
k	sample number	1
T_s	sampling interval	T
T	time delay	T
NMP	non minimum phase	1
G	steady-state gain	1
z^{-1}	backward shift operator	1
δ_{kj}	Kronecker delta	1
J	cost function	1
EVN	order identification criterion	1
R_T	goodness-of-fit criterion	1
u_k	input	1
y_k	output	1
V (6.9)	volume	L^3
S	salinity in well mixed compartment	ML^{-3}
S_i	salinity input	ML^{-3}
Q_i	inflow	$L^3 T^{-1}$
h	waterdepth	L
A	surface area of compartment	L^2
e	effective evaporation	LT^{-1}
R	"reservoir coefficient"	TL^{-2}
$\underline{X}_k$ (6.11)	vector of state variables	1

$\underline{\mu}_k$	white noise vector	1
$\underline{\nu}_k$	vector of known inputs	1
T_k	trend component	1
S_k	seasonal component	1

APPENDIX 1

THE MICROCAPTAIN COMPUTER PROGRAM PACKAGE

All of the numerical results presented in this chapter were obtained directly from the MICROCAPTAIN computer program package. A main-frame, interactive computer program CAPTAIN (Computer Aided Program for Time Series Analysis and Identification of Noisy Systems) was developed by the author in the early nineteen seventies and has been utilised success-fully in the modelling of many different types of system, from well defined, man-made engineering processes to poorly defined natural envir-onmental and socio-economic systems. The MICROCAPTAIN package is a microcomputer version of CAPTAIN developed by the author for use on the APPLE II and IIe computers. It is an easy to use, interactive program which makes considerable use of computer graphics with hard copy from a dot matrix printer. The program is menu-driven and consists of a number of sub-programs for: file preparation and editing, data pre-processing, optimal recursive smoothing, ordinary and simple refined IV estimation (including time-varying parametric estimation), and high order auto-regressive spectrum estimation. An IBM-PC version of the package is under development.

REFERENCES

Beck, M.B. and Young, P.C. (1976) 'Systematic identification of DO-BOD model structure', Proc. A.S.C.E., Jnl. Env. Eng. Div., 102, EE5, 909.
Beck, M.B. (1978) 'A Comparative study of dynamic models for DO-BOD-Algae interaction in a freshwater river', International Inst. of Applied Syst. Analysis, Rep. No. RR-78-19.
Beer, T. and Young, P.C. (1983) 'Longitudinal dispersion in natural streams', American Society of Civil Engineers Jnl. Env. Eng., 109, 1049-1067.
Box, G.E.P. and Jenkins, G.M. (1970) Time Series Analysis, Forecasting and Control, Holden Day: San Francisco.
Bryson, A.E. and Ho., Y.C. (1969) Applied Optimal Control, Blaisdell: Mass.
Chatfield, C. (1975) The Analysis of Time-Series: Theory and Practice, Chapman and Hall: London.
Ellis, J., Kanamori, S. and Laird, P.G. (1977) 'Water pollution studies on Lake Illawarra', Aust. Jnl. Mar. Freshwater Res., 28, 467-477.
Gauss, K.F. (1821, 1823, 1826) Theoria combinationis observationum erroribus minimis obnoxiae, Parts 1, 2 and Supplement, Werke 4, 1-108.
Humphries, R.B., Young, P.C. and Beer, T. (1981) 'Systems analysis of an estuary: the Peel-Harvey estuarine study (1976-1980)'. Dept. of

Conservation and Environment, Western Australia, Bulletin No. 100.

Jakeman, A.J. and Young, P.C. (1979) 'Refined instrumental variable methods of recursive time-series analysis, Part II; multivariable systems', Int. Jnl. of Control, 29, 621–644.

Kalman, R.E. (1960) 'A New approach to linear filtering and prediction problems', ASME Trans., Jnl. Basic Eng., 83D, 95–108.

Kendall, M.G. and Stuart, A. (1961) The Advanced Theory of Statistics, Vol. 2., Griffin: London.

Ljung, L. (1979) 'Convergence of recursive estimators', in R. Isermann (ed.) Identification and System Parameter Estimation, Pergamon: Oxford, 131–144.

Mehra, R.K. and Tyler, J.S. (1973) 'Case studies in aircraft parameter identification', in P. Eykhoff (ed.) Identification and System Parameter Estimation, North Holland/American Elsevier: Amsterdam/New York.

Meijer, O.H. (1941) Simplified flood routing, Civil Eng., 11, 306–307.

Moll, J.R. (1984) 'Analysis of hydrological data using microCAPTAIN'. Waterschapsbelangen, 69, 410–415.

NERC (1975) Flood Studies Report, Natural Environment Research Council: Wallingford, UK.

Ng, C. and Young, P.C. (1985) Recursive estimation and forecasting of nonstationary time-series, in preparation.

Plackett, R.L. (1950) 'Some theorems in least squares', Biometrika, 37, 149–157.

Smith, R. (1980) 'Buoyancy effects upon longitudinal dispersions in wide well mixed estuaries', Phil. Trans. Royal Soc., A296, 467–496.

Soderstrom, T. (1973) 'An on-line algorithm for approximate maximum likelihood identification of linear dynamic systems', Lund Inst. of Tech., Div. Auto Control., Rep. No. 7308.

Soderstrom, T. and Stoica, P. (1983) Instrumental Variable Methods of System Identification, Springer-Verlag: Berlin.

Todini, E. (1978) 'Mutually interactive state/parametric (MISP) estimation in hydrological applications', in G.C. Vansteenkiste (ed.) Modelling, Identification and Control in Environmental Systems, North Holland: Amsterdam.

Truxal, T.G. (1955) Control System Synthesis, McGraw Hill: New York.

Unny, T.E., and McBean, E.A. (1982) Decision Making for Hydrosystems: Forecasting and Operation, Water Resource Publications: Colorado.

Whitehead, P.G. and Young, P.C. (1975) 'A dynamic-stochastic model for water quality in part of the Bedford-Ouse River system', in G.C. Vansteenkiste (ed.) Modelling and Simulation of Water Resource Systems, North Holland: Amsterdam.

Whitehead, P.C., Young, P.C. and Hornberger, G. (1979) 'A systems model of stream flow and water quality in the Bedford-Ouse River, I: Stream flow modelling', Water Res., 13, 1155–1169.

Wiener, N. (1949) The extrapolation, interpolation and smoothing of stationary time-series, Wiley: New York.

Wilde, D.J. (1964) Optimum Seeking Methods, Prentice-Hall: N.J.

Wong, K.Y. and Polak, E. (1967) 'Identification of linear discrete-time systems using instrumental variables', IEEE Trans. Auto. Control, AC-12, 707.

Young, P.C. (1965) 'Process parameter estimation and self adaptive

control', Proc. IFAC Symp. Teddington; appears in P.H. Hammond (ed.)
  Theory of Self Adaptive Control Systems. Plenum Press: New York, 1966.
Young, P.C. (1970) 'An instrumental variable method for real-time iden-
  tification of a noisy process'. Automatica, 6, 271-287.
Young, P.C. (1974) 'Recursive approaches to time-series analysis', Bull.
  Inst. Maths. Appl., 10, 209-224.
Young, P.C. (1976) 'Some observations on instrumental variable methods
  of time-series analysis', Int. Jnl. of Control, 23, 593-612.
Young, P.C. (1979) 'Self adaptive Kalman filter', Electronics Letters,
  15, 358.
Young, P.C. (1981) 'Parameter estimation for continuous-time models -
  a survey'. Automatica, 17, 23-39.
Young, P.C. (1982) 'The validity and credibility of models for badly
  defined systems', in M.B. Beck and G. van Straten (eds.) Uncertainty
  and Forecasting of Water Quality, Springer Verlag: Berlin.
Young, P.C. (1984) Recursive Estimation and Time-Series Analysis: An
  Introduction, Springer Verlag: Berlin.
Young, P.C. (1985) 'Recursive identification, estimation and control'.
  Chapter 8 of Handbook of Statistics Vol. 5, E.J. Hannan, P.R. Krishnaiah
  and M.M. Rao (eds.) Elsevier: Amsterdam.
Young, P.C. and Jakeman, A.J. (1979) 'Refined instrumental variable
  methods of recursive time-series analysis, Part I: single input,
  single output systems', Int. Jnl. of Control, 29, 1-30.
Young, P.C. and Jakeman, A.J. (1980) 'Refined instrumental variable
  methods of recursive time-series analysis, Part III: extensions'
  Int. Jnl. of Control, 31, 741-764.
Young, P.C. and Wallis, S. (1985) 'Recursive estimation: a unified
  approach to the identification estimation and forecasting of hydrological systems'; to appear in Applied Mathematics and Computation.
Young, P.C., Jakeman, A.J. and McMurtrie, R. (1980 'An instrumental
  variable method for model order identification', Automatica, 16, 281-
  294.

7. RELATIONSHIP BETWEEN THEORY AND PRACTICE OF REAL-TIME RIVER FLOW FORECASTING

G.A. Schultz

Ruhr University
P.O. Box 102148
4630 Bochum
F.R.G.

7.1 Link between theoretical chapters and case studies

At this stage a number of theoretical tools have been handed to the reader and it seems appropriate to make a brief summary of what has been presented as theoretical background and how this theory will be applied in the case studies that will be presented in the following chapters.

Theory was demonstrated for the following 5 fields:
1. Deterministic Catchment Models (Chapter 2 by O'Donnell)
2. Flood Routing Models (Chapter 3 by Dooge)
3. Low Flow Models (Chapter 4 by Mull)
4. Meltwater Runoff Models (Chapter 5 by Lang)
5. Time Series and Recursive Estimation Models (Chapter 6 by Young).
   For each of these 5 model types several mathematical models were presented. The choice of the adequate forecasting model depends on several features, such as (see also Chapter 12, section 12.3 by Němec):
- accuracy of input data and their resolution in time and space (instrumentation)
- availability of data in real-time depending on the data transmission system
- qualification and training of staff for the application of more or less sophisticated models
- performance of available data processing equipment
- purpose of forecast
- consequences of over-/under-estimation of forecast river flows.
   This incomplete enumeration of items influencing the choice of forecasting technique reveals that it is impossible to give general rules and that it is only possible to choose a combination of model elements tailored for the special project, conditions and purpose. Thus, the best way to learn about the appropriate model choice is by means of examples. Such examples are given by selected case studies presented in the following chapters of this volume.
   For didactical purposes it seems reasonable to present only one forecasting element in each chapter: e.g. Chapter 3 deals only with

D.A. Kraijenhoff and J.R. Moll (eds.), River Flow Modelling and Forecasting, 181-193
© 1986 by D. Reidel Publishing Company.

flood routing, Chapter 2 only with rainfall-runoff models. In practice, however, the forecast can seldom be based on the determination of one element only. Usually a combination of various elements is required. A frequently used combination is that of deterministic catchment models with flood routing models. The river Dee model presented by O'Connell in Chapter 8 shows the combination of a rainfall-runoff type model with flood routing models. This model is extended such that besides the linear reservoirs of the hydrologic models non-linear reservoirs representing dams built in the Dee catchment are also incorporated.

A further example of the combination of a deterministic catchment model, i.e. the Stanford Watershed Model, with a flood routing technique, i.e. the kinematic wave model, will be demonstrated by Fleming in Chapter 13 for the Orchy River in Scotland.

A combination of deterministic models (St. Venant equations of flood routing) with stochastic components (ARIMA model and Kalman filter) is presented by Moll in Chapter 11 for short range flood forecasts for the river Rhine.

An example of interesting model element combinations is given by Němec in Chapter 12 for the Indus River (Pakistan).

In Chapters 8 to 13 many more model applications to real world rivers are presented for single element forecasting (e.g. the Rijkswaterstaat forecasting and warning system for the Rhine river using a multiple regression model) as well as for multi-element forecasting. It is, however, not the purpose of this chapter to mention all but rather to highlight certain examples of the application of the techniques, presented in the first part of this volume, on real world river systems as demonstrated in the case studies to be presented in the following chapters.

7.2 Model input fields

One topic not dealt with so far, which is most important for real-world, river flow forecasts, is the acquisition of the relevant model input data (rainfall, runoff, groundwater, snowpack, energy, etc.) and the real-time data transmission to the forecast center. This general topic comprises several subtopics which will be dealt with the in the following paragraphs.

7.2.1 Instrumentation/Data collection

For simple river flow forecasting techniques only flow data of one or more river gauges further upstream are needed. More complex data acquisition systems require river flows at several gauges, water levels in reservoirs and their actual releases, rainfall data at various points within the catchment and possibly data about snowmelt, interflow and groundwater flow into the river. The case studies show several of such data collection systems, one of the most advanced of which is the River Dee system in Wales (Chapter 8).

7.2.2 Data transmission and processing

A further important point is the data transmission in real-time which

requires in many cases rather complex telecommunication systems. Furthermore these data coming from various sources distributed over the catchment have to be collected and filed in a data processing system in the
computer center calculating and issuing the forecast. The forecasting
center has to fulfil special hardware and software requirements.

These problems of data collection, transmission and processing are
described along with the presentation of case studies. As examples
only a few will be mentioned: for the River Dee System (Wales) Chapter
8 subsection 8.2.3; for the Haddington System (Scotland) subsection 8.3.3;
for the Bedford Ouse System (England) which monitors water quality,
subsection 8.4.3; for the River Rhine system Chapter 10 section 10.5;
for the Derwent River System (England) Chapter 13 section 13.3.

More general statements about data collection systems are given by
Nêmec in subsection 12.2.1, about data transmission systems in subsection
12.2.2 and about data processing, filing and retrieving systems in subsection 12.2.3.

7.2.3 Remote sensing techniques

The increased infrastructure in the river valleys gives rise to increased
damage due to floods, low flow periods or rapid deterioration of water
quality. This damage may be alleviated by early and precise river flow
forecasts. Thus there is a demand for improved flood forecasts which
in turn requires data acquisition networks giving information with a
high resolution in time and space. This can be achieved either by a
very dense network of conventional data acquisition systems (e.g. raingauges) or by new techniques such as remote sensing. Among the several
remote sensing techniques available are two of particular relevance for
hydrological purposes: radar and satellites.

Radar

A radar used for the detection of rain in the atmosphere emits a narrow
beam with a width of one to two degrees. As the distance from the radar
increases, it will reach a higher level and will eventually attain the
0° level, where ice formation occurs. At the 0° level the signal will
become very distorted thus forming a practical limit to the distance at
which radar can be used. Clearly, the curvature of the earth is also a
limiting factor to the distance.

In Germany and the Netherlands, the maximum distance covered by
radar is 70 to 100 kilometers. In England six radars cover the entire
country. Rainfall radars are used in Pakistan for distances up to 200 km.
It may be assumed that this is possible because the 0° level there is
much higher. In Germany, a 5 cm. wave length was found optimal for
rainfall measurements. The number of terrestrial rainfall gauges needed
for the calibration of the radar signal depends on climatological conditions. In southern Germany, 50 raingauges, covering an area of 20,000
km^2, were used to calibrate the radar echo. A sensitivity analysis
showed that only 3 of the 50 raingauges were necessary. Research is now
being conducted on the effects of various drop spectra on radar measurements. One of the results of this research may be that in the future no

raingauges will be needed for calibration of the radar signal.

Weather radars are capable of measuring rainfall over an area (e.g. a catchment area) with very high resolution in time and space. So-called C-band radar can procure rainfall intensities at a rate of e.g. one information per minute and for each square km in area (WMO 1975, Attmannspacher and Schultz 1981, Klatt and Schultz 1983).

This characteristic can be used e.g. for flood forecasts if the radar rainfall measurements are used as input for a deterministic catch-ment model. Figure 7.2.1 shows an example of a flood forecast for a river in Germany based on radar rainfall measurements. The model para-meters are adaptively improved each time (e.g. every hour) a new forecast is computed. This forecast is rather good - it should be noted, however, that the forecast was made after the rainfall terminated. Forecasts calculated while it is still raining are of inferior quality (see the next subsection).

The radar technique still has some problems but it seems feasible to use radar for several forecasting purposes. Examples are given in Chapter 8, for the Dee catchment, in Chapter 12, Annex 2, for the Indus River in Pakistan.

Satellite based methods

Satellites produce images in various spectral bands. Particularly the infrared imagery can be related to cloud top temperature which, in turn, can be converted into estimates of rainfall volumes from convective cells (Krüger, Harboe and Schultz 1983). Although these efforts are still in an experimental stage, it is hoped that in future such satellite information can be used as input into rainfall-runoff models for real-time river flow forecasts (see also Chapter 12, subsection 12.2.1).

7.2.4 Rainfall forecasts as basis for flood forecasts

The flood forecasting techniques discussed so far all use measured run-off and/or measured rainfall as model input. Thus in smaller catchments the lead-time of the forecast is not very long because the flood rapidly follows the rainfall that causes the flood. Therefore it becomes necessary in many cases to issue a flood forecast before the end of rainfall, i.e. during a storm event. This implies that some kind of forecast of the rainfall to be expected during the coming hours has to be made. Quantitative rainfall forecasting, however, is an unsolved prob-lem in meteorology. For hydrological purposes an exact quantitative rain forecast may not be necessary; an estimate that the rainfall during the remainder of a storm will not exceed a certain quantity with a probability of p (e.g. 90) per cent may suffice.

Several research efforts are being made to try to achieve such a stochastic rainfall forecast. A model using conditional probabilities for volume and duration of the remainder of a storm (while it is still raining) is currently being investigated in Bochum. After rainfall volume and duration are calculated in this way, a simple assumption about the time distribution of the intensity during the duration is made. Figure 7.2.2 shows an example of a flood forecast based on radar-observed

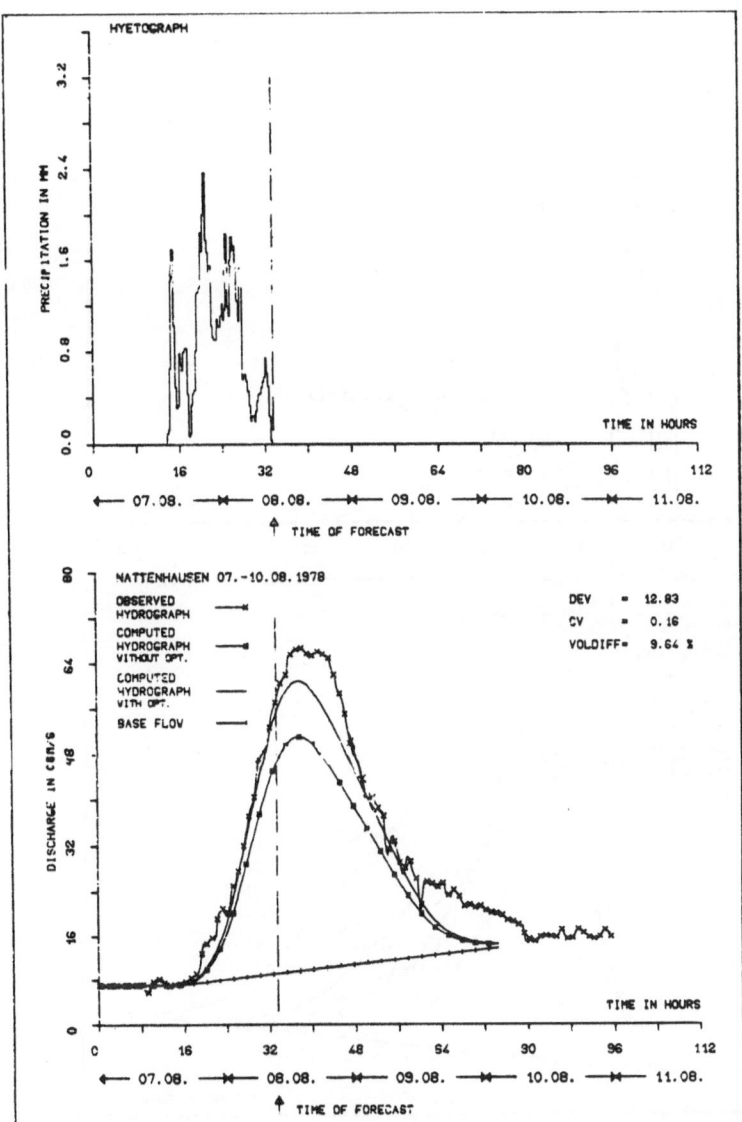

Fig. 7.2.1. Flood forecast based on radar rainfall measurements

plus forecast rainfall where the assumption of gradually decreasing
rainfall intensity for the remainder of the storm was made (Klatt and
Schultz 1983). As can be seen from Figure 7.2.2 the result depends
strongly on the rainfall forecast. In any case the results for probabilities of non-exceedance p between 60 and 90% are much better than they
would be if no forecast had been made (p = 0%).

This example is presented in order to stress the necessity also of
making rainfall forecasts in order to gain a long lead-time of a flood

forecast for small and medium size catchments.

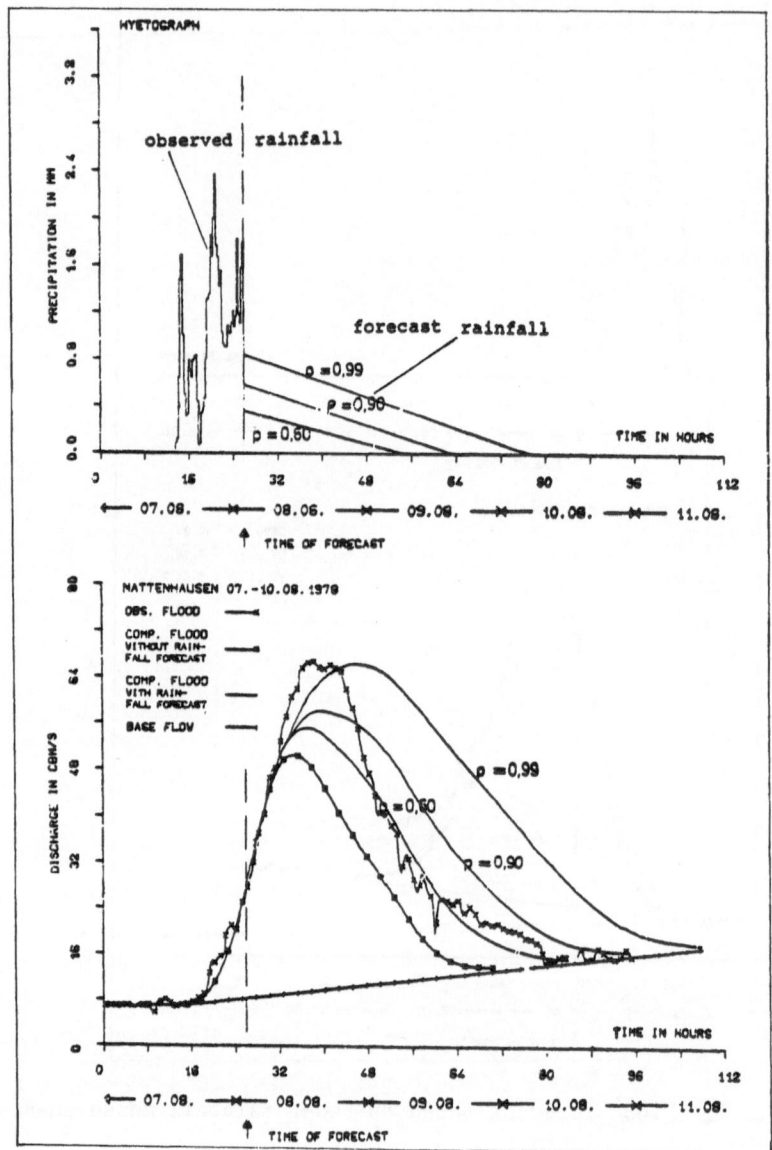

Fig. 7.2.2.   Flood forecast based on radar rainfall measurement
              and rainfall forecast

7.3   <u>Theory versus practice in real-time river flow forecasting</u>

During the last decade there has been an extensive production of theoretical techniques for real-time flow forecasting. Many can be used for

the development of good - or even optimal under the prevailing conditions
- forecasts but only a few are suitable for real-time forecasts of real-
world systems. Therefore a distinction seems necessary between two
phases as shown in Figure 7.3.1.

The specifications of Phase I are described in subsection 7.3.1,
those of Phase II in subsection 7.3.2. Both are compared in Figure
7.3.2 and in section 7.4. It should be noted that the points made in
the following two subsections are not restricted to forecasts only but
are also valid for the objectives of such forecasts, i.e. the issuing of
warnings (flood warnings, low flow warnings or warnings of water quality
deterioration) and the operation of reservoirs with the purpose of
alleviating danger and damage caused by such special river flow conditions.

7.3.1 Phase I

By preparatory measures (Fig. 7.3.1) is meant the specification of the

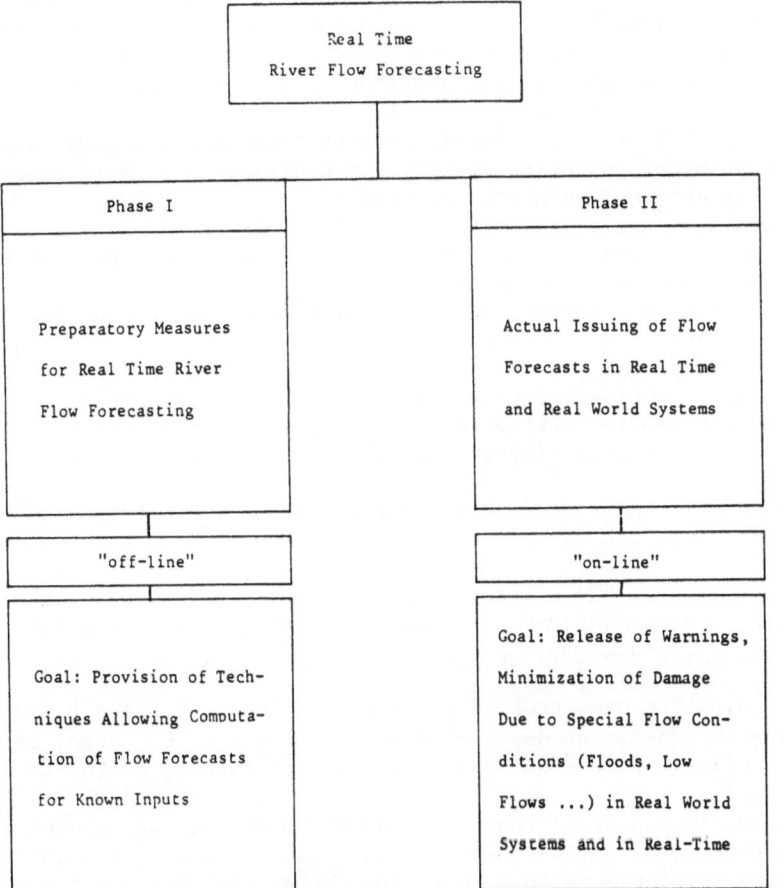

Fig. 7.3.1. The two phases of real-time river flow forecasts

application of theoretical methods for the development of more or less
general techniques which hopefully could eventually be employed in real-
time forecasting for real world river systems. During the development
of such techniques there is usually little or no direct contact with the
real world system and its operation, i.e. the development of these
techniques occurs "off-line", e.g. on a computer.

People active here are mostly theoreticians, "model builders" - in
the terminology of Chapter 1 they are "other world people" (OWP). For
their work the following criteria usually are valid (see also Fig. 7.3.2,
left part):

1. Most models are developed in universities or other institutions which
   are not directly responsible for issuing forecasts or the operation
   of hydrosystems.

2. Techniques applied stem from the field of deterministic and stochastic
   hydrology, decision theory, operations research, systems analysis,
   control theory, etc. They are not adapted to the peculiarities of a
   specific real world system.

3. They are neither concerned with, nor responsible for, real-world
   systems data collection, data transmission and mass data processing.

4. The quality of real-time forecasting techniques is highly important
   for success or failure of thus developed forecasts, warnings or
   reservoir operating policies.

5. Risk and uncertainty imply simply numbers between 0 and 100% following some prior or posterior probability distribution. Best results
   are usually obtained with risk proneness.

6. Unwanted results or mistakes do not matter. In such a case a new
   computer run will be started with other parameters, another systems
   functions and other decision parameters.

7. Human and political factors are of no importance as long as they
   cannot be quantified in numbers.

8. Objections to acceptance of the most modern techniques are considered:
   unqualified, conservative (= old fashioned), "risk averse".

9. Unfortunately people working in Phase I are frequently neither will-
   ing nor capable of explaining their results and the criteria on which
   their "optimum forecast" is based to people outside their field of
   work who do not understand their jargon. At a recent symposium on
   "Real-Time Operation of Hydro-systems" you could find quotations like
   this:
   "In practice ..... the following question (will be) posed to the
   user: The system management informs you that in the coming interval
   you will receive "a" m^3/s of water with the probability of 50% or
   "b" m^3/s with the same probability assuming that a > b and b < a < p
   (p, being the user's needs). What amount of water "c", supplied with
   the probability of 100% would satisfy you identically as the uncertain offer?".

The author is afraid that the "users" he knows, e.g. bosses of
water authorities, would answer this question in a rather unpolite
way, like asking him to go back to the "other world" and not further
waste his time. The user wants to know what he can get and not
answer complicated questions of which the sense is not clear to him.

Although some of the above statements may sound somewhat negative, there can be no doubt that the work and results of the OWP are of great relevance for real-time flow forecasts, warnings and operation of real-world hydrosystems.

7.3.2 Phase II

Phase II is concerned with the practical application of techniques, developed in Phase I, that is in a real world system and in real-time (see also Fig. 7.3.2, right hand part):

1. The hydrological models to be applied must always function; also under unusual conditions. It must be clear what must be done if some data are highly erroneous or missing entirely or if the model cannot be operated (e.g., if the computer breaks down).
2. The techniques to be applied must exist in the form of perfectly documented computer programs tested for all possible conditions which could occur in a specific real world system.
3. Since the "forecaster" is responsible for all consequences of his systems operation he must develop a "fool-proof" data collection, transmission and processing system, the organization of which also works if some of the field data do not reach the central computer.
4. Although flood forecasts are highly relevant in practice, their potential cannot be fully used due to the necessary risk aversion of the forecaster or operator of a reservoir system.
5. Risk and uncertainty are not just numbers. In each event the probability of that very event (e.g., flood) is 100%. This fact leads to the usual risk aversion of the real-time decision maker. The consequences are usually lower benefits but also less damages.
6. Unwanted results and mistakes do matter; in fact they may be crucial: erroneous forecasts and consequently false system operation during floods may cause epidemic diseases, loss of goods or even lives, and with them loss of the operator's job. They can also have political consequences.
7. Human and political factors are of highest importance. Example: theory recommends in the case of the expected occurrence of a flood to release a great amount of water from a reservoir on the basis of a flood forecast in order to use an increased storage capacity for catching the actual flood peak. Since the forecast is always imperfect information due to its uncertainty, the system response (operation) cannot be optimal for a single event. Thus it may happen that pre-release will cause some flood damage downstream of a dam while later, due to overestimation of the forecast, the reservoir will not become full. This could lead to a court case where the cost of the damage would be claimed from the river authority (operator).
    Therefore pre-release of water in the case of flood is politically not feasible.
8. A reluctant attitude towards the newest techniques is necessary as the consequences of all changes in an operating policy must be considered in advance; the trade-off between the expected pros and cons must be evaluated.

9. If certain new techiques, particularly their theoretical basis, the
    incorporated decision criteria and the obtained results cannot be
    explained to the user in a brief and clear way, these methods are
    completely useless and serve only to irritate potential users with
    the consequence that <u>all</u> new techniques will be rejected!

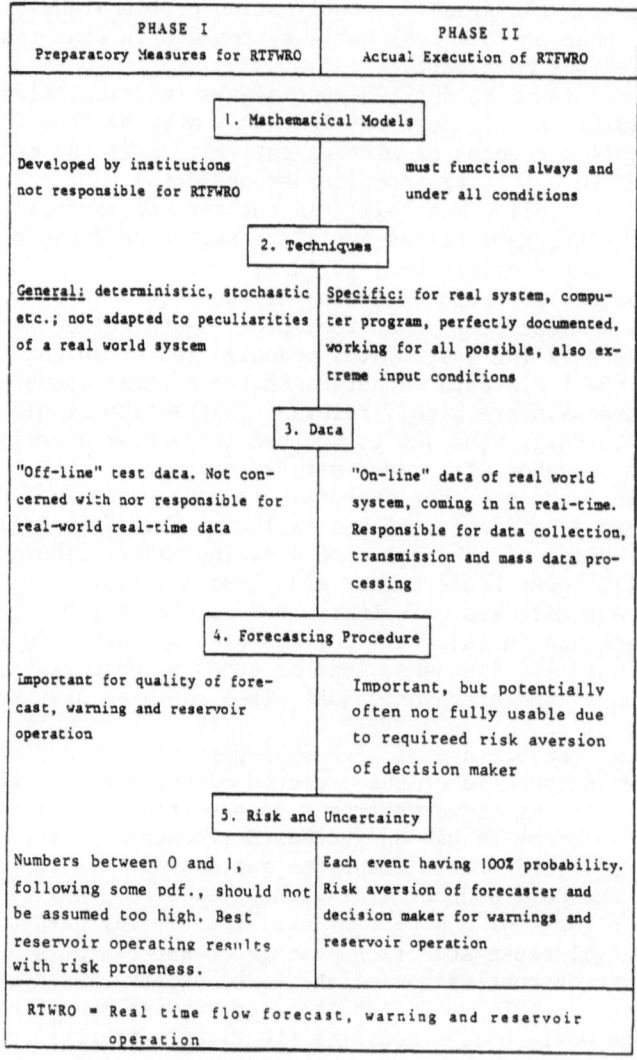

PHASE I Preparatory Measures for RTFWRO	PHASE II Actual Execution of RTFWRO
1. Mathematical Models	
Developed by institutions not responsible for RTFWRO	must function always and under all conditions
2. Techniques	
<u>General:</u> deterministic, stochastic etc.; not adapted to peculiarities of a real world system	<u>Specific:</u> for real system, compu- ter program, perfectly documented, working for all possible, also ex- treme input conditions
3. Data	
"Off-line" test data. Not con- cerned with nor responsible for real-world real-time data	"On-line" data of real world system, coming in in real-time. Responsible for data collection, transmission and mass data pro- cessing
4. Forecasting Procedure	
Important for quality of fore- cast, warning and reservoir operation	Important, but potentially often not fully usable due to requireed risk aversion of decision maker
5. Risk and Uncertainty	
Numbers between 0 and 1, following some pdf., should not be assumed too high. Best reservoir operating results with risk proneness.	Each event having 100% probability. Risk aversion of forecaster and decision maker for warnings and reservoir operation
RTWRO = Real time flow forecast, warning and reservoir operation	

Fig. 7.3.2.   Relevant aspects of Phase I and II for real-time flow
              forecasts, warnings and reservoir operation.

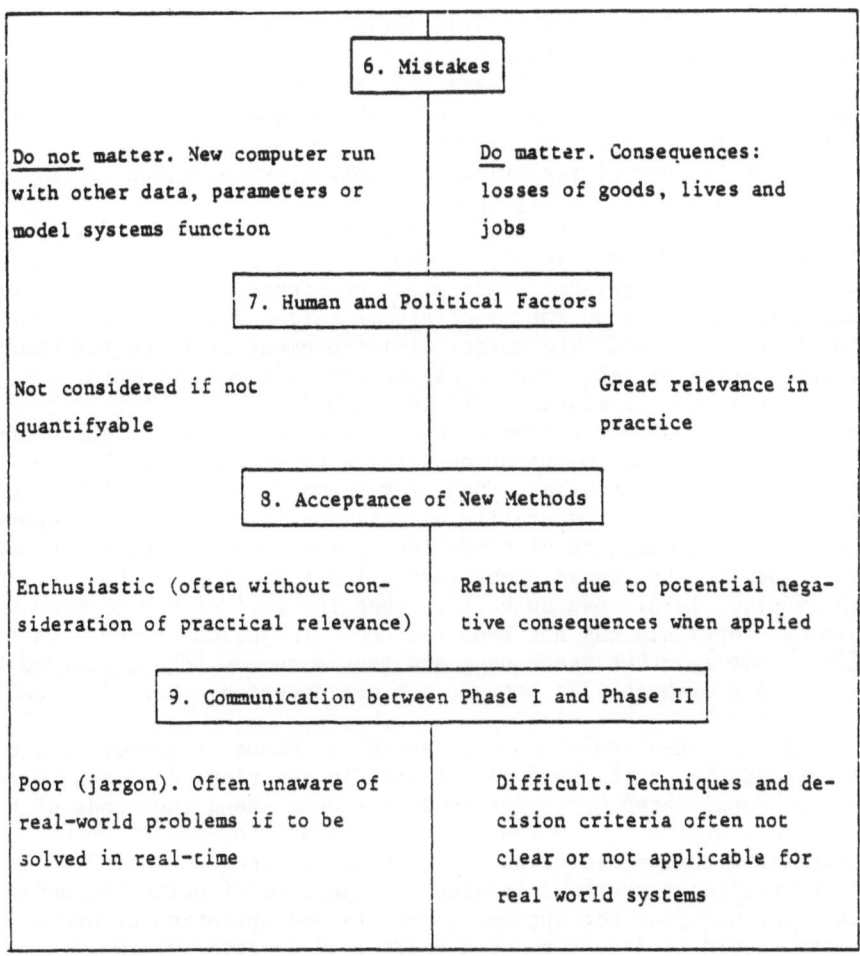

Fig. 7.3.2.   Relevant aspects of Phase I and II for real-time
              (continued) flow forecasts, warnings and reservoir
              operation.

7.4 Conclusions

After having looked at certain aspects of the real-time flow forecasts,
warnings and operation of hydrosystems the author would like to draw
some conclusions concerning future work in the field of both Phase I
and Phase II.

1. The majority of publications dealing with forecasting techniques stem
    from Phase I. This is unsound. The practitioners say that theoret-
    icians just play around with their models in the ivory tower, present
    non-understandable papers and leave the practitioner on his own,
    with his unsolved problems.

2. When developing mathematical models for real-time flow forecasting
    in real world systems they should be designed such that they can

really be used in practice. This implies that models should require only input data fields which exist, that they should be flexible, such that they also work if parts of the required input data are not transmitted; that consideration of risk and uncertainty should be realistic and not spectacular, etc.

3. Success or failure of real-time forecasting, warning and operation of hydrosystems depend greatly on the quality of forecasts. Although the imperfect information of forecasts can be incorporated in decision models (Duckstein 1981), attempts should be made to improve short-term and long-term forecasting techniques not only by improving known methods based on Markov-models or Kalman filters, etc., but also by new methods. The potential improvement of flood forecasts by application of radar and satellite rainfall measurements as input for rainfall-runoff models is to be evaluated.

4. Communication between Phases I and II must be improved! There are already examples of co-operation between them; there is, however, still much to be improved. When, for example, at an IIASA Workshop (IIASA 1982) the representative of a famous river authority reported about modern techniques of real-time operation of hydrosystems and had to admit, when questioned, that all his new methods have never been applied in his own authority, then the goal of a book such as this one certainly has not been reached. Yevjevich observed there: "Experience in other areas suggests that when new ideas, claimed to be an order-of-magnitude better, are not accepted, there is usually a good cause".

    Some of these causes must be sought in Phase I: papers are not understandable, techniques do not work in practice, decision criteria are too complicated, researchers do not understand the needs of the practice. Others lie in Phase II: inertia ("the system works perfectly with our old techniques, we do not need new ones"); lack of theoretical background leading to rejection of better methods; fear that previous non-optimal forecasts and operation of hydrosystems could be discovered and could lead to legal claims.

    The Ruhr University Institute for Water Resources and Hydrology has cooperated with a German river authority for many years. First they were very reluctant but gradually became convinced of the advantages of new methods. Eventually they gave the first consulting contract and after the first results were (seemingly) convincing, another was obtained. Meanwhile, the standard was raised and increasingly more complex techniques applied. Thus psychological factors and time are tools to be used to improve the standards.

5. The human factor and its role in real-time forecasting and operation of hydrosystems must be clarified. Until recently it was considered as ideal to develop a fully automatic real-time operation of hydrosystems on the basis of automatic flow forecasts. In the meantime many doubts have occurred. At the above mentioned IIASA Workshop, all experts from many countries agreed that no real-time operation of hydrosystems should ever be made without a competent person supervising it. A similar agreement was reached, when Hamlin conducted an opinion poll at the IAHS Symposium at Oxford, 1980: "What is most important for flood forecasts: good data, a good model or a

good forecaster?" Almost 100% of the participants said: "a good forecaster". This role of the human factor must be considered when designing data transmission and processing systems.

The author wishes to make it quite clear that practical implementation of real-time forecasting techniques in real world systems is a rather complex business with many problems. Participation in a course on "Real-time River Flow Forecasting" is certainly not enough to solve these problems. There are, however, colleagues who have successfully solved such problems in various projects in several countries, from which we can learn much about special solutions for these problems for particular projects. The idea behind the case studies to be presented in the following chapters is to use know-how and experience in order to stimulate the imagination and fantasy of the reader in his search for original solutions for his own real-time river flow forecasting problems.

REFERENCES

Attmannspacher, W. and Schultz, G.A. 1981, Wasserwirtschaft Vol. 1, 1981. 'Möglichkeiten und potentieller Nutzen eines bundesdeutschen Niederschlags-Radar-Verbundsystems' ('Possibilities and potential benefits of a West German coupled rainfall-radar system').

Duckstein, L. 1981 Proceedings of the NATO Advanced Study Institute, Operation of Complex Water Resources Systems, Erice, Italy, Application of Decision Theory to Operation.

Kaczmarek, Z. and Kindler, J. 1982, IIASA Collaborative Proceedings Series, IIASA, Laxenburg, Austria. 'The Operation of Multiple Reservoir Systems'.

Klatt, P. and Schultz, G.A., 1983 Proceedings of the IAHS Symposium on Remote Sensing and Remote Data Transmission, Hamburg. 'Flood Forecasting on the Basis of Radar Rainfall Measurement and Rainfall Forecasting'.

Krüger, L.R., Harboe, R. and Schultz, G.A. 1983, IAHS Symposium on Remote Sensing and Remote Data Transmission, Hamburg. 'Estimation of Convective Rainfall Volumes with the Aid of Satellite Data'.

WMO 1975 'Intercomparison of Conceptual Models Used in Operational Hydrological Forecasting', WMO-Operational Hydrology Report No. 7, WMO, Geneva, No. 429.

8. CASE STUDIES IN REAL-TIME HYDROLOGICAL FORECASTING FROM THE UK

P.E. O'Connell, G.P. Brunsdon, D.W. Reed and P.G. Whitehead

Institute of Hydrology
MacLean Building
Crowmarsh Gifford
Wallingford, Oxon.
U.K.

8.1 Introduction

The developments which have taken place in the fields of electronic
engineering and hydrological modelling over the past decade are now
generating considerable operational benefits in the areas of flood
warning, reservoir management and pollution control. On-line monitoring
of rainfall, flow and water quality variables can now be achieved with
reliable instrumentation and telemetry at modest cost: ever increasing
advances in the performance of on-line monitoring schemes are being
achieved through the exploitation of microprocessor technology. Dedicated low-cost microcomputers can be programmed to control telemetry
schemes automatically while also providing the necessary computing power
to run real-time flow and water quality forecasting models. Thus the
capacity of the water engineer to respond efficiently and effectively to
emergency situations created by flood or pollution events has been
greatly enhanced. As more operational experience with this new technology is acquired, it is important that case studies are reported in the
literature (and particularly presented in post-experience courses)
to assist the practising engineer in choosing the instrumentation, telemetry, computer, and forecasting models appropriate for his particular
problem.

The three case studies reported here will help to give an impression
of how real-time forecasting has developed into an operational tool in
the UK while also covering three completely different forecasting problems. The first of the three systems to be developed and implemented
was that for the river Dee basin in North Wales; this system is of
special interest since its commissioning in 1975 marked the culmination
of an extensive research programme undertaken to explore the potential
of weather radar and real-time hydrological forecasting in multipurpose
reservoir management and control. The case study description presented
here will focus more on the operational experience which has been
acquired since implementation rather than on the details of the actual
system implemented.

The second case study describes a real-time flood warning scheme
installed in 1981 in a small catchment area east of Edinburgh in Scotland.

D.A. Kraijenhoff and J.R. Moll (eds.), River Flow Modelling and Forecasting, 195–241.
© 1986 by D. Reidel Publishing Company.

This case study illustrates how modern microprocessor-based technology
and simple rainfall-runoff modelling can be combined to produce a flood
warning scheme which is inexpensive, automatic, flexible and reliable.

The third case study has as its focal point the real-time forecasting
and control of water quality rather than water quantity. For the
Bedford Ouse River in East Anglia, the objective of the real-time forecasting scheme installed in 1980 is to help protect water supply abstraction intakes along the river from accidental upstream discharges of
pollutants. This forecasting system also represents the culmination of
a previous extensive research programme to develop dynamic models of
flow and water quality for use in water quality management.

8.2 Real-time flow forecasting system for the river Dee

8.2.1 Background

Commissioned in 1975, the river Dee real-time flow forecasting system
represented the culmination of a collaborative research programme initiated in 1966 to develop new methods for multipurpose reservoir management; real-time flow forecasting and control (for both high and low flow
regimes) occupied a central role in this research programme together
with a project aimed at developing the potential of weather radar for
the quantitative measurement of rainfall. The Dee Weather Radar and
Real-time Hydrological Forecasting Project was coordinated by a Steering
Committee and its report (Dee Steering Committee 1977) presents a complete account of the various phases of the research programme together
with its major conclusions and an extensive bibliography of many related
scientific papers. The forecasting model and much of the control software was developed by the Institute of Hydrology. Considerations underlying the choice of real-time forecasting model of the river Dee are
discussed by Lambert and Lowing (1980) who also describe how the original
model has been modified in the light of operational experience.

8.2.2 The catchment

Topographically, the river Dee basin (area 1816 km^2 at Chester) divides
naturally into upper and lower catchments (Figure 8.2.1); the upper
catchment has an area of 1018 km^2 at Manley Hall. The tributary sub-
catchments in the upper catchment are steep, with only a thin covering
of soil overlying the impermeable bed rock, resulting in a rapid response
of runoff to rainfall. Rainfall over the basin ranges from 2200 mm per
annum in the west to 700 mm per annum in the east. There are four
reservoirs located in the upper part of the basin (Figure 8.2.1); these
are listed together with their contributing catchment areas, storage
volumes and principal uses in Table I. Regulation of the Dee supports
abstractions to supply totalling 10.3 m^3/s. The major abstraction
points are sited between Eccleston and Chester and, during low flow regulation periods, the operational requirement is to maintain a flow of
about 13 m^3/s at Eccleston. Releases from storage take up to 36 hours
to reach this point, giving rise to problems of forecasting release
requirements.

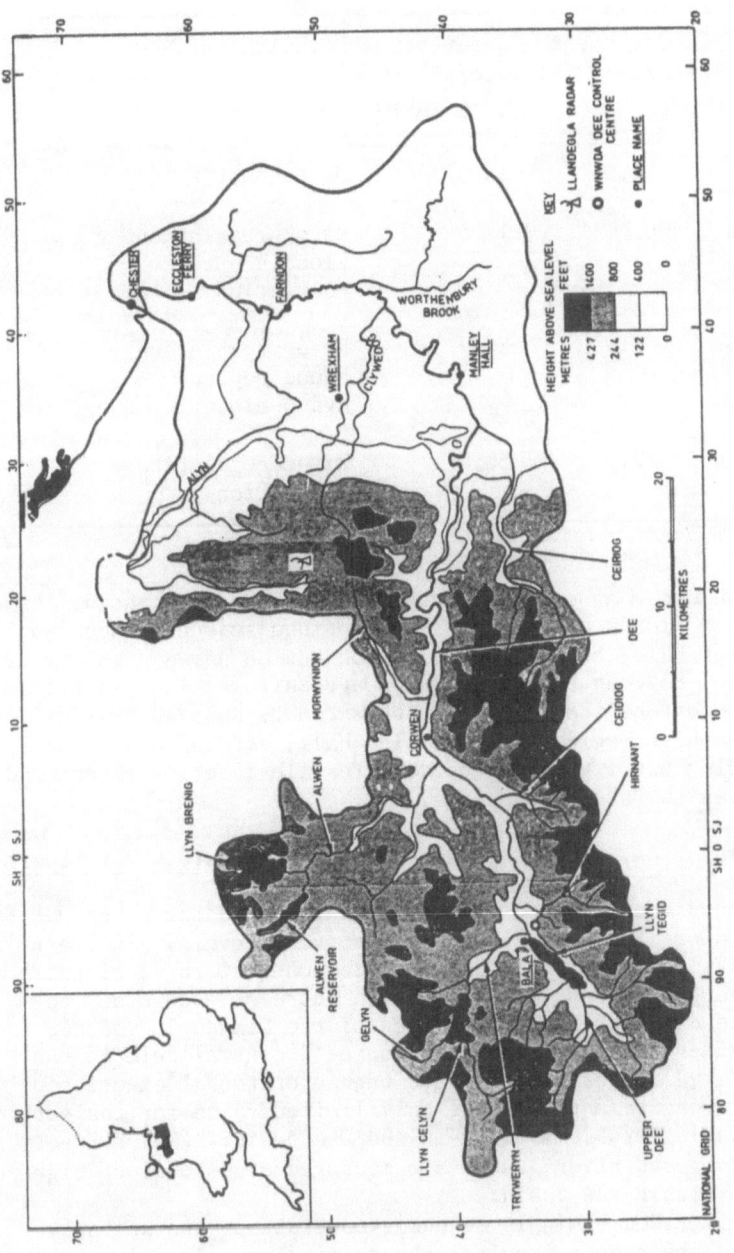

Fig. 8.2.1. Topographic map of the river Dee basin showing main tributaries and reservoirs.

Table I. River regulating and direct supply reservoirs in the Dee
          Basin

Reservoir	Catchment area km^2	Reservoir storage $m^3 \times 10^6$	Principal uses
Alwen	25.5	14.5	Direct water supply
Llyn Tegid	262.0	17.6	River regulation Flood control Recreation
Llyn Celyn	59.9	80.7	River regulation Flood control Hydro-electric power
Llyn Brenig	22.0	68.0	River regulation Recreation

Flooding can occur at several places in the river Dee basin. Land
around Llyn Tegid floods easily and in exceptional circumstances so
would the town of Bala, but flood mitigation action gives priority to
preventing this. Between Bala and Corwen agricultural land is flooded
when the river overtops its banks, but this can be delayed by limiting
the release from Llyn Tegid. Below Manley Hall, agricultural land is
liable to flooding but little can be done to alleviate its effect and
priority is given to issuing warnings.

8.2.3 Instrumentation, telemetry, and computer hardware and software

Figure 8.2.2 shows the locations of the oustations in the river Dee basin
for measuring rainfall, river flows and reservoir levels. Twelve raingauges were installed at a total of six locations; at three of these
locations, three gauges were sited within a metre of each other to
provide reliable data for the calibration of the Llandegla weather radar.
Although the latter installation had occupied a central role in the
research programme, it was discontinued when operational experience
showed rainfall measurement to be of only limited value for the particular flow forecasting problems found in the Dee basin. (See subsection
8.2.5). Reservoir and river levels are measured at a total of eighteen
different sites within the basin.
    The upper river Dee basin is somewhat mountainous and sparsely
populated with sites remote from telephone lines or mains electricity
supplies. To overcome these difficulties, three different data transmission techniques are used, namely privately laid land lines, perman-
ently rented British Telecom lines and UHF radio, with repeater stations
strategically placed to obtain near line of sight paths between or over

hills. The distances between sites, their remoteness and lack of facilities raised serious maintenance problems, especially during bad weather,
and so the telemetry scheme was designed to minimize these difficulties.

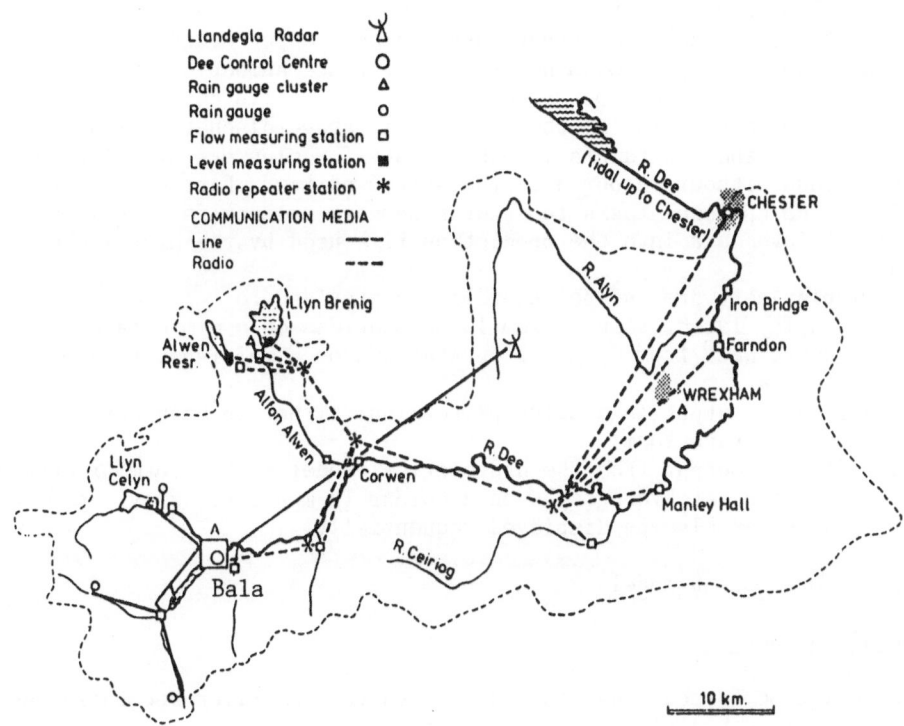

Fig. 8.2.2.   Map of the river Dee basin showing the telemetry
              system.

At the Dee Control Centre located at Bala, a PDP 11/40 processor
with 56K words of core storage controls the telemetry, receives and
processes the data, runs the real-time forecasting model. The software
in the Bala computer is under the control of a proven real-time systems
executive (RSX-11M), which looks after:
a. disc-accessing (for both data and programs);
b. scheduling of programs at specific times or under specific circum-
    stances;
c. core and processor utilization so that the computer does not waste
    time waiting for some event to occur when there is other work to be
    done.
Wherever possible, the programs for the computer at Bala were written
in the high level FORTRAN language so that the operating staff could
appreciate more easily the functions and ramifications of any individual
program within the system. Excluding the system software, the software
in the Bala computer can be divided into three main components; telemetry
software, hydrological model software and display software.

In normal operation, a control utility program initiates a scan of outstations at half-hourly intervals. Appropriate conversions are carried out before the telemetered values are passed on to the data base update task. Thereafter, control is passed to the hydrological modelling task. (See subsection 8.2.4).

There are a number of other tasks which are performed as the need arises. Once a day at 0900 hours the reference number of tips made by a raingauge is changed so that daily rainfall totals can be calculated. Interactive tasks exist so that changes in stage-discharge relationships and the tolerance in rainfall measurements for clusters of raingauges can be made without recourse to the system editor. Particularly powerful are the optional tasks to enter reservoir release forecasts and rainfall forecasts into the prediction file used by the hydrological model.

Display software and peripherals are provided to:
(i)   display latest values from 10 outstations on a mimic panel;
(ii)   record data from up to 12 outstations on a multi-point chart recorder;
(iii) provide output on a teletype terminal either in hard copy or punched tape form;
(iv)   display output from the hydrological model in the form of observed and forecast hydrographs on a colour TV monitor. Previous forecasts can also be displayed sequentially.

8.2.4 Forecasting model

Overall structure

The real-time forecasting model developed for the river Dee basin has three distinct hydrological components:
a.   the rainfall-runoff model for the sub-catchments;
b.   the model for main channel routing between Bala and Manley Hall;
c.   the channel model for the lower Dee between Manley Hall and Eccleston.
    A schematic representation of the full model is shown in Figure 8.2.3; keys to the labels for the various sub-catchments and outstations are given in Table II. The separation of the sub-catchments is dictated by the location of the reservoirs and the telemetering gauging stations.

Sub-catchment model

The Lambert model (Lambert 1972) is used to provide real-time forecasts of sub-catchment outflow from rainfall in real-time. This model is a member of the inflow-storage-outflow (ISO) function class of models described in Chapter 2. The relationship between storage and discharge Q is assumed to be of the form

$$S = k \log_e Q + C \qquad\qquad (8.2.1)$$

which, when combined with the continuity equation gives the following analytical solution for Q:

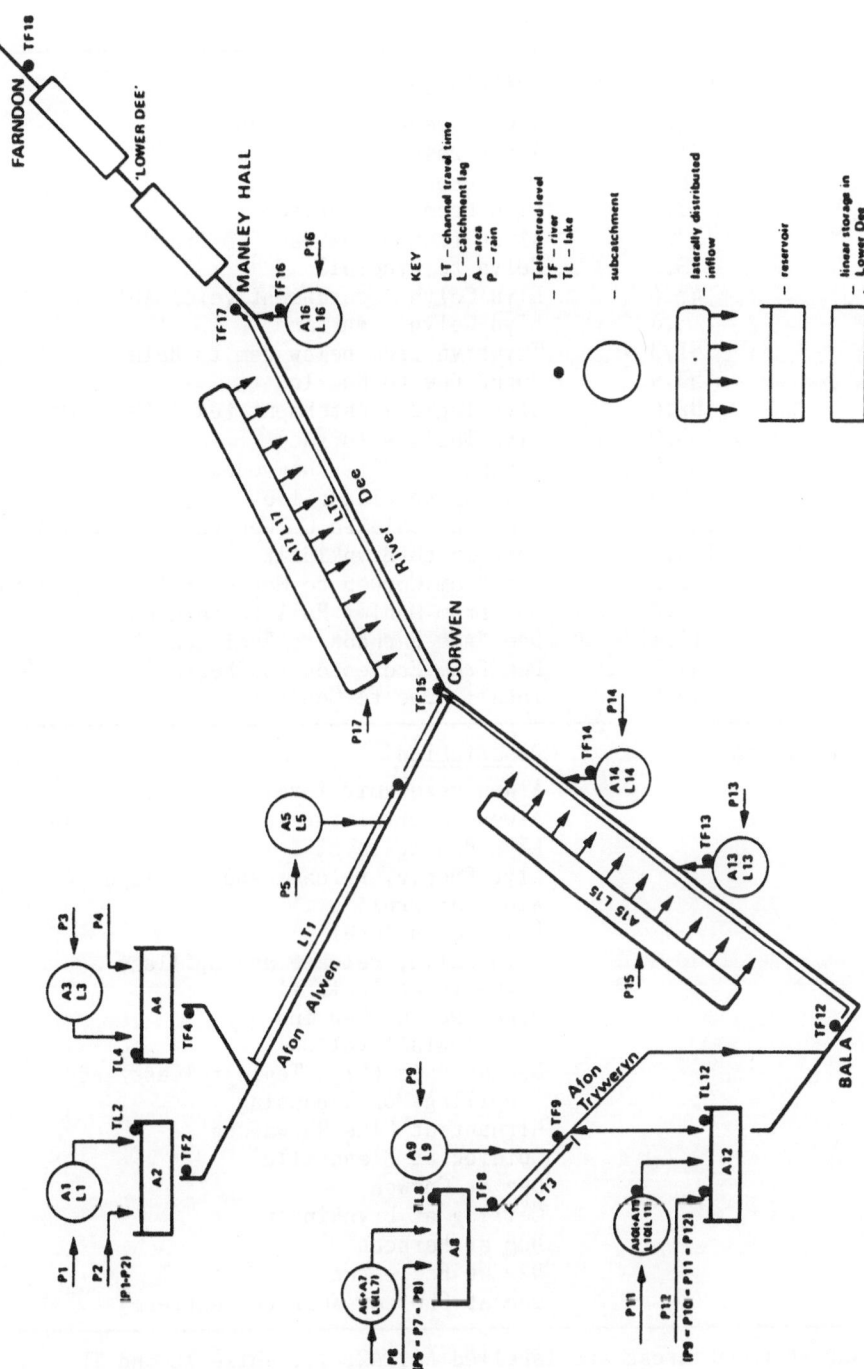

Fig. 8.2.3. Schematic representation of the river Dee forecasting model

Table II. Reference system for schematization of the Dee basin shown
           in Figure 8.2.3.

Sub-catchment	Area (km^2)	Description
A1	24.0	Alwen reservoir – catchment
A2	1.5	Alwen reservoir – surface
A3	18.3	Llyn Brenig – catchment
A4	3.7	Llyn Brenig – surface
A5	137.2	Alwen from below dams to Druid
A6	13.1	Celyn to Cynefail
A7	43.4	Llyn Celyn – catchment (excl A6)
A8	3.4	Llyn Celyn – surface
A9	51.3	Tryweryn from below dam to Bala
A10	53.9	Upper Dee to New Inn
A11	92.6	Llyn Tegid – catchment (excl A9, A10)
A12	3.9	Llyn Tegid – surface
A13	33.9	Hirnant to Plas Rhiwaedog
A14	36.5	Ceidiog to Llandrillo
A15	137.4	Dee from Bala to Corwen (excl A13, A14)
A16	113.7	Ceiriog to Brynkinalt
A17	250.2	Dee from Corwen to Manley Hall (excl A16)
A18	340.7	Dee from Manley Hall to Farndon
A19	314.1	Dee from Farndon to Eccleston
A20	142.7	Dee from Eccleston to Chester
	1815.5	Total: Dee to Chester

Telemetered variable	Description
TL2	Alwen reservoir level
TF2	Alwen reservoir, release and spillage
TL4	Llyn Brenig level
TF4	Llyn Brenig, release and spillage
TF5	Alwen at Druid
TL8	Llyn Celyn level
TF8	Llyn Celyn, release and spillage
TF9	Tryweryn at 'weir x'
TF10	Upper Dee at New Inn
TL12	Llyn Tegid level
TF12	Dee at Bala (Llyn Tegid release and spillage or bypassing)
TF13	Hirnant at Plas Rhiwaedog
TF14	Ceidiog at Llandrillo
TF15	Dee at Corwen
TF16	Ceiriog at Brynkinalt
TF18	Dee at Farndon
TF19	Dee at Eccleston
TF21	Dee at Chester Weir (downstream)

Note: Subcatchment areas are labelled A1, A2, ... while TL and TF
numbers refer to telemetered reservoir and river levels respectively.

(i) <u>for zero rainfall (P = 0)</u>

$$Q_{t+\ell} = Q_t/(1 + Q_t \cdot \ell/k);$$ (8.2.2a)

(ii) <u>for non-zero rainfall P > 0</u>

$$Q_{t+\ell} = Q_t/\{e^{-(P \cdot \ell/k)} + (1 - e^{-(P \cdot \ell/k)})Q_t/P\}$$ (8.2.2b)

The flow at time $t + \ell$ is thus defined as a function of the flow at time t, the storage parameter k and the gross rainfall P during the period $t - b$ to $t + \ell - b$ where b is the pure time delay between rainfall and streamflow response. Because $Q_{t+\ell}$ is obtained as an explicit function of Q_t, the observed value of flow can be used continually to update model forecasts by locating the origin of each forecast on the correct part of the storage-outflow relationship.

There are thus two parameters to be estimated for the sub-catchment model: the storage parameter, k, and the pure time delay parameter, b. These parameters were estimated off-line using an objective optimization procedure (Rosenbrock 1960) whereby a function F^2 defined as

$$F^2 = \sum_{i=1}^{n} (Q_i - \hat{Q}_i)^2$$ (8.2.3)

was minimized, where Q_i denotes observed discharge and $\hat{Q}_i$ denotes simulated model discharge (i.e. discharge generated by the model as a function of preceding model discharge and gross rainfall). To take some account of losses due to evaporation, separate sets of parameters were derived for summer (May-October) and winter (November-April) data in this way, the effect of 'losses' is implicitly accounted for through the derived values of the model parameters. Details of the calibration of the sub-catchment models are given in McKerchar (1975).

Historical half hourly flow data were available to calibrate the above model for five subcatchments and a sixth was subsequently calibrated from telemetered data. Values of the pure time delay parameter, b, for the above sub-catchments range from 0.5 to 2.5 hours; thus, using measured rainfall, the above model can at best provide flow forecasts up to 2.5 hours ahead. This factor, together with the particular disposition of the reservoirs and management problems found in the Dee basin, led to the decision not to retain the experimental weather radar station at Llandegla, since measured rainfall could generate only relatively minor improvements in resource management and flood warning.

The responses of ungauged areas in the upper Dee above Manley Hall are modelled by scaling from adjacent gauged catchments; these flow contributions are fed into the channel routing model either as tributary inputs or lateral inflow distributed along the main channel reach, as indicated in Figure 8.2.3.

Channel routing model

The travel time from Bala to Manley Hall for low flow regulation releases
is about 19 hours while, for flood peaks, it is about 9 hours. Since the
model was required to route both high and low flows accurately, it was
important that this variation of travel time with discharge be incorporated into the model. This led to the use of the Variable Parameter
Diffusion (VPD) model developed by Price (1973) in which the parameters
of the diffusion equation (derived in Chapter 3 as an approximation to
the St. Venant equations) are allowed to vary with discharge Q: the
resulting model is

$$\frac{\partial Q}{\partial t} + \bar{c}(Q)\,\frac{\partial Q}{\partial x} = Q.\bar{a}(Q)\,\frac{\partial^2 Q}{\partial x^2} + \bar{c}(Q).q', \qquad (8.2.4)$$

where

$\bar{a}(Q)$ = attenuation parameter (m^{-1})
$\bar{c}(Q)$ = celerity of flood wave (m/s)
q' = lateral inflow per unit length of river (m^2/s).

Because $\bar{c}$ and $\bar{a}$ vary with discharge, iterative finite difference
methods are needed to give the most accurate solution to Equation (8.2.4).
A computationally more efficient method is to regard $\bar{c}$ and $Q.\bar{a}$ as being
constant throughout a time step, taking values calculated for the beginning of the step; the resulting finite difference equation is then linear
and can be solved by Gaussian elimination.
    To apply the VPD model, the main river Dee channel between Bala and
Manley Hall was divided into two reaches, Bala to Corwen and Corwen to
Manley Hall; the Bala-Corwen reach is flatter, with a small flood plain,
while the Corwen-Manley Hall reach is steeper and twists through a deep
narrow valley. Calibration consisted of deriving appropriate $a(Q)$ and
$\bar{c}(Q)$ functions for the two reaches. The $\bar{a}(Q)$ function was calculated
from channel topography for each reach (Hydraulics Research Station
1975). The $\bar{c}(Q)$ function was obtained by fitting to the propagation
speeds of recorded peak flows. Figure 8.2.4 shows the wave celerity
curves for the two reaches; there is a pronounced variation in $\bar{c}(Q)$ with
discharge.
    Routing of flows is carried out through a number of sub-reaches with
gauged tributary flows added in between. Each sub-reach contains a
number of computational nodes; the distance between nodes is about 2000
metres. Figure 8.2.5 shows a comparison of observed and simulated flows
at Manley Hall for January 1974; the latter were obtained by routing
measured tributary flows, and so the comparison is a test of the routing
method and the ungauged lateral inflow assumptions.
    When the channel routing model is used in real-time, telemetered
flows are available from Corwen and Manley Hall, and so the question
arose as to how this information should be used to update model forecasts in real-time. Forecasting errors derive from two main sources:
(i) lateral inflow, which is estimated from the nearest gauged tributary

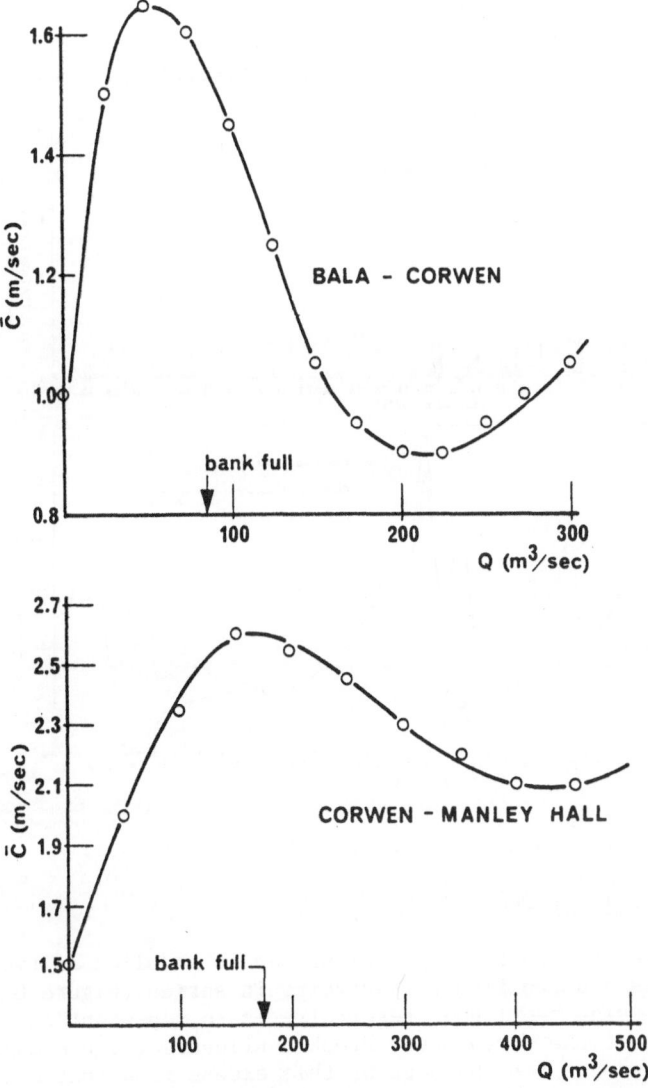

Fig. 8.2.4. Variation of wave speed with discharge for the Bala
to Corwen and Corwen to Manley Hall reaches of the
river Dee.

with adjustments in terms of basin area and annual effective rainfall,
may be incorrect; (ii) the main channel routing model may be in error.
These two sources of uncertainty cause errors mainly in terms of volume
and timing respectively and a simple ordinate disparity, which is the
symptom, gives no clues as to the underlying cause. Simply to make a
correction equal to this disparity to the whole of the forecast would be
counter-productive if the problem was one of timing. In the case of
Manley Hall, trial and error showed when, in terms of what was happening

upstream, it was best to make no correction or a partial correction. The
end result, programmed into the model, is a practical but inelegant solution to what is a central problem in real-time forecasting.

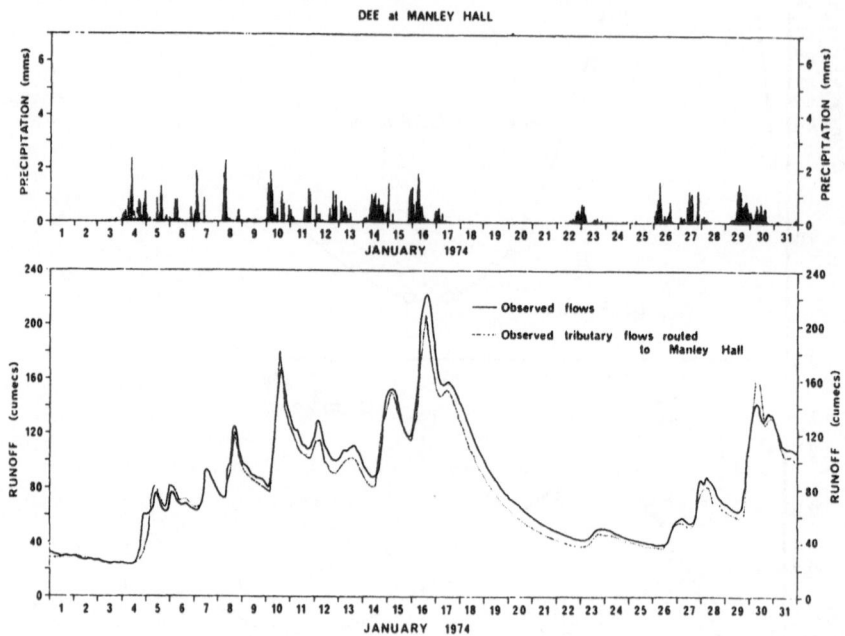

Fig. 8.2.5.   Observed and simulated flows at Manley Hall.

Channel model for the lower Dee

The Manley Hall to Farndon reach of the river Dee is modelled as two
quasi-linear channels and two linear reservoirs in series (Figure 8.2.6).
The main reason for using two linear reservoirs is to represent spillage
onto the flood plain at the Worthenbury Brook confluence (Figure 8.2.1).
The model does not account for the part of this excess flow that eventually returns to the river. The resulting error is small. Between
Manley Hall and the confluence there is assumed to be a linear channel
with pure time delay b hours and a linear reservoir with a storage
constant K hours; between the confluence and Farndon an identical combination is assumed. Forecasts of flow are produced for Eccleston but
the time of travel of a flood wave from Farndon to Eccleston is ignored.
    Values of K and b have been evolved by trial and error under various
conditions. This has produced more acceptable results than objective
optimization using a minimum sum of squares criterion. The aim of the
fit was to match observed attenuation and travel times while making
allowance for the varying and unknown inflow below Manley Hall. The
storage parameter K is constant for each reservoir and equal to 3 hours.
The pure time delay is allowed to vary for each channel according to

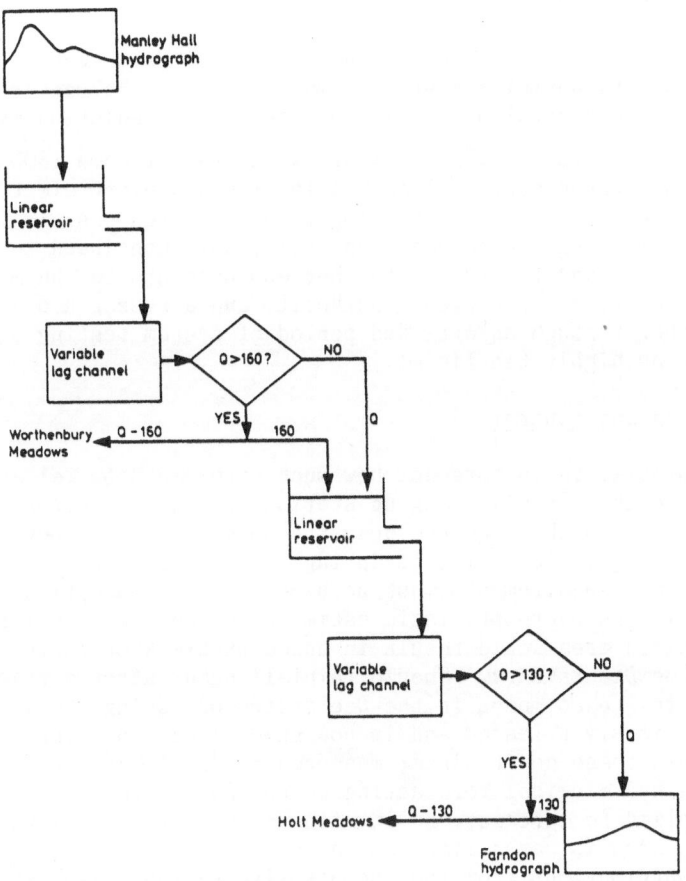

Fig. 8.2.6.   Schematic representation of lower river Dee model.

$$b = \frac{140}{21 + \bar{Q}_{29}} \qquad\qquad (8.2.5)$$

where $\bar{Q}_{29}$ is the average flow at Manley Hall for time $(t - 5)$ to time $(t + 24)$ hours based on observed and forecast flows. The pure time delay b does not change during the calculation of a particular forecast but it may change before the next time a forecast of Farndon flow is made in order that the observed variation in translation time with discharge may be matched more closely. The average value $\bar{Q}_{29}$ in Equation (8.2.5) is used rather than the Manley Hall flow at one instant because the relation was derived for the time of travel of reservoir releases between Manley Hall and Farndon and not for peak flows. It was felt, therefore, that the use of peak values in (8.2.5) would lead to errors.

8.2.5 Operational performance

Since 1976, the forecasting system has been used by the Welsh Water
Authority as part of its normal operational management of the Dee, to
assist in the effective control of a multipurpose river regulation system
of 160×10^6 m^3 total storage set in a river basin area of some 1800 km^2,
and used for regional water supply, flood mitigation and hydro-electric
power generation. The Institute of Hydrology, who were responsible for
developing and implementing the forecasting model, have continued to
monitor its performance and to try out further enhancements to the model.
This arrangement whereby an operational authority and a research organization work together through an extended period of system testing and
refinement is seen as highly beneficial.

Low flow forecasting and control

The hydrological problem is to forecast how much water must be released
at Bala (and, in drought, Brenig) gauging station so that Eccleston
flows will not fall below the required quantity some 36 hours later,
the travel time for regulation releases during low flow conditions.
Predictions of release requirements must be based on the assumption of
zero future rainfall, as over-optimistic estimates of future flow from
the uncontrolled basin area would result in unacceptable shortfalls in
residual flow to the Dee Estuary. Should rainfall occur after a release
has been made, there are no means in the Dee system of saving the water
which has been previously released and is now superfluous to river
requirements. Given these constraints, measured rainfall has minimal
value for low flow hydrological forecasting on the Dee system.

The method applied to forecast the low flow contribution from the
uncontrolled basin area is based simply on a master recession extrapolation of present contributions from the uncontrolled basins. Two cases
are considered: contributions upstream and downstream of Manley Hall.
Present values of these are obtained by continually carrying out 'route
and lag' calculations; e.g. the Manley Hall flow is 'routed and lagged'
to Eccleston, where it is compared with the observed flows; the difference between the two values (after smoothing) is taken as the present
value of uncontrolled inflow below Manley Hall. Similar calculations
of routed and lagged Bala/Brenig releases are compared with current
Manley Hall flows to give uncontrolled inflows between Bala and Manley
Hall.

The simplicity of approach to the problem of low flow forecasting
requires a fundamental commitment to highly accurate hydrometric instal-
lations on the main river, as the forecasting is based on 'differences'
between successive main river gauging stations, and requires reliable
modelling of channel routing processes. Despite unsteady flow conditions
between Farndon and Chester caused by tidal effects for 30 per cent of
the time in summer, the operational results are impressive in severe
droughts, with an average regulation over-release in dry summer months
of only 0.35 m^3/s at Eccleston; in wet summers with frequent rainfall,
regulation losses are obviously greater but this is not operationally

significant on the Dee as Llyn Celyn and Llyn Tegid refill by gravity
and are not drawn down significantly in such years, while Llyn Brenig
would not be used at all. It is considered that greater operational
efficiency on the Dee is more likely to be achieved by improved engineering management techniques than through improved hydrological forecasting.

Flood forecasting and control

As discussed in subsection 8.2.4, the extra lead time gained through sub-
catchment modelling in the upper Dee has little impact on the accuracy
of forecasts in the lower Dee catchment and on reservoir operation for
flood control. However, it is important to realize that the value of
rainfall data, and a radar installation, for flood forecasting might
have been much greater had (a) the shape and channel network of the Dee
basin and (b) the locations of the multipurpose reservoirs, been different. Consider a hypothetical situation in which a substantial tributary
(X) entered the Dee just downstream of Manley Hall; that the agricultural
flood plain between Manley Hall and Farndon had been developed into a
large city protected by flood embankments; that a multipurpose reservoir
(incorporating flood control storage) was to be built at Manley Hall to
increase the degree of flood protection, and that there were no other
flood control reservoirs in the Dee basin area. Logic would suggest that
the 5 hour warning time available using Corwen gauging station would be
insufficient for optimal utilization of flood storage in the hypothetical
Manley Hall reservoir, but measurements of areal rainfall by radar in
the Dee basin upstream of Corwen would enable reliable predictions of
Corwen flow to be made perhaps 6 hours in advance, and would also enable
accurate flow forecasts of the flow of tributary X to be made up to, say,
4 hours ahead. In such circumstances, the radar would almost certainly
have proved an essential part of the control and forecasting system.
    In the River Dee project, short-time rainfall forecasts from the
Meteorological Service are combined with information from radars and from
the geostationary satellite.

Model refinements

As part of the research programme of the Institute of Hydrology, some
further enhancements to the Dee model have been made in pursuit of
greater forecasting accuracy and, although they may generate only margi-
nal operational benefits on the Dee, they are likely to be of broader
interest in flood forecasting. The method of model fitting described in
subsection 8.2.4 is geared to long-term flow simulation, and does not
guarantee that the model performance will be optimal for short-term
forecasting. Green (1979) has investigated this problem for a specific
lead time by assuming knowledge of $Q_{t+\ell}$ and then inverting Equations
(8.2.2) to yield a value of k which would reproduce $Q_{t+\ell}$ exactly. Green
then fitted curves relating k and Q; he also related k' to Q where

$$S_t = k'(Q) \log_e Q_t \qquad\qquad (8.2.6)$$

and found it easier to construct the curvilinear relationships when this log-linear form was used. In either case, a significant improvement in forecasting performance was obtained if separate relationships were used on the rising and falling limbs of the hydrograph.

The question of a fixed pure time delay b in the model has also been the subject of further study; a fixed lag implies no attenuation due to channel routing as is conventionally provided by the time-area diagram or other rainfall smoothing functions. Various simple smoothing functions were tried but the sensitivity of the forecast hydrograph to these changes was found to be rather small, although a general improvement in hydrograph shape was obtained.

Quite apart from the above investigations which were specific to the Dee forecasting model, the data base generated by the Dee Research Programme has been the subject of a number of research studies at the Institute of Hydrology and elsewhere; some of this research is reported in O'Connell (1980). In particular, the use of the Kalman filter in real-time rainfall-runoff modelling and channel routing has been investigated, and a number of other real-time models have been compared with the river Dee sub-catchment model; in general, the latter performed as well as models employing more sophisticated updating procedures.

Simple channel routing models incorporating forecast updating have been developed and tested on the Bala-Corwen-Manley Hall reaches of the river Dee (Wood 1980; Moore and Jones 1980; Jones and Moore 1980) and have been shown to perform well on historical data. They have not, however, been tested operationally.

8.3 The Haddington flood warning system

8.3.1 Background

For large river basins with major flooding problems, the installation of a comprehensive real-time forecasting system can frequently be justified in cost/benefit terms; for such systems, significant capital (instrumentation, telemetry, computer hardware and software) and operating costs (specialist manpower to operate and maintain the system, frequently on a 24 hour basis) will be involved but the benefits will be spread widely. However, there are many smaller catchments, particularly in remote upland areas, with relatively localised flooding problems affecting small centres of population for which a proper flood forecasting system would, in the past, have been too expensive to install and operate. However, with the advent of the microprocessor, new horizons have been opened up whereby the flood warning requirements of such small flood-prone communities can be met through the judicious combination of microprocessor based technology and relatively simple flood forecasting models. The prototype flood warning system developed for the Haddington area in East Scotland exemplifies the capabilities of this low-cost, robust, and flexible technology. The system devised to do this fits into the administrative framework obtaining in Scotland where both financial and manpower resources are severely limited. The system automatically notifies the police in the flood-prone area using messages which do not require interpretation by skilled hydrologists.

Engineers and hydrologists may however monitor the system's performance using remote terminals connected via standard British Telecom telephone lines.

The system description presented here is based largely on that of
Brunsdon and Sargent (1982).

8.3.2 The catchment

The catchment, situated in East Lothian, Scotland, some 25 km east of
Edinburgh (Figure 8.3.1) has an area of 270 km^2. Here the river Tyne
rises in the steep Lammermuir Hills and flows through the town of
Haddington causing periodic inundation. The runoff is concentrated by
two tributary systems, one of which enters the main branch of the Tyne
immediately upstream of the town. This, combined with the steepness and
configuration of the tributaries, prevents the use of upstream river
monitoring to provide flood warnings. In such cases, flood forecasting
has to be based largely on rainfall-runoff modelling.

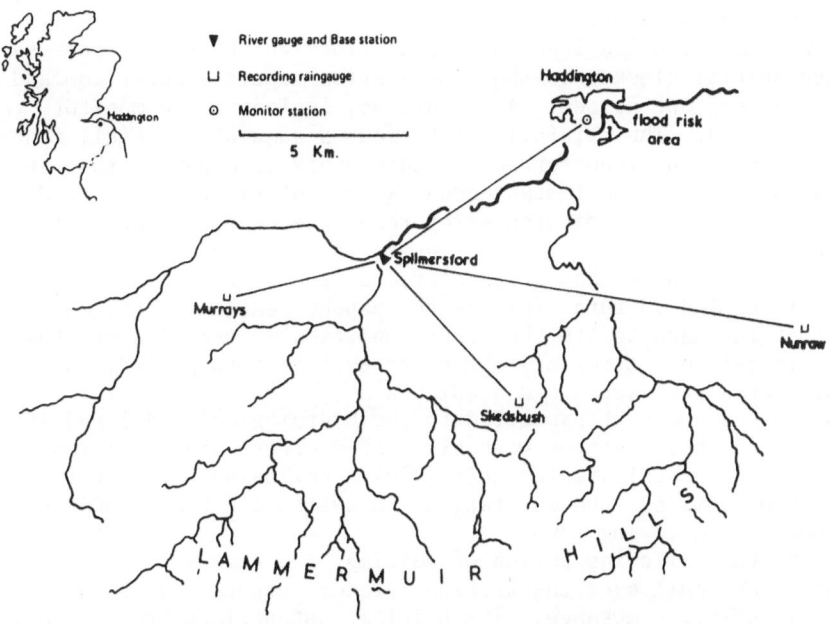

Fig. 8.3.1. The Haddington catchment

8.3.3 Instrumentation, telemetry and computer hardware and software

The telemetering network

It was decided to install the base station, which provides the central
intelligence and control functions for the scheme, at Spilmersford

(Figure 8.3.1) where there is good accommodation with mains power and
telephone access and an existing river flow gauge which could be simply
hard-wired into the base station. For reasons of economy and reliability
a simple half-duplex radio telemetry system was chosen to link the
scheme's other gauges to the base station. Three raingauges were deployed
at altitudes mid-way between the low-lying main valley and the tops of
the Lammermuir Hills. The sites at Nunraw, Skedbush and the Murrays
were chosen to provide potentially useful data on direction of movement
of storm cells as well as rainfall amounts at the sites (Figure 8.3.1).
This arrangement of stations was shown to be feasible by radio path
surveys which promised acceptable communication links between the raingauge outstations and the base station. A further link was necessary to
transmit information from the base station to a site with 24-hour manning,
so it was decided to provide a monitor at Haddington police station where
arrangements for action on flood warnings could be implemented.

The base station, which is a small, desk-top, microcomputer, gathers
information from outstations over radio links using a polling technique.
The microcomputer interrogates the outstations and monitors there outputs, carries out checks on the data and reports system faults, calcu-
lates and displays flood forecasts when necessary, and relays information
to the outlying monitor.

Each outstation has its own crystal clock which is synchronised with
the base station clock when the base station issues a reset command
during start-up operations. All stations, including the monitor, respond
to this command. The rainfall outstations accumulate rainfall continuously, values being transferred to store every 30 minutes; the river
level outstation stores instantaneous values of river level at 30-minute
intervals. Thirty-one minutes after reset the base station calls an
outstation and thereafter calls it every thirty minutes, so requesting
data one minute later than it is stored at the outstation. Each outstation is called in turn; failure to respond results in two more
attempts being made before the error condition is recorded and the next
station is called. After all the stations have been polled, the data
thus collected are stored and displayed at the base station.

These data are next presented to the hydrological model and its
forecasts, if any, together with all system errors are displayed on the
base station's visual display unit (VDU). Following interrogation, the
base station commands the outstations to switch off their radio equipment thereby conserving power.

During the following period of waiting, the base station will
respond to any call over the British Telecom network and acknowledge it
with an identifying message. It solicits instructions from the caller
which may take the form of requests for data, test procedures or re-
booting following a power failure of the base station. The outstations
switch themselves on again from their own clocks when it is time to
update their stores and then listen for a call from base. The base
station will call them again as previously described.

However, if the base station receives a reply from an outstation
which was previously in error, it sends a further command to that outstation requesting all its stored data. These are received by the base
station and the gaps resulting from previous loss of data for that

station are filled in (the capacity of the outstations' stores is 24 hours).

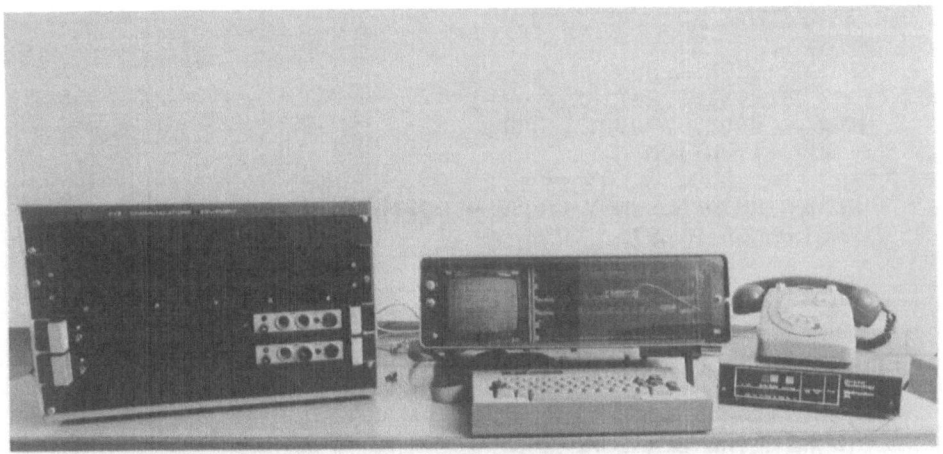

Fig. 8.3.2.   Complete base station arrangement; radio on left, microcomputer in centre and telephone modem at right.

The base station

The base station microcomputer, shown in Figure 8.3.2, uses the M6800 microprocessor from Motorola, and wherever possible, micro-modules have been used in its construction. These modules, which are developed and tested printed circuit boards, include a central processor with one parallel and two serial input/output ports, random access memory for data use, read-only memory for program storage, a display control module and a serial input/output module for connection via a modem to the radio telemetry system. The visual display unit has no intelligence of its own so it is managed by the central processor. The system runs auto-matically; it gives alarms, warnings, displays predictions and notes failures. Examples of the information printed out are given in Figure 8.3.3.

The computer may be accessed remotely over the telephone network which allows the operator to intervene and modify, or start remotely, the complete system. Commands are provided to list data from the base station databank with retrieval of up to five days of data possible (see example in Figure 8.3.3(c)). The screen display may also be reproduced over the telephone network. There is a command to test the hydrological model; requests for manual entry of rainfall and river level data are made and the resulting flood forecasts appears on the VDU screen, the contents of which may be sent back to the operator. A further command sends the data and results to the monitor station if required.

Real-time operations are fed into the computer via the interrupt line. Suspension of the normal program loop, which waits for one of the telephone commands (see Figure 8.3.4(a)), is forced by a real-time

```
    TIME 1858.
 +:D
      INSTITUTE OF HYDROLOGY
      HADDINGTON FLOOD WARNING
      TIME-DATE 1846.02.10.1981.
      DATA:
      SPILMR-SKSBSH-NUNRAW-MURRYS
      R-L MM.R-TIPS.R-TIPS.R-TIPS
      1526.  0002.  ?000.  ?000.
      FLOOD SITUATION
 0       +++    BLUE ALERT  +++
      FURTHER RAINFALL MAY CAUSE FLOODING
      MAINTENANCE ALERT    CODE  6
```

Fig. 8.3.3(a) showing BLUE alert at 18'46 on 2 Oct. 1981

```
    60875 340 436,INPUT YOUR INSTRUCTIONS
    :D
    ;)       INSTITUTE OF HYDROLOGY
      HADDINGTON FLOOD WARNING
      TIME-DATE 2116.02.10.1981.
      DATA:
      SPILMR-SKSBSH-NUNRAW-MURRYS
      R-L MM.R-TIPS.R-TIPS.R-TIPS
      1831.  0000.  ?000.  ?000.
      FLOOD SITUATION
    )       +++ AMBER ALERT  +++
      MINOR FLOODING EXPECTED TO
                COMMENCE IN    2.00   HR
                CONTINUE FOR   0.50   HR
      PEAK RIVER LEVEL OF      1.580  M
                EXPECTED IN    2.00   HR
      MAINTENANCE ALERT    CODE  6
```

Fig. 8.3.3(b) showing AMBER alert at 21'16 on 2 Oct. 1981

interrupt and the interrupt service routine (Figure 8.3.4(b)) is entered.
As interrupt activities have top priority any telephone call in progress
to the computer is halted temporarily in order to permit calls to outstations at the arranged interrogation times. Figure 8.3.4(c) shows
the logic of the subroutine which deals with data received over a radio
link and is part of the interrupt service routine.

The outstations

The outstations are powered by a 12-volt D.C. battery charged by solar
panels. Rainfall stations use a tipping bucket system whereas the river
level station employs a float recorder with optical shaft encoder.
Figure 8.3.5 shows a raingauge outstation.
    The raingauge outstation counts bucket tips in a buffer counter and
each 30 minutes transfers the total into that location in its 48 byte

```
+:RIME 2130.
RAW DATA


TIME & DATE 2130 02 10 1981
SPILMR-SKDBSH-NUNRAW-MURRYS
R-L.MM.R-TIPS.R-TIPS.R-TIPS

1831.  0000.  ?000.  ?000.
1776.  0000.  ?000.  ?000.
1699.  0000.  ?000.  ?000.
1645.  0005.  ?000.  ?000.
1606.  0004.  ?000.  ?000.
1526.  0002.  ?000.  ?000.
1458.  0003.  ?000.  ?000.
1407.  0001.  ?000.  ?000.
1384.  0002.  ?000.  ?000.
```

Fig. 8.3.3(c) showing raw data for the period 1930–2130 on 2 Oct. 1981 (river level in mm and rainfall from gauge 1 in ½ mm tips).

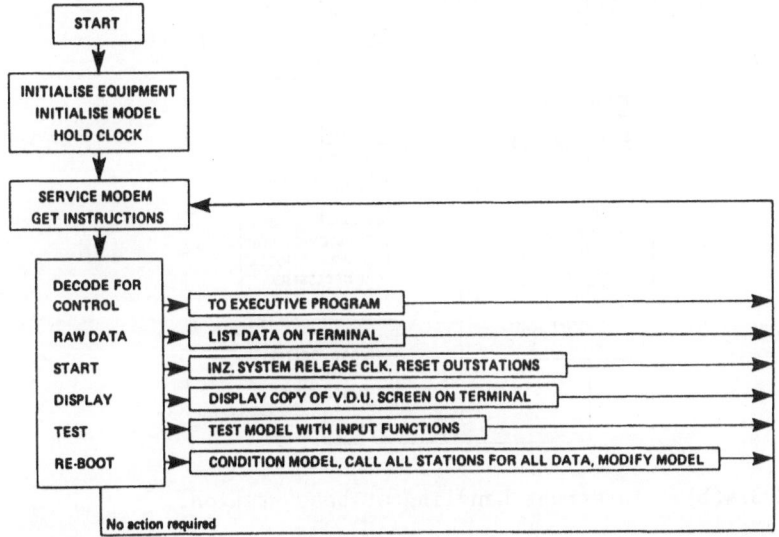

Fig. 8.3.4(a). Main program in base station

store which is pointed to by an address register. This register is then incremented and the buffer counter reset to zero, ready for the next accumulation period. These operations continue until the store is full, when the address counter is reset to zero and further data overwrite data from the beginning of the store. This provides a data capacity of 24 hours which matches the recovery period of the model.

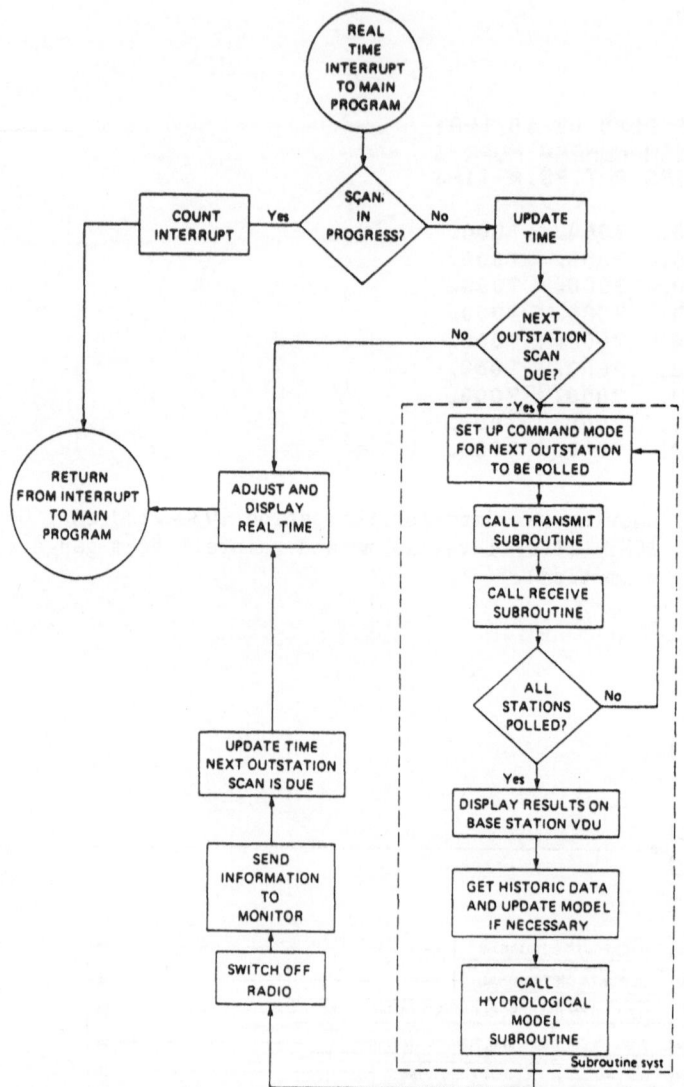

Fig. 8.3.4(b).   Interrupt handling at base station.

River level data are stored in a similar sequence except that the sensor provides its output directly in digital form for storage every 30 minutes.

In the early outstations the logic was hard wired, but now the CMOS version of the M6805 microprocessor is incorporated. This processor improves the station's noise immunity, simplifies maintenance, provides flexibility and opens possibilities for distributed intelligence, yet maintains the low power consumption (less than 1 mW when on standby) which is essential if the stations are to be sited at remote locations.

Its software features are similar to the M6800.

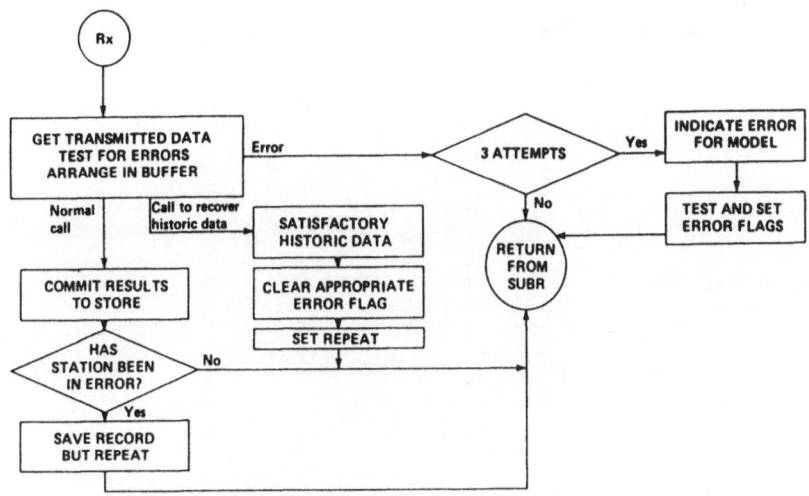

Fig. 8.3.4(c). Receive data control sub-routine

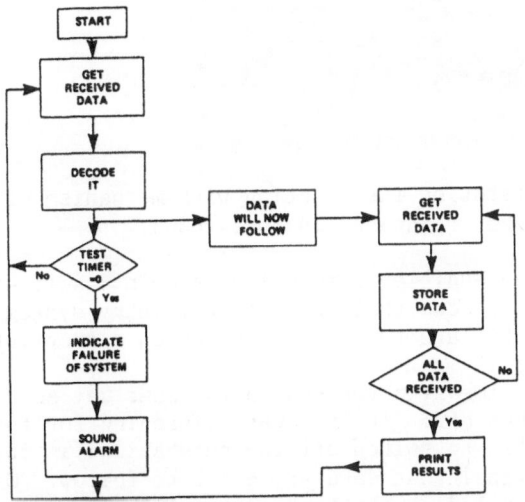

Fig. 8.3.4(d). Monitor program.

The monitor station

The monitor station consists of a radio receiver interfaced to electronic hardware with an attached printer. This provides a hardcopy output of the base station display. The unit is fully automatic, providing visual warnings of floods, together with transmission and system errors. All

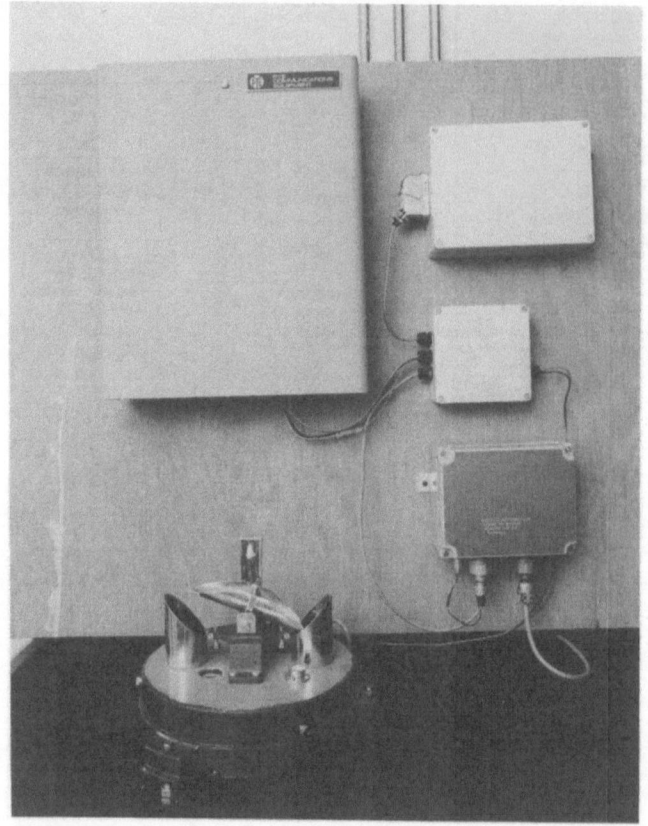

Fig. 8.3.5. Typical outstation installation with mechanism of
tipping-bucket raingauge in foreground.

errors and warnings actuate an audible alarm which cancels itself after
one minute. The monitor also checks that the main telemetry system has
responded every 30 minutes. Failure to do so actuates an appropriate
warning.

The monitor is reset and time-synchronised in the same way as the
outstations, but never switches off: it is always listening for further
commands. The code transmitted to switch off the outstations is interpreted by the monitor as a signal that data are about to follow. On
receipt, the complete data, including the base station's screen display,
are checked and stored. Providing a correct transmission has been
received, the data are printed. Failure results in an appropriate alarm
being set, accompanied by a typewritten message on the printer. The
monitor completes its cycle of actions by resetting the time and awaiting
a further call from the base station. These operations are controlled
by a program which runs in the monitor's own microprocessor and associated random access and read-only memories. The program logic is summarised in Figure 8.3.4(d).

8.3.4 Forecasting model

A simple four parameter catchment response model for simulating isolated
storm events (the Isolated Event Model) has been developed and proved in
yielding realistic flood hydrographs for design purposes (Natural Environment Research Council 1975; Mandeville 1975); this model has been
adapted for real-time forecasting and implemented on the base station
microcomputer. The model contains a non-linear storage component from
the inflow-storage-outflow class of models described in Section 2.4. It
is a 'lag and route' model, effective rainfall being delayed prior to
routing through storage. The storage-discharge relationship is assumed
to be of the form

$$S = \alpha Q^{1/2} \tag{8.3.1}$$

where α is a routing coefficient. Combined with the equation of continuity, (8.3.1) yields the differential equation

$$\frac{dQ}{dt} = \frac{2}{\alpha} (P' - Q)Q^{1/2}. \tag{8.3.2}$$

Here P' denotes the lagged effective rainfall and is defined in terms of
the pure time delay, b, and the runoff coefficient, c:

$$P'_t = c.P_{t-b}. \tag{8.3.3}$$

   In the model developed by Mandeville (1975), the runoff coefficient
was estimated as a function of the prevailing soil moisture deficit.
However, in the absence of such information in real-time on the Tyne
catchment, c was defined in terms of the flow, Q_o, in the river prior to
the storm, through the relationship:

$$c = \gamma + \delta \ln(Q_o) \tag{8.3.4}$$

Values of α and b were estimated by trial and error while γ and δ were
estimated by regression using historical data for ten flood events
recorded on the Tyne; a further eight events were used to test the calibrated model.
   The model thus developed was for the Tyne catchment to East Linton,
a site some 12 km downstream of Haddington. In order to provide effective flood warnings for Haddington, the model was adjusted by reducing
the pure time delay parameter, b, to represent the faster response of
the smaller catchment. In addition, Equation (8.3.4) was amended to
estimate c from telemetered flows at Spilmersford, some 8 km upstream of
Haddington.
   The above model structure has the advantage of requiring only one
value of flow, Q_o, to run the model in simulation to forecast the flood
hydrograph. No attempt is made to update model forecasts using telemetered flow data since these refer to a different site; the model is,

however, re-run at half hourly intervals as new rainfall data become
available in real-time.

The computer program for the model, written in FORTRAN, occupies
about 16 Kb of storage on the base station microcomputer and typically
takes 11 seconds to compute a forecast of river levels at Haddington
for up to 36 hours ahead. Incoming rainfall data from the outstations
are checked for spurious values (rainfalls greater than 100 mm in half
an hour are assumed invalid) and a weighted mean of the acceptable
catches returned from working outstations is calculated for input to the
model. If the rainfall exceeds a predetermined threshold an event is
deemed to have begun. Then the river level measured at Spilmersford is
converted to flow: levels less than 0.42 m or greater than 10 m are
assumed to be invalid and the previous flow value issued. If no historic
value is available the median flow at Spilmersford is assumed. From
this initial flow, the runoff coefficient c is calculated and this, with
the rainfall data, is fed through the model to produce forecasts of level
at Haddington. When the latter reach preset values, flood alerts are
relayed to the local police via the remote monitor. The alerts are in
the form of a series progressing from a warning of possibile flooding if
rainfall continues (BLUE alert), through minor inundation in the flood-
way (AMBER alert) to major flooding of domestic property (RED alert).
In each case the time of commencement of flooding, time of peak flooding
and the predicted flood duration are relayed to the remote monitor.

As more rainfall data become available at each half hour interval,
fresh forecasts are made and the monitor output is updated. As the
flood progresses and the peak flow passes, the expected time of flood
cessation is relayed to the monitor, updated at half hourly intervals,
and the flood alarms recede down to the initial one of possible flooding
if further rainfall occurs. Finally, 12 hours after significant rainfall
ceases, the event is considered to have ended. An "all clear" is
relayed to the monitor and the operating system cuts out the model and
reverts to watching for significant rainfall.

If contact is re-established with a previously incapacitated raingauge the data stored at the outstation are accessed and the model
recalculates its flow forecasts. Similarly the system is programmed to
make an orderly recovery from a mains failure at the base station.

8.3.5 Operational performance

The system was installed in the summer of 1981 following two years
design, construction and testing at the Institute of Hydrology, and is
working satisfactorily with the radio links, base station and monitor
all proving to be reliable. The outstations, as already mentioned,
have been modified and better reliability and easier maintenance are
expected.

The scheme has already successfully predicted flooding during a
complex double-peaked event with a 12-year return period which occurred
during 2nd to 4th October 1981. It also demonstrated robustness by
recovering from a mains failure at the base station during the event
and by making forecasts when two of the rain outstations were out of
commission.

This flood event was monitored remotely on a terminal at the Institute of Hydrology in Wallingford and three of the print-outs obtained are shown in Figures 8.3.3. Figure 8.3.3(a) shows a copy of the VDU screen contents obtaining at 18'46 hr on 2nd October when a BLUE alert existed. By 21'16 hr an AMBER alert (Fig. 8.3.3(b)) existed, predicting flooding at Haddington in 2 hours time. The listing (Fig. 8.3.3(c)) of the data collected at 30 minute intervals (most recent first) shows that the river level was still rising at Spilmersford but that rainfall had stopped. The maintenance alert code 6 indicates that the two raingauge outstations with code values 2 and 4 were inoperative. This condition is reflected in the listed data by a question mark preceding the missing data items.

There are a number of further improvements which are foreseen for the present system. The incorporation of 'intelligence' in the outstations through microprocessors permits easy modification for use with other sensors and allows data validation on-site and distributed modelling. The model itself can be improved by allowing a variation in c during an event and by making better use of the recorded variations in rainfall at the three gauges. Improvements in event definition are also under consideration. All these modifications exploit the ease with which the system may be programmed. Each modification can be incorporated into a new PROM (programmable read-only memory) and simply plugged into the processor. Similarly changes or additions to the network of gauges can be accommodated, and other models (including those for ungauged catchments) may be installed to give the system great flexibility. Power consumption by the outstations can be reduced to a level where the solar panels and large batteries could be replaced by small cells which would need replacing every six months. The average costs of an outstation in the U.K. are £2000. A monitor or base station costs £3000. It was found in practice that solar batteries for outstations perform well.

8.4   An on-line monitoring, data management and water quality forecasting system for the Bedford Ouse river basin

8.4.1 Background

In the operational management of water resource systems, there is frequently a major requirement for information on the present state of the system and on likely future changes in the system state, particularly with respect to water quality. Operational managers must be able to respond quickly to emergency situations in order to protect and conserve water quality and maintain adequate water supplies for public use. Moreover, the costs of water treatment and bankside storage are particularly high and there are therefore considerable benefits to be gained from the efficient operational management of river systems with respect to water quality (Beck 1981).

In recent years there has been some progress towards providing more efficient operational management of river systems by the installation of automatic river level recorders and continuous water quality monitors. The latter measure river levels and water quality variables such as

dissolved oxygen, ammonia, and temperature and, if combined with a telemetry scheme relaying information to a central location, provide immediate information on the state of the river for pollution officers.
Advances in technology now provide an opportunity to use this information
in conjunction with mathematical models for making real-time forecasts
of water quality. The practical problems associated with the continuous
field measurement and telemetering of water quality have largely limited
the application of on-line forecasting and control schemes. The continuous flow of water past sensors for measuring water quality gives rise
to severe fouling of optical and membrane surfaces, thereby drastically
reducing the accuracy of the data produced. In recent years, however,
there have been several studies and applications of continuous water
quality monitors (Briggs 1975; Kohonen et al. 1978). Most UK water
authorities have established monitoring and telemetry schemes (Hinge and
Stott 1975; Cooke 1975; Caddy and Akielan 1978; Wallwork 1979) and report
reasonable reliability provided the monitors are regularly maintained.

The application of particular interest here is an extensive water
quality monitoring and telemetry scheme which has been developed by the
Anglian Water Authority along the Bedford Ouse river system in eastern
England. As indicated in Figure 8.4.1 automatic water quality monitors
have been installed at several sites along the river and data on such
variables as dissolved oxygen, pH, ammonia, and temperature are telemetered at regular intervals to the central control station located in
Cambridge. This telemetry scheme has been extended to include information on flow and to use a mini-computer located in Cambridge to analyse
the data on-line. The system provides rapid information on the present
state of the river and incorporates a dynamic water quality model for
making real-time forecasts of flow and quality at key locations along
the river system (Whitehead et al. 1979, 1981).

The Bedford Ouse river system had previously been the subject of an
extensive research programme aimed at developing mathematical models for
use in water quality planning and operational management (Fawcett 1975;
Whitehead 1975). The dynamic water quality model developed as part of
this research programme has formed the basis of the model now used for
operational forecasting on the Bedford Ouse. The forecasting system is
described in detail by Whitehead et al. (1982).

8.4.2 The catchment

The Great Ouse basin is one of the largest river catchments in England
and is located in the low lying eastern part of the country. It
receives one of the lowest annual rainfalls in the country which,
combined with a lack of gradient, produces a generally extremely slug-
gish river.

For simplicity, the Great Ouse basin can be considered as consisting
of four sections: the Bedford Ouse, the Ely Ouse, the Middle Level, and
the Tidal River. The Bedford Ouse area is the most upstream section of
the river and has the highest population density - over 50% of the
population is in the upper basin which represents only 35% of the total
area. Figure 8.4.2 is a schematic map of the Bedford Ouse area, showing
River Ouse catchment and its major centres of population. As might be

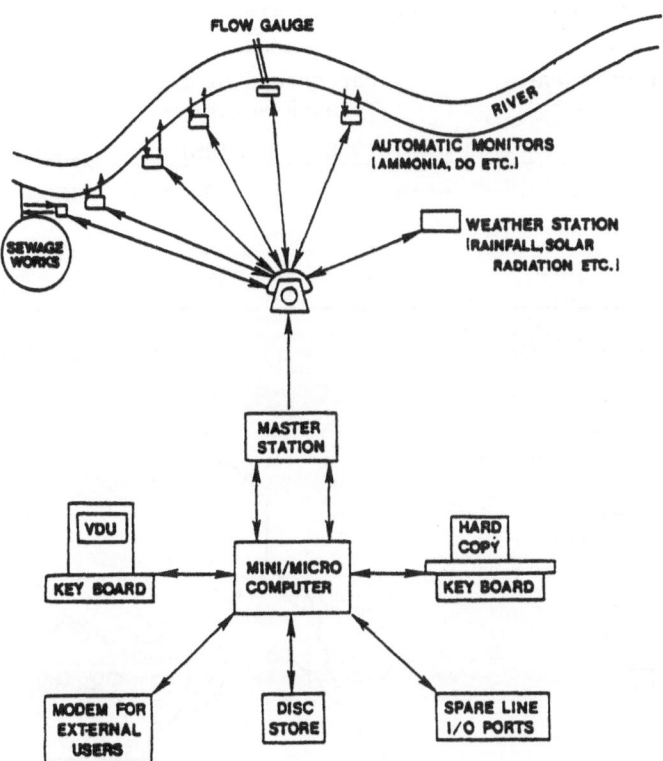

Fig. 8.4.1.   Schematic representation of water quality monitoring,
              telemetry and data processing system for the Bedford
              Ouse.

expected in a predominantly rural area, there are few large sewage
treatment works and very few direct discharges of industrial effluent.
Nevertheless, effluent discharges can represent a significant proportion
of the river flow under low flow conditions as is demonstrated in Table
III which underlines the deteriorating situation in the future due to
the expected growth in population of the New City of Milton Keynes.   It
was largely for this reason, and its associated implications in terms
of volumes of water required for public supply and volumes of effluent
requiring disposal, that this study has concentrated upon the Bedford
Ouse area.
     There are three major abstractions for potable supply within the
area – at Foxcote, Clapham (Bedford) and Offord; and these are also
shown in Figure 8.4.2. A further abstraction at Brownshill may be
considered in the future as a means of increasing the yield of the public
water supply reservoir at Grafham (which is at present supplied by the
abstraction at Offord).

Table III.   Proportion of Effluent in the River Ouse

Location (in order down-stream)	Percentage of Effluent in Total River Flow		
	1968	1981	2001
Foxcote	20	30	30
Clapham	35	60	70
Offord	45	60	85
Brownshill	25	35	45

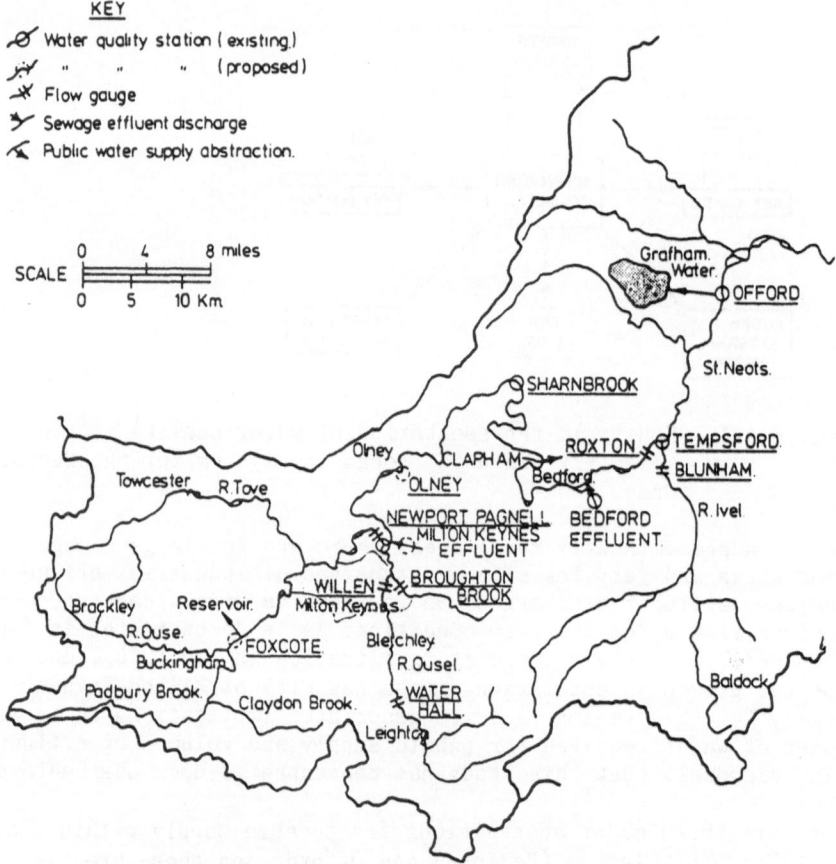

Fig. 8.4.2.   Bedford Ouse system showing the deployment of water quality monitors and flow gauging stations.

8.4.3 Instrumentation, telemetry and computer hardware and software

System benefits

The benefits to be gained from the installation of a water quality
monitoring and forecasting system on the Bedford Ouse are the following:
(a) Improved operational management. The system provides real-time (i.e.
immediate) information on the quality of the river thus enabling pollution officers to detect pollutions at an early stage and to respond
quickly and efficiently;
(b) Dissemination of information. Information on the present quality of
the river in the form of daily, weekly and monthly summaries can be made
available for management reports;
(c) Optimization of capital and revenue expenditure. A significant
saving in capital expenditure and in laboratory and other running costs
should be possible through the more efficient operational management of
the system. For example, it would be unnecessary to expand the water
storage facilities at Bedford (currently 18 hours hold-up time) as a
reliable forecast of pollutant loads allows sufficient time to switch in
alternative supplies from groundwater or reservoirs. Also savings in
operational costs can result if the time of a pollution load can be forecast and treatment such as chlorination increased to match the arrival
time rather than be increased well in advance of the arrival;
(d) Improved understanding of water quality. Information from the water
quality data would improve the understanding of the river system and the
effect of water quality on the river ecology;
(e) Accurate predictions of time of travel and dilution factors. Given
a model of the river system it is possible to predict the time of travel
of a pollution load and the dilution factors along a particular stretch
of the river. This assists the pollution inspectorate in assessing the
significance of pollution under given conditions.
(f) Forecasts of water quality. A model of the system can provide forecasts of key water quality variables (nitrate, dissolved oxygen, ammonia,
etc.) at critical points along the river.

Outstations and telemetry

The Bedford Ouse automatic water quality monitoring stations have been
installed to monitor the impact of effluent discharges on river water
quality and to protect river abstraction sources for potable supply. In
choosing which variables to measure, the philosophy has been to use only
sensors which are reliable for long-term duty; variables such as ammonia
and nitrate are monitored using specific ion electrodes while standard
sensors are used for dissolved oxygen, conductivity, temperature and pH.
In Figure 8.4.3, an example of a DO record shows clearly the daily
oscillation of dissolved oxygen production and consumption processes and
the longer term fluctuations which are due to other variables such as
temperature and streamflow.
     Automation of the outstations is achieved using electronic timers
which activate various processes in a predetermined sequence. These
control processes are:

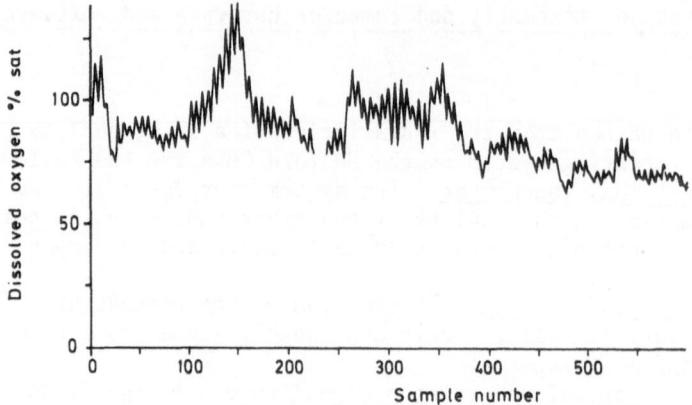

Fig. 8.4.3. Dissolved oxygen levels on the Bedford Ouse (Tempsford
             Monitor), July to December, 1978, measured at 4 hourly
             intervals.

(a) the automatic cleaning of pipelines and sensors using compressed air;
(b) the automatic correction of sensors using standard solutions;
(c) the automatic selection of river water to be monitored e.g. at
    Tempsford (Figure 8.4.2) two rivers can be monitored: two hours
    monitoring on the River Ivel is followed by monitoring on the River
    Ouse. At Newport Pagnell, a similar monitoring scheme applies to
    the Rivers Ouse and Ouzel;
(d) the automatic suspension of telemetry to keep the station off-line
    when automatic cleaning and calibration events occur.
    Further details of the outstations are given in Caddy and Whitehead
(1981).
    The telemetry scheme used for flow and water quality monitoring is a
Dynamic Logic TELTEL system which employs the public switched network
telephone system. Outstations are interrogated by the mini-computer
controlling the system through the Dynamic Logic TELTEL dedicated master
station.
    Flow stations included in the telemetry scheme are located at the
following sites (Figure 8.4.2):
(i)   Newport Pagnell - River Ouse
(ii)   Willen   - River Ouzel
(iii)   Broughton   - Broughton Brook
(iv)   Roxton   - River Ouse
(v)   Blunham   - River Ivel
    Water quality stations are located at the following sites:
(i)   Newport Pagnell - River Ouse and River Ouzel
(ii)   Sharnbrook Mill - River Ouse
(iii)   Bedford   - Sewage Effluent
(iv)   Tempsford   - River Ouse and River Ivel
(v)   Offord   - River Ouse
(vi)   Barrington   - River Rhee
(vii)   Bottisham   - River Cam
    Additional stations are to be installed by the Anglian Water Authority

at Buckingham and at Olney, both on the River Ouse.

Computer hardware and software

The success of the monitoring and forecasting system depends on the reliability of the instrumentation and telemetry hardware and the way in which the data are processed. At the most basic level of analysis it is necessary to convert the instrument signals to meaningful concentration units. With the use of a mini-computer it is possible to make these conversions conveniently and to provide data listings, summaries and graphical output for operational management. On a more sophisticated level the computer can be programmed to forecast water quality conditions using a mathematical model and hence provide advance warning of the effects of pollution events.

In order to meet these requirements a DEC MINC computer system has been installed at the Cambridge office of the Great Ouse Area Office of the Anglian Water Authority. The DEC MINC system incorporates a PDP 11/23 (16 Bit) computer with 128k of store, a storage system based on 2 floppy discs (1 Megabyte of store), a visual display screen, together with a keyboard, extensive input-output facilities and a software system capable of supporting a FORTRAN compiler. In addition to the basic system a DEC writer was also required for listing data and a plotter for producing graphical output.

Data are relayed to the control centre at Cambridge where equipment is situated for telemetry control, data processing and report production (Figure 8.4.1). The scanning of the flow gauges and water quality monitors is computer controlled with the outstation telephone numbers, which are held in computer memory, being fed to the master station (autodialler/receiver). After making contact with an outstation, data are transmitted and checked before being stored in the computer memory for further processing. Information can be presented to operating staff in several forms, and data summaries, alarm messages and graphical displays are available on either the printer or on the visual display unit. The computer system can be operated by staff not qualified in computer science and the Institute of Hydrology have devised an interactive system which prompts the user into selecting options which appear on the visual display unit.

An important function of the mini-computer is the storage of data and this is achieved using the two double density floppy discs capable of storing several weeks' data for all the monitoring stations. Data stored on the floppy discs can be retrieved at any time for direct listing or for the calculation of summary statistics such as daily mean, maximum, minimum and 95 percentiles. The logging program on the micro-computer allows the acquisition of both analogue data (i.e. measurements from water quality instruments) and digital signals related to equipment status (e.g. pump failure). Comparison of data with preset alarm levels enables warning messages to be given in the event of a pollution incident. Given such warnings the situation can be investigated using the mathematical model to obtain forecasts of flow and water quality up to 80 hours ahead, given certain assumptions on the upstream conditions during the forecast period.

The options available on the mini-computer system are selected using a simple interactive system by which the operator is prompted to answer questions displayed on the screen. The options available may be summarised as follows:

Start logging: allows the operator to start the collection and storage of data from the outstations at predetermined times. The logging program also converts the instrument signals to river flow or concentration units, allows for calibration factors and prints out alarm level messages, equipment status and daily statistics.

Interrogate outstation: allows the operator to interrogate any outstations of his/her choice immediately. This is particularly useful if a pollution incident has occurred, or when installing and maintaining instruments.

Plot data: provides the operator with a choice of various graphical presentations of data on the visual display screen. Plots may be obtained for the data collected at any outstation.

Edit master file: allows the operator to create or edit the masterfile which holds all the fixed parameters describing each outstation's configuration and other attributes such as site name, telephone number, type of instruments, equations defining data conversions, calibration factors, alarm levels and warning messages and equipment reliability.

Print master file: allows the operator to list the contents of the masterfile on the DEC printer.

Print dial out statistics: reports the number of successful and failed dial out attempts for each outstation.

Print daily, weekly, monthly and annual summaries: prints the selected summary (mean, maximum, minimum standard deviation, 95 percentile) on the DEC printers (see Table IV).

Table IV. Summary of daily data (Current data at Barrington on Friday 21 August 1981)

Instrument	'n'	Minimum	Maximum	Mean	Standard deviation	95%'ile	Units
Dissolved oxygen	7	48.8	105.1	75.9	22.70	113.3	% sat
Temperature	7	14.0	15.5	14.6	0.51	15.5	centigrade
Conductivity	7	869.1	957.0	927.7	33.35	982.7	μ siemens
Ammonia (as N)	7	0.04	0.04	0.04	0.00	0.04	(mg/1)
pH	7	7.7	8.3	8.0	0.21	8.3	-

Run flow and water quality model: asks the system to run the flow and water quality model using data held on disc.

Run impulse model: asks the system to run the impulse model using data supplied by the operator on the type of pollutant (BOD, ammonia or conservative pollutant), pollutant flow rate, concentration and location of discharge.

List station data: prints the output data for a specified station (see Table V).

Table V. List of station data (Data logged to data for Barrington on Tuesday 22 September 1981 11-09-39).

Date	Time	Dissolved Oxygen (% saturation)	Temperature (centigrade)	Conductivity (μ siemens)	Ammonia (as N) (mg/1)	pH
19.8.81	0900	50.00	15.60	947.21	0.04	7.73
19.8.81	1120	64.06	15.79	966.74	0.04	7.83
19.8.21	1149	66.79	15.79	966.74	0.04	7.86
19.8.81	1216	71.48	15.79	956.97	0.04	7.88
19.8.81	1219	72.26	15.89	956.97	0.04	7.89
19.8.81	1224	73.43	15.89	956.97	0.04	7.89
19.8.81	1234	75.78	15.89	956.97	0.04	7.93
19.8.81	1239	76.95	15.89	956.13	0.04	7.83
19.8.81	1244	76.56	15.89	956.97	0.04	7.99
19.8.81	1249	77.73	15.89	956.97	0.04	7.95

The programs for all these options have been written in standard FORTRAN so that the software can be easily modified as water authority requirements change or can be transferred to other computer systems.

8.4.4 Forecasting model

General structure

The model developed for the Bedford Ouse is designed to use time varying input (upstream) measures of flow and water quality to compute time varying output (downstream) responses. The model characterizes the short term (hourly) system behaviour and provides a mathematical approx- imation to the physico-chemical changes occurring in the river system. The structure of the model is shown in Figure 8.4.4. A multi-reach flow model is linked with the water quality model so that flow-quality inter- actions are incorporated directly. A detailed description of the flow and quality model is provided elsewhere (Whitehead et al. 1979, 1981); a summary is included here.

Flow routing model

The flow routing model is based on a 20-reach representation of the Bedford Ouse river (see Table VI), in which each reach is characterized by a number of compartments. The model for flow variations in each compartment is based on an analogy with the mass balance equations for the variations in concentration of a conservative pollutant under the

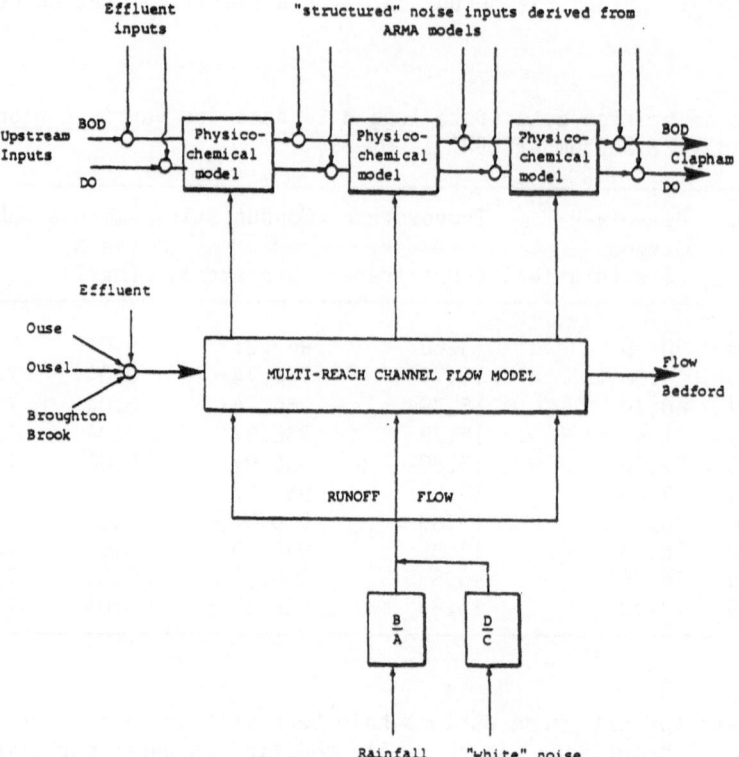

Fig. 8.4.4. Schematic representation of dynamic flow and water
           quality model

assumption of uniform mixing over the compartment. Alternatively, the
model may be viewed in hydrological flow routing terms as one in which
the relationship between inflow I, outflow Q and storage S in each
compartment is represented by the equation:

$$\frac{d(KQ)}{dt} = (I - Q) \qquad\qquad (8.4.1)$$

where K is a travel time or residence time parameter which varies as a
function of stream velocity as follows:

$$K = \frac{L}{UN} \qquad\qquad (8.4.2)$$

where N is the number of compartments in the reach, L is the reach
length, and U, the mean flow velocity in the reach, is related to
discharge through:

$$U = a\, Q^b. \qquad\qquad (8.4.3)$$

Table VI. List of reach boundaries

Reach boundaries	length of reach (m)
Newport Pagnell	
to 5200	
Tyringham Bridge	
to 3000	
Ravenstone Weir	
to 4800	
Olney Weir	
to 4000	
Lavendon Weir	
to 4800	
Turvey	
to 5300	
Harold Weir	
to 8040	
Felmersham	
to 8850	
Stafford Bridge	
to 6260	
Oakley Weir	
to 3370	
Clapham (AWA Abstraction)	
to 3220	
Bromham Weir	
to 4000	
Bell End Weir	
to 5150	
Cardington	
to 2080	
Castle Mills	
to 3680	
Willington	
to 2240	
Barford	
to 4320	
Roxton	
to 7200	
Eaton Socon	
to 4000	
St. Neots	
to 2080	
Offord (AWA Abstraction)	

Thus in the terminology of Chapter 3, the flow routing model for each reach constitutes a cascade of non-linear storage reservoirs in which the non-linearity is introduced by allowing the storage parameter K to vary as a function of discharge Q, thus achieving the observed variation in travel time with flow.

In order to estimate the parameters of the Bedford Ouse velocity-flow relationships (Equation 8.4.3), a series of tracer experiments has been conducted on the river. A known mass of iodide was injected into the river and the iodide concentration determined at one or more selected locations downstream either continuously using selective ion detection equipment or by sampling the river water and subsequent analysis of the samples at the Institute of Hydrology (see Fig. 8.4.5). Using information on velocity and flow rate from these experiments and on earlier experiments conducted in 1975, the parameter values in equation (8.4.3) have been determined as:

$a = 0.045$ and $b = 0.67$.

Given information on upstream and tributary inputs, the flow routing model simulates downstream flow by solving Equation (8.4.1) with K defined through Equations (8.4.2) and (8.4.3).

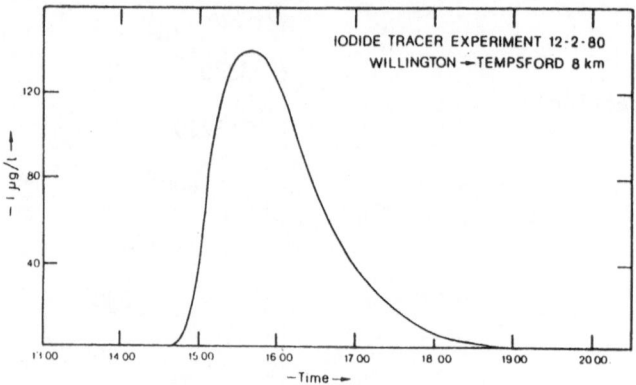

Fig. 8.4.5.   Tracer curve obtained from iodide tracer experiment conducted on 12.2.1980 for the Willington to Tempsford reach of the Bedford Ouse.

Water quality model

The water quality model for the Bedford Ouse is similarly based on a mass balance principle but includes factors to allow for the non-conservative nature of water quality variables. For example, dissolved oxygen in the river is a balance between the various sources and sinks of oxygen. On the one hand, there is oxygen supplied by the re-aeration from the atmosphere and photosynthetic oxygen produced by plants and algae and, on the other hand, oxygen is being consumed by respiration processes and the removal of oxygen during the bacterial breakdown of organic material and effluents. The river is assumed to consist of a large number of compartments. Flow through these compartments is described by Equation (8.4.1). Each compartment represents a continuously stirred tank. The mass balance equations developed to simulate water quality behaviour are as follows:

Chloride

$$\frac{dx_1(t)}{dt} = \frac{Q_i(t)}{V} u_1(t) - \frac{Q_o(t)}{V} x_1(t) + S_1(t);$$

(8.4.4)

Nitrate

$$\frac{dx_2(t)}{dt} = \frac{Q_i(t)}{V} u_2(t) - \frac{Q_o(t)}{V} x_2(t) - k_1 x_2(t) + S_2(t);$$

(8.4.5)

Ammonia

$$\frac{dx_3(t)}{dt} = \frac{Q_i(t)}{V} u_3(t) - \frac{Q_o(t)}{V} x_3(t) - k_2 \left[\frac{1}{Q_o(t)}\right] x_3(t) + S_3(t);$$

(8.4.6)

Dissolved oxygen (DO)

$$\frac{dx_4(t)}{dt} = \frac{Q_i(t)}{V} u_4(t) - \frac{Q_o(t)}{V} x_4(t) - 4.33 \, k_2 \left[\frac{1}{Q_o(t)}\right] x_3(t)$$

$$- k_3 x_5(t) + k_4 \left(C_s(t) - x_4(t)\right) + S_4(t);$$

(8.4.7)

Biochemical Oxygen Demand (BOD)

$$\frac{dx_5(t)}{dt} = \frac{Q_i(t)}{V} u_5(t) - \frac{Q_o(t)}{V} x_5(t) - k_3 x_5(t) + S_5(t),$$

(8.4.8)

where
$x_j(t)$ refers to the downstream (reach output) concentration (mg/1) for the jth determinand $j = 1,2,\ldots,5$;
$u_j(t)$ refers to the upstream (reach input) concentration (mg/1) for the jth determinand, $j = 1,2,\ldots,5$.
$Q_i(t)$ is the upstream (input) flow rate.
$Q_o(t)$ is the downstream (output) flow rate (determined from the flow model);
V is the reach volume;
$S_j(t)$ refers to the additional sources and sinks affecting the jth determinand such as the net rate of addition of DO in the reach by the photosynthetic/respiration activity of plants,

$j = 1,2,\ldots5.$

$C_s(t)$ is the saturation concentration of dissolved oxygen.

The model has been programmed to run in two modes:

(1) Normal operating mode, in which information on upstream input flows and quality are taken from the outstations and forecasts produced for all downstream reach boundaries up to 80 hours ahead. Because of the considerable travel time on the river, downstream forecasts are based on measured upstream flow and quality. It is necessary, however, to forecast upstream conditions to provide reasonable forecasts for the upper reaches. Techniques for this are given in Whitehead and Young (1979).

(2) Impulse mode, in which the operator can supply information on an upstream impulse discharge of a conservative pollutant, ammonia or BOD. The resultant simulation uses the current flow and river quality data and simulates the slug of pollutant moving down the river system. Again forecasts up to 80 hours ahead are available at all of the 20 reach boundaries.

It is assumed that weather conditions remain stable during this forecast period. It would be possible however to link existing rainfall gauges into the telemetry scheme to forecast flow and quality.

Figures 8.4.6 and 8.4.7 show a typical simulation in impulse mode. The output data from the model can be plotted either as a profile down the river at a specific time or else as a function of time at any selected reach boundary. Thus if information on say the location of the minimum is required the river profile would be plotted. If, however, it is required to know the likely time of arrival of a slug of pollutant at a given point the time-concentration curves would be plotted.

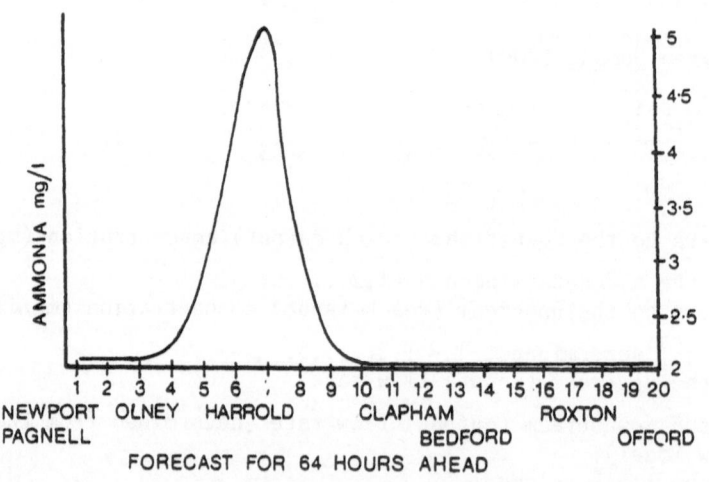

Fig. 8.4.6. Forecast of ammonia profile along the Bedford Ouse.

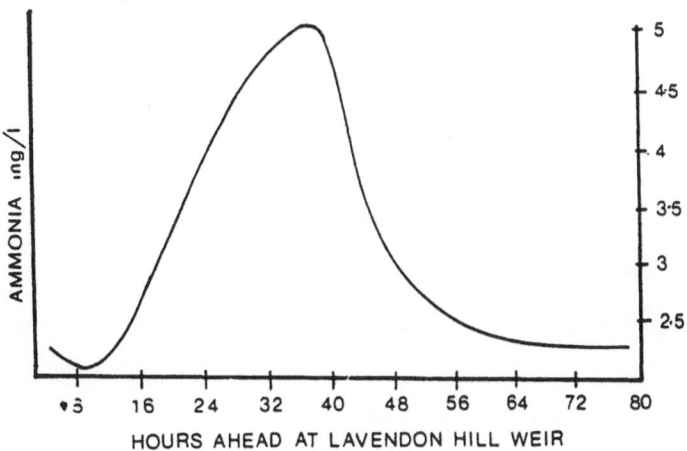

Fig. 8.4.7. Forecast of ammonia for the Lavendon Hill Weir site
on the Bedford Ouse.

8.4.5 Operational performance

The system has performed well operationally during a number of pollution
events; two examples are given here. Following a loss of oxidative
treatment at the Cotton Valley Sewage Treatment works, effluent containing elevated levels of ammonia was discharged into the Bedford Ouse
river at Newport Pagnell. In order to protect the water supply at
Bedford (Clapham abstraction) information was required on the likely
ammonia concentrations at Clapham and the time of arrival of the pollutant. Simulating ammonia using the model indicated an arrival time of
four days with concentrations of ammonia of 1.2 mg/l at Clapham. The
observed levels at Clapham were 1.12 mg/l and the arrival time was 4
days 2 hours. In this situation the model provided valuable information
for the management at the Clapham water abstraction plant and gave them
effectively a four day warning.

The second example relates to the release of unsatisfactory effluent
from Bedford sewage works which resulted in a significant pulse of
ammonia being discharged to the river. Again running the model in an
impulse mode gave reasonable forecasts of ammonia and dissolved oxygen
concentrations downstream at Tempsford (see Fig. 8.4.8).

Considerable experience has been acquired over a number of years by
the Great Ouse River Division of the Anglian Water Authority in the
operation of automatic continuous water quality monitors. This experience has revealed the difficulties associated with operating existing
monitors in the field, namely the fouling of electrode membranes with
consequent signal drifting, difficulties in calibration and electrical
interference. As a result, it has proved difficult to obtain from outstations continuous series of short-term water quality data which would
allow further improvements to be made to the model. As a number of
existing outstations are due for replacement and further outstations are
required, a study has been initiated to develop a new microprocessor

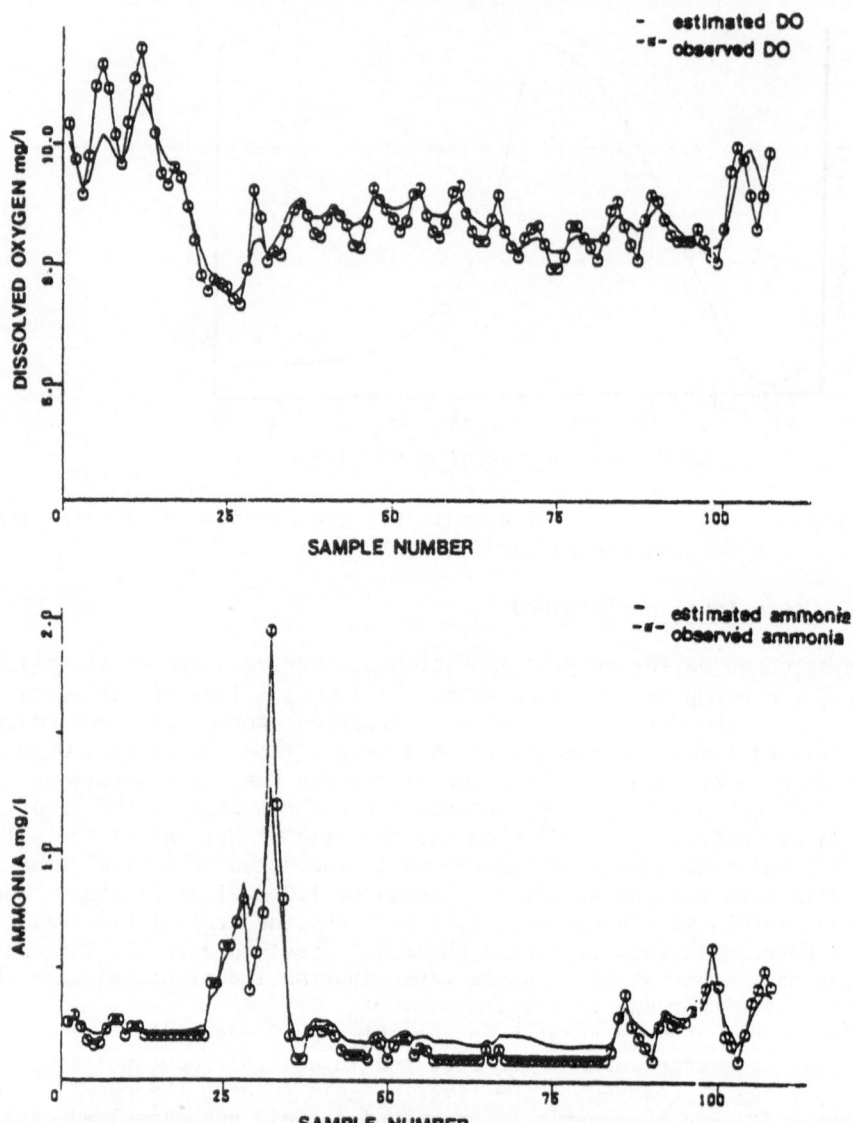

Fig. 8.4.8.   Dissolved oxygen and ammonia forecasts downstream at
             Bedford.

controlled pollution monitor.

Incorporation of a microprocessor provides distributed intelligence
which can be used locally to provide facilities such as data logging,
automatic sensor calibration and cleaning, local processing of data and
control of pumps. Thus, outstation processes would be controlled
directly from a master station, thus achieving considerable savings in
maintenance costs over the present system. The new monitor is currently
undergoing development at the Institute of Hydrology and should be

available for detailed testing in 1983. Once proven the monitor should
open the way for new advances in the field of water quality monitoring,
forecasting and control, and will provide the data necessary to allow
further improvements to be made to the present Bedford Ouse forecasting
model.

8.5 Discussion

The three case studies presented above allow comparisons to be made under
a number of headings. The Dee and Haddington case studies represent
good contrasts hydrologically; on the Dee the value of measured rainfall
data in flood forecasting is relatively small whereas in the case of
Haddington, rainfall data and rainfall-runoff modelling are indispensible
for flood forecasting. This is largely because the gauging of headwater
flows in the upper Tyne is not feasible due to the rather diffuse river
network and the difficulties of accessing outstations for maintenance
purposes. On the Dee, accurate flood forecasts at sites on the lower
Dee which are prone to flooding can be made by routing upstream main
channel and tributary flows, while the locations of the reservoirs are
such that rainfall-runoff modelling can provide little benefit to short-
term reservoir operation. These case studies illustrate that the relative merits of measuring rainfall and flow must be weighed carefully
when designing a telemetering network.

As far as instrumentation, telemetry and computer hardware are
concerned, the rapid microprocessor-driven developments which are
currently taking place in these areas offer exciting prospects for the
future. Although installed in 1975, the Dee system hardware is now
technologically obsolete; the relatively large and cumbersome computer
installed then can now be outperformed by a relatively small mini-microcomputer and at a fraction of the cost. The possibilities for utilizing
microprocessor intelligence in outstations also promises major advances,
particularly in providing more reliable methods of monitoring water
quality, and in reducing both capital and maintenance costs. The
unanticipated nature of accidental pollution events and the potentially
disastrous effects which they can have on water supplies underline the
need for water quality monitoring and forecasting systems of the kind
installed on the Bedford Ouse.

From the modelling standpoint, the developments which have taken
place in the telecommunications and computing fields have opened up new
horizons for mathematical modelling. Rainfall-runoff, flow routing and
water quality models can now be put to the test operationally, allowing
the research hydrologist to assess where he should direct his efforts to
increase the operational efficiency of his models. One factor to emerge
from the case studies considered here is that simple models can satisfy
the requirements of the practising engineer; they can also be readily
understood by him. The case for using a more complex model should be
established in terms of the additional benefits which can accrue from
its use in preference to a simpler model, rather than accepting a priori
that the more complex model is best.

SYMBOLS

S	storage	L^3
Q	flow	L^3T^{-1}
P	rainfall	L^3T^{-1}
K,k	storage parameter	T, L^3
b	time delay	T
F^2	goodness of fit criterion	1
$\bar{a}(Q)$	attenuation parameter	L^{-1}
$\bar{c}(Q)$	celerity of flood wave	LT^{-1}
q^1	lateral inflow per length of river	L^2T^{-1}
p^1	lagged effective rainfall	L^3T^{-1}
U	mean flow velocity	LT^{-1}
N	number of compartments in a reach	1
$x_j(t)$	downstream concentration jth determinand	ML^{-3}
$u_j(t)$	upstream concentration jth determinand	ML^{-3}
$Q_i(t)$	upstream flow rate	L^3T^{-1}
$Q_o(t)$	downstream flow rate	L^3T^{-1}
V	reach volume	L^3
$C_s(t)$	saturation concentration dissolved oxygen	ML^{-3}
$S_j(t)$	sources and sinks affecting jth determinand	$ML^{-3}T^{-1}$

REFERENCES

Beck, M.B. 1981 'Operational Water Quality Management: Beyond Planning and Design', Executive Report 7, IIASA, Laxenburg, Vienna.

Briggs, R. 1975 J. Soc. Wat. Treatment and Examination, 24, 23, 'Instrumentation for monitoring water quality'.

Brunsdon, G.P. and Sargent, R.J. 1982 Proc. of the Exeter Symposium Publ. No. 134, 257, 'The Haddington Flood Warning System'.

Caddy, D.E. and Akielan, A.W. 1978 Int. Environ. Safety, 1, 18, 'Management of river water quality'.

Caddy, D.E. and Whitehead, P.G. 1981 Effl. and Water treat. Jour., 21, 9, 407, 'Practical river quality monitoring and pollution forecasting: Part I - Continuous quality monitoring and pollution forecasting'.

Cook, G.H. 1975 Instruments and Control Systems Conference: Water Research Centre, Medmenham, 'Water quality monitoring in the River Trent system'.

Dee Steering Committee 1977 'Dee Weather Radar and Real Time Hydrological Forecasting Project', Central Water Planning Unit, Reading Bridge House, Reading, Berkshire, UK.

Fawcett, A. 1975 Proc. Symp. on "Water Quality Modelling of the Bedford

Ouse, Anglian Water Authority, Huntingdon, UK, 29, 'A management model for river water quality'.

Green, C.S. 1979 'An Improved Subcatchment Model for the River Dee', Internal Report No. 58, Institute of Hydrology, Wallingford, Oxfordshire, UK.

Hinge, D.C. and Stott, D.A. 1975 Proc. Instruments and Control Systems Conference: Water Research Centre, Medmenham, Buckinghamshire, UK, 'Experience in the continuous monitoring of river water quality'.

Hydraulics Research Station 1975 'A Flow Routing Model for the River Dee', Report EX712, Wallingford, UK.

Jones, D.A. and Moore, R.J. 1980 Proc. of the Oxford Symposium on Hydrological Forecasting, IAHS Publ. No. 129, 397, 'A simple channel flow routing model for real-time use'.

Kohonen, T., Hell, P., Muhonen, J. and Vuolas, E. 1978 'Automatic Water Quality Monitoring Systems in Finland', Report of the National Board of Waters, Helsinki, Finland, no. 153.

Lambert, A.O. 1972 J. Instn. Wat. Enginrs. 26, 413, 'Catchment models based on ISO-functions'.

Lambert, A.O. and Lowing, M.J. 1980 Proc. of the Oxford Symposium on Hydrological Forecasting, IAHS Publ. No. 129, 525, 'Flow forecasting and control on the River Dee'.

Mandeville, A.N. 1975 'Non-Linear Conceptual Catchment Modelling of Isolated Storm Events', Ph.D. Thesis, Univ. of Lancaster, Lancaster, UK.

McKerchar, A.I. 1975 'Subcatchment Modelling for Dee River Forecasting', Internal Report no. 30, Institute of Hydrology, Wallingford, Oxfordshire, UK.

Moore, R.J. and Jones, D.A. 1980 Proc. Int. Workshop on Realtime Hydrological Forecasting and Control, Institute of Hydrology, Wallingford, 160, 'A simple adaptive finite difference flow routing model'.

NERC 1975 'Flood Studies Report', Material Environment Research Council, London, Great Britain.

O'Connell, P.E. 1980 (Ed.) 'Real-time Hydrological Forecasting and Control', Proc. 1st International Workshop, Institute of Hydrology, Wallingford.

Price, R.K. 1973 Proc. Instn. Civ. Engrs., 55, 913, 'Flood routing methods for British rivers'.

Rosenbrock, H.H. 1960 Comp. Jour. 3, 175, 'An automatic method of finding the greatest or least value of a function'.

Wellwork, J.F. 1979 Proc. River Pollution Control Conference: Water Research Centre, Medmenham, Buckinghamshire, UK, 'Protecting a water supply intake - river water data collection and pollution monitoring'.

Whitehead, P.G. 1975 Proc. Symp. on Water Quality Modelling of the Bedford Ouse, Anglian Water Authority, Huntingdon, UK, 49, 'A dynamic-stochastic model for a non-tidal river'.

Whitehead, P.G. and Young, P.C. 1979 Water Resources Research, 15, 451, 'Water quality in river systems: Monte Carlo analysis'.

Whitehead, P.G., Young, P.C. and Hornberger, G.E. 1979 Water Research, 13, 1155, 'A systems model of flow and water quality in the Bedford Ouse river system: Part I - Streamflow modelling'.

Whitehead, P.G., Beck, M.B. and O'Connell, P.E. 1981 Water Research, 15, 1157, 'A systems model of flow and water quality in the Bedford Ouse river system; Part II – Water quality modelling'.

Whitehead, P.G., Williams, R.J., O'Connell, P.E. and Black, K.B., 'Operational Management of Water Quality in River Systems', Final Report on Contract ENV 400–80 UK(B) to the Commission of the European Communities.

Wood, E.F. 1980 <u>Proc. Int. Workshop on Real-time Hydrological Forecasting and Control, Institute of Hydrology, Wallingford</u>, 183, 'An adaptive input–output flow routing model'.

9. RIVER FLOW SIMULATION

J.G. Grijsen

'De Voorst' Laboratory
Postbus 152
8300 AD Emmeloord
Netherlands

9.1 Introduction

River flow simulation and forecasting and the study of consequences of
projected engineering works may well be performed on the basis of
mathematical models. An adequate quantitative mathematical description
of the physical processes involved is a fundamental pre-requisite of
such models. This description often comprises a system of ordinary or
partial differential equations, together with suitable boundary conditions and other data. Equations for fluid flow are frequently so complicated that they can be solved only by computers. Moreover, the equations usually include empirical factors associated with turbulence.
These two properties determine to a large extent the feasibility of
applying a mathematical model.

   Although more facilities are rapidly becoming available, numerical
three-dimensional, time dependent calculations will not be feasible for
most practical purposes in the near future. Solving two-dimensional
time-dependent problems is within the capabilities of existing computers,
although this still requires a considerable computational effort. One-
dimensional calculations, however, are more easily performed, and satisfactory for many purposes. They are the main subject of this chapter.

   Chapter 3 gives a brief account of various numerical methods for
river flow simulation, including a presentation of the Saint-Venant
equations for one-dimensional, non-steady flow in open channels. The
present chapter deals in more detail with one group of these methods,
i.e. finite difference methods. The concept underlying these methods
is presented in Section 9.2; the numerical properties of various types
of finite difference schemes are dealt with in Section 9.3. In the next
section this is examined in more detail for a specific scheme currently
used at the Delft Hydraulics Laboratory. In Section 9.4 attention is
also given to other aspects such as boundary conditions, the solution of
the finite-difference equations and the treatment of special flow problems. Various aspects of the practical application of numerical methods
for river flow simulation are dealt with in Section 9.5, and Section
9.6 gives a brief account of a case study, i.e. flood control for the
rivers Parana and Paraguay. Finally a strategy for the use of the above

D.A. Kraijenhoff and J.R. Moll (eds.), River Flow Modelling and Forecasting, 241-272.
© 1986 by D. Reidel Publishing Company.

type of model in a river flow forecasting system is outlined in Section 9.7.

The present chapter is merely meant to introduce the reader in the subject of river flow simulation by mathematical modelling. For more details reference is made to textbooks such as Cunge et al. (1980) and Jansen (1979). Grateful acknowledgement is made to Dr. C.B. Vreugdenhil of the Delft Hydraulics Laboratory for his approval for using his lecture notes on the subject in the writing of this chapter.

9.2 Finite difference methods

In order to clarify some of the concepts introduced by Dooge in Chapter 3, the kinematic-wave equation is used as a starting point

$$\frac{\partial h}{\partial t} + c\frac{\partial h}{\partial x} = 0 \qquad\qquad (9.2.1)$$

in which the propagation velocity c is considered a constant.

The equation must be discretized in order to obtain a numerical solution, in a similar way a function is usually characterized by a finite number of function values in a grid (Figure 9.2.1).

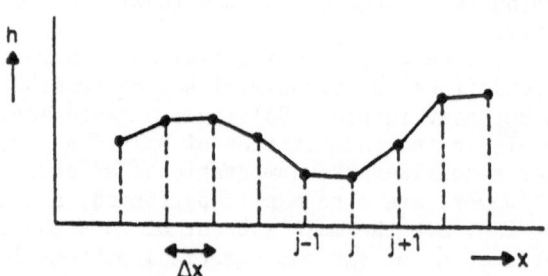

Fig. 9.2.1. Discretizing a function

Interpolations can be made between the grid points. Depending on the interpolation technique used, several mathematical expressions are possible for the derivatives in (9.2.1), for example:

$$\frac{\partial h}{\partial x}(x = x_j) = \frac{h_j - h_{j-1}}{\Delta x} + o(\Delta x) \quad \text{(backward difference)}$$

$$= \frac{h_{j+1} - h_j}{\Delta x} + o(\Delta x) \quad \text{(forward difference)}$$

$$= \frac{h_{j+1} - h_{j-1}}{2\Delta x} + o(\Delta x^2) \quad \text{(central difference)} \qquad (9.2.2)$$

For the derivative with respect to time similar possibilities exist.
The function therefore must be discretized on a two-dimensional grid
(Fig. 9.2.2).

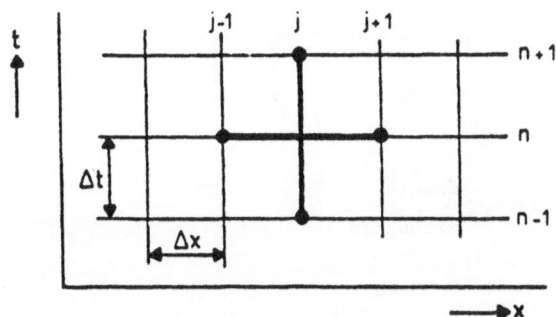

Fig. 9.2.2. Two-dimensional grid with leap-frog scheme

In (9.2.2) the truncation error, indicating the accuracy, is also
shown. The use of central differences, both for time- and space-deriv-
atives, produces a difference method known as the leap-frog method (see
Fig. 9.2.2).

$$\frac{h_j^{n+1} - h_j^{n-1}}{2\Delta t} + c \frac{h_{j+1}^n - h_{j-1}^n}{2\Delta x} = 0 \qquad (9.2.3)$$

in which $h_j^n = h(j\Delta x, n\Delta t)$. There are however still various other possibilities, e.g.
-    modified Lax scheme

$$\frac{h_j^{n+1} - h_j^n}{\Delta t} + c \frac{h_{j+1}^n - h_{j-1}^n}{2\Delta x} - \alpha \frac{h_{j+1}^n - 2h_j^n + h_{j-1}^n}{2\Delta t} = 0 \qquad (9.2.4)$$

-    Crank-Nicholson scheme

$$\frac{h_j^{n+1} - h_j^n}{\Delta t} + \theta c \frac{h_{j+1}^{n+1} - h_{j-1}^{n+1}}{2\Delta x} + (1 - \theta) c \frac{h_{j+1}^n - h_{j-1}^n}{2\Delta x} = 0 \qquad (9.2.5)$$

-    four-point scheme (Preismann, 1960)

$$\frac{1}{2\Delta t}\left(h_{j+1}^{n+1} - h_{j+1}^n + h_j^{n+1} - h_j^n\right) + \frac{\theta c}{\Delta x}\left(h_{j+1}^{n+1} - h_j^{n+1}\right) +$$

$$+ (1 - \theta) \frac{\theta c}{\Delta x}\left(h_{j+1}^n - h_j^n\right) = 0 \qquad (9.2.6)$$

These schemes all approach the same equation in various ways and with varying degrees of accuracy. The "difference molecules" are illustrated in Fig. 9.2.3.

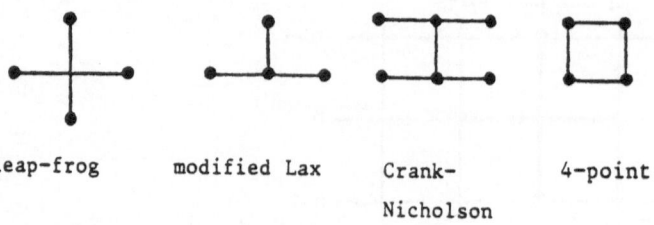

| leap-frog | modified Lax | Crank-Nicholson | 4-point |

Fig. 9.2.3. Some difference schemes

The schemes (9.2.3) and (9.2.4) allow the calculation of one new point from the given values at time level n. They are therefore called explicit schemes (Section 3.7). In (9.2.5) and (9.2.6) however all values h_j^{n+1} are related through the difference equations and must be solved together (with the boundary conditions). Hence these methods are called implicit (Section 3.8).

Clearly, this procedure is more complicated than the explicit one, but this is outweighed by other advantages (Section 9.3). The systems of equations, produced with implicit methods, often have a band structure and can therefore be solved efficiently. Various implicit finite difference schemes were discussed by Liggett and Cunge (1975).

9.3 Numerical properties

A suitable selection of numerical methods from the large number available requires some knowledge of their properties.

9.3.1 Consistency and truncation error

The difference equations are intended as an approximation of the differential equation. Both equations are said to be consistent if they approach each other as the grid sizes go to zero. The error in the approximation may be analyzed by transforming the difference equation via a Taylor series back into a continuous structure. For the modified Lax scheme (9.2.4) this produces:

$$\frac{\partial h}{\partial t} + c\,\frac{\partial h}{\partial x} = \alpha\,\frac{\Delta x^2}{2\Delta t}\,\frac{\partial^2 h}{\partial x^2} - \frac{1}{2}\,\Delta t\,\frac{\partial^2 h}{\partial t^2} + \dots \qquad (9.3.1)$$

The difference with the differential equations (9.2.1) is in the right-hand side of (9.3.1); this is called the truncation error. If Δx and $\Delta t \to 0$, the truncation error also goes to zero, provided that the second derivatives of h behave normally. This implies that the difference equation is <u>consistent</u> with the differential equation. Furthermore, the truncation error is proportional to Δt and (if Δx and Δt are in a fixed relation to each other) to Δx. As a result the truncation error decreases linearly with Δx and Δt, if the step size is reduced. The scheme is therefore said to be of the first order in Δx and Δt. In practice, Δx and Δt are not made equivalent to zero; in that case (9.3.1) would be solved instead of (9.2.1). Differentiation of (9.2.1) leads to the following equation.

$$\frac{\partial^2 h}{\partial t^2} = -c \frac{\partial^2 h}{\partial x \partial t} = c^2 \frac{\partial^2 h}{\partial x^2} \tag{9.3.2}$$

Thus (9.3.1) can also be written as follows:

$$\frac{\partial h}{\partial t} + c \frac{\partial h}{\partial x} = \frac{\Delta x^2}{2\Delta t} (\alpha - \sigma^2) \frac{\partial^2 h}{\partial x^2} + \ldots \tag{9.3.3}$$

in which $\sigma = c \, \Delta t / \Delta x$ represents the <u>Courant number</u>. The right-hand member is in the form of a diffusion term. Due to discretization, a numerical diffusion has been introduced. In fact a kind of diffusion approach is applied, in which the diffusion coefficient (and thus the influence on the solution) is determined by c, α, Δx and Δt. The effectively solved Equation (9.3.3) is also called the modified equation. Its properties give some insight in the behaviour of the solution of the difference equation.

From (9.3.3) it appears that the first term of the truncation error becomes zero if $\alpha = \sigma^2$. The next term (not mentioned here) then becomes important. It is one order higher and gives a second-order scheme, known as the Lax-Wendroff scheme.

For the Crank-Nicholson scheme (9.2.5) the following equation is similarly derived

$$\frac{\partial h}{\partial t} + c \frac{\partial h}{\partial x} = \frac{\Delta x^2}{2\Delta t} \sigma^2 (2\Theta - 1) \frac{\partial^2 h}{\partial x^2} + \ldots \tag{9.3.4}$$

For $\Theta = 0.5$ a second-order scheme is obtained.

9.3.2 <u>Accuracy</u>

Possibly of greater importance is the error in the solution of the difference equation (discretization error). Unfortunately it is rather difficult to discuss this in general terms. Under certain conditions the equivalence theorem applies; for linear equations the discretization error is of the same order in Δx and Δt as the truncation error, provided the difference equation is stable. For the example of (9.2.4) this implies that for the numerical solution h_j^n in relation to the analytical solution $h(x,t)$:

$$h_j^n - h(j\Delta x, n\Delta t) = o(\Delta x, \Delta t) \tag{9.3.5}$$

One advantage is that the method is <u>convergent</u>, because the discretization error goes to zero if Δx and $\Delta t \rightarrow 0$, and hence the values of the finite difference solution approach the values of the differential equation solution. This, however, does not give any indication of the actual magnitude of the error, as for this purpose, the coefficients of Δx and Δt in (9.3.5) should be known. It is quite possible that a second-order method is less accurate than a first-order method, due to these coefficients in the discretization error.

A completely different method to estimate the inaccuracy is indicated in subsection 9.4.5.

9.3.3 <u>Stability</u>

A difference scheme may produce unrealistic results despite an approximation of a certain order, as the following example of the leap-frog equation (9.2.3) with $\sigma = 2$ demonstrates. At time level $t = \Delta t$ a unit error has been introduced at $x = 5\Delta x$.

$x/\Delta x =$	1	2	3	4	5	6	7	8	9
$t/\Delta t$									
0	0	0	0	0	0	0	0	0	0
1	0	0	0	0	1	0	0	0	0
2	0	0	0	-2	0	2	0	0	0
3	0	0	4	0	-7	0	4	0	0
4	0	-8	0	20	0	-20	0	8	0

Fig. 9.3.1.   Numerical example: instability leap-frog scheme.

This phenomenon of explosive growth of errors is called instability. For each differential scheme, not only the truncation error, but also the <u>stability</u> must be analyzed. A method is given in Jansen (1979). In most cases, a rigorous stability analysis is difficult. Here a physical explanation is given of the possibility that such an instability occurs (see Fig. 9.3.2). The influence and dependence regions of the characteristics theory are involved (Sections 3.3 and 3.4).

The influence region of the points $j - 1$, j, $j + 1$ at time level n lies below the (single) characteristic through $(j - 1, n)$. If the new point to be calculated $(j, n + 1)$ falls outside that region (as in fig. b), it cannot be adequately determined; in fact also points to the left of $(j - 1, n)$ would have to be taken into account. This then leads to instability. If the new point lies within the influence region, it will, at best, receive too much information, which is not necessarily bad. This then leads to the Courant-Friedrichs-Lewy (CFL) criterion for <u>explicit</u> schemes:

$$\sigma = c \frac{\Delta t}{\Delta x} \leq 1 \tag{9.3.6}$$

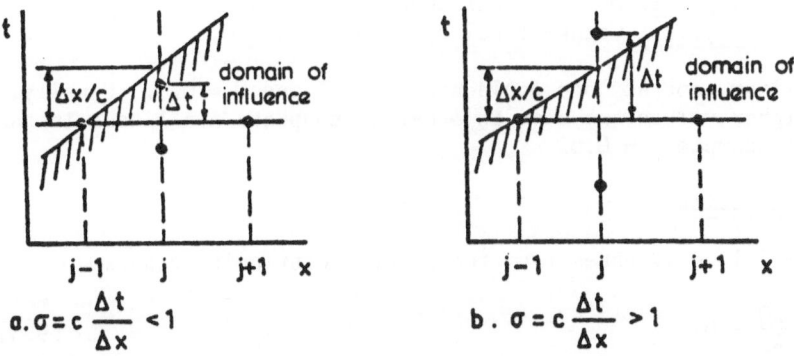

Fig. 9.3.2. CFL criterion for stability.

This is a necessary, but not a sufficient condition! For _implicit_ schemes this condition generally does not apply, because there the whole physical influence region is taken into account simultaneously (Fig. 9.3.3), irrespective of Δt. This is one of the major advantages of implicit schemes: large time steps can be used without instability. The effects on accuracy, however, will have to be closely watched.

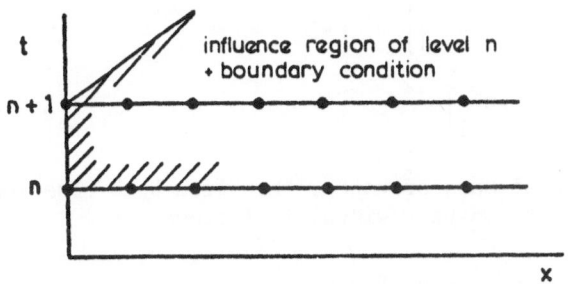

Fig. 9.3.3. CFL criterion does not apply to implicit schemes

In the event that the propagation velocity c varies, the CFL criterion must be valid everywhere, hence also at the worst point. For example, if one makes a tidal computation for a river where, at one point, a very deep hole exists, this determines the maximum possible time step there as $c \approx (gh)^{\frac{1}{2}}$ is maximal. Also in the remainder of the field one must use a small time step (hence a Courant number appreciably smaller than 1), although not necessarily from the point of view of accuracy.

Stability criteria can also be derived from the numerical diffusion coefficients, as given in the right hand members of (9.3.3) and (9.3.4). These coefficients should be positive, as negative values produce an amplification of errors, and therefore instabilities. Together with (9.3.6) this results in the following stability criteria:

for the explicit modified Lax scheme: $\sigma^2 \leq \alpha \leq 1$
for the implicit Crank-Nicholson scheme: $\overline{\Theta} \geq \overline{0.5}$ (9.3.7)

For maximum accuracy one would need $\alpha = \sigma^2$ and $\Theta = 0.5$. Usually, slightly higher values are used in order to suppress non-linear instabilities, for example $\Theta = 0.55$.

9.3.4 Conservation

Equation (9.2.1) originates from the general continuity equation

$$\frac{\partial A}{\partial t} + \frac{\partial Q}{\partial x} = 0 \tag{9.3.8}$$

combined with the momentum equation for uniform, steady flow, which leads to

$$Q = f(h) \tag{9.3.9}$$

In continuous form (9.3.8) and (9.2.1) are equal. In difference form this often is no longer the case, particularly if the coefficient c is not constant. The leap-frog scheme (9.2.3) applied to both equations gives

$$h_j^{n+1} - h_j^{n-1} = c_j \frac{\Delta t}{\Delta x} \left(h_{j-1}^n - h_{j+1}^n \right) \tag{9.3.10}$$

$$A_j^{n+1} - A_j^{n-1} = \frac{\Delta t}{\Delta x} \left(Q_{j-1}^n - Q_{j+1}^n \right) \tag{9.3.11}$$

If we now observe the total mass of water, then from both equations (for the sake of simplicity at constant width) it follows

$$\Delta x \sum_j B_j h_j^{n+1} - \Delta x \sum_j B_j h_j^{n-1} = \Delta t \sum_j B_j c_j \left(h_{j-1}^n - h_{j+1}^n \right) \tag{9.3.12}$$

$$\Delta x \sum_j A_j^{n+1} - \Delta x \sum_j A_j^{n-1} = \Delta t \sum_j \left(Q_{j-1}^n - Q_{j+1}^n \right) \tag{9.3.13}$$

It can now be seen clearly that the right-hand member of (9.3.13) produces zero, except at the boundaries. This agrees with (a discrete form of) the mass balance. The terms in the right member of (9.3.12) do not cancel, so therefore, in this case, the discrete mass balance condition is not exactly fulfilled. Although a small error in the mass balance condition can be accepted as long as the solution of the difference equation is accurate, it still seems advisable to meet the discrete mass balance. For example it provides an opportunity to check the numerical results. This requires that the equation be presented in a conservative form (like (9.3.8)). Similar considerations are valid for the conservation of momentum (important for shock waves, hydraulic jumps, bores, dambreak floods, etc.) and energy (important for stability).

9.4 The Delft Hydraulics Laboratory method

In this section the Delft Hydraulics Laboratory method is discussed as
an example of a mathematical model for river flow simulation (see also
Grijsen et al., 1979).

9.4.1 Equations and difference scheme

One-dimensional flow in open channels is described by the following two
equations (see also definition sketch in Fig. 9.4.1 and Sections 3.1 and
3.2).
Equation of continuity:

$$\frac{\partial A}{\partial t} + \frac{\partial Q}{\partial x} + q_1 + q_3 = 0 \tag{9.4.1}$$

Equation of motion, written in conservative form:

$$\frac{\partial Q}{\partial t} + \frac{\partial}{\partial x}\left(\alpha\, Q^2/A_s\right) + u_1 q_1 + u_2 q_2 + g\, A_s \frac{\partial h}{\partial x} - B_s \frac{\tau_s}{\rho} + \frac{g Q|Q|}{C^2 R A_s} = 0 \tag{9.4.2}$$

It is important to use the above conservative form of the equation
of motion, rather than, for example, (3.2.4), in order to prevent numerical problems in case hydraulic jumps or super-crical flow may occur
locally in a channel system to be modelled.

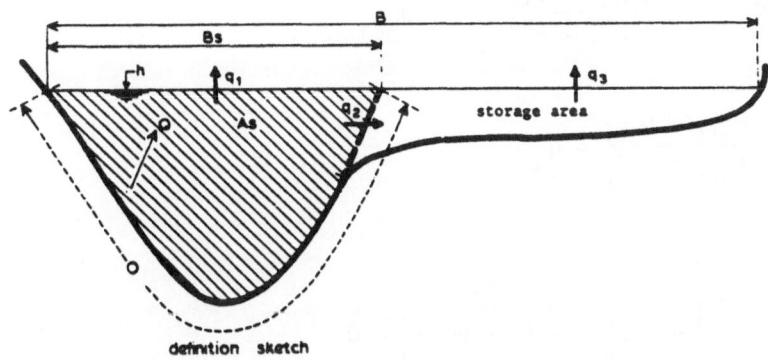

definition sketch

Fig. 9.4.1. Definition sketch

It is assumed as usual that the longitudinal velocities $u_{1,2}$ of the
lateral fluxes $q_{1,2}$ in the momentum equation are equal to zero. The

effects of density gradients are also not taken into account in (9.4.2).

If the flow is more or less uniformly distributed over the cross-section, the coefficient $\alpha = 1$, the C-value is the one defined for the whole flow area, and the hydraulic radius is defined as A_s/O.

If the distribution of velocity in a cross-section is clearly non-uniform the term C^2R and the coefficient α can be determined by means of a special algorithm. This is discussed in subsection 9.4.6.

The Equations (9.4.1) and (9.4.2), together with boundary conditions as discussed in subsection 9.4.3, are solved by means of an implicit finite difference method of the Crank-Nicholson type (Section 9.2). A space-staggered grid is used, i.e. depths and discharges are defined on alternating grid points, except at the boundaries where both are defined. The equations are written in a conservative form where possible. In the following, only one in/outflow term q will be mentioned; the remaining one is treated similarly.

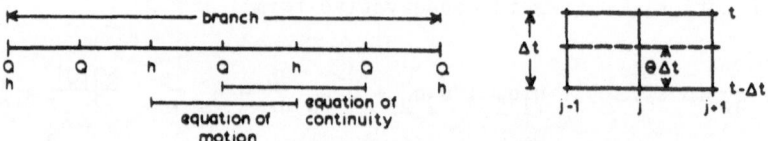

Fig. 9.4.2. Space staggered finite-difference scheme

The finite difference equations read, in the absence of wind shear stress:

$$\frac{A_j^{n+1} - A_j^n}{\Delta t} + \Theta \, M_j^{n+1} + (1 - \Theta) \, M_j^n = 0 \qquad (9.4.3)$$

with

$$M_j = \frac{Q_{j+1} - Q_{j-1}}{x_{j+1} - x_{j-1}} + q_j$$

and

$$\frac{Q_j^{n+1} - Q_j^n}{\Delta t} + \Theta \, N_j^{n+1} + (1 - \Theta) \, N_j^n = 0 \qquad (9.4.4)$$

with

$$N_j = \frac{1}{x_{j+1} - x_{j-1}} \left\{ \alpha_{j+1} \frac{\bar{Q}^2_{j+1}}{A_{s_{j+1}}} - \alpha_{j-1} \frac{\bar{Q}^2_{j-1}}{A_{s_{j-1}}} + g\,\bar{A}_{s_j} \left(h_{j+1} - h_{j-1} \right) \right\} +$$

$$+ \frac{g Q_j |Q_j|}{\bar{C}^2 \bar{R} \bar{A}_{s_j}}$$

The quantities with an overbar are averaged between adjacent grid points.

9.4.2 Network schematization

Flow in a river system is determined by the equations of motion and continuity and external boundary conditions (subsection 9.4.3). It is, however, useful to extend the idea of boundary conditions to internal boundaries, which may be formed by conditions such as:
- a river junction
- a relatively sudden (almost discontinuous) change in river cross-sections
- a weir or dam or other construction, separating the upstream and downstream flow regime
- a storage basin not included in the flood plain schematization.

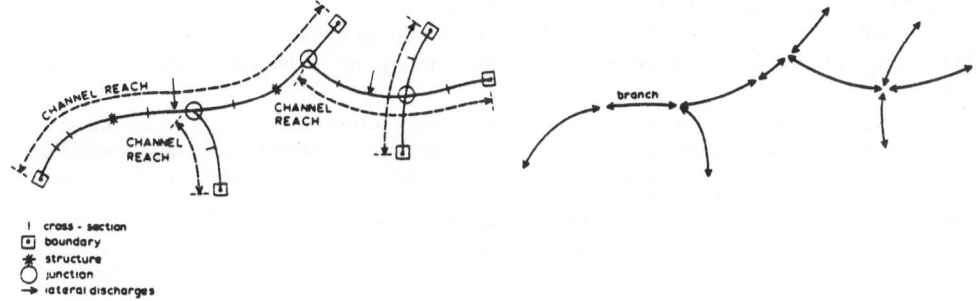

I   cross - section
[□] boundary
✳ structure
◯ junction
→ lateral discharges

Fig. 9.4.3. Definition sketch for model components

All such points are considered as nodes of a network of river branches. Flow in the connecting branches is governed by the equations of motion and continuity and by the internal boundary conditions at the nodes, and, of course, by relevant external boundary conditions. The flow in flood plains can also be approximated by a network of equivalent channels, provided the flow directions are more or less fixed by topographical conditions. If a river reach consists of several rather distinct channels, these may be included either as separate branches, connected by nodes at regular intervals, or be schematized into one equivalent branch with the same effective parameters (cross-section, resistance, etc.). Schematization of a river system requires considerable care and due consideration of anticipated flow patterns.

9.4.3 Boundary conditions

For subcritical flow (the only case considered here) one boundary condition is needed at each end point of a river reach or branch in terms of
a network schematization (Section 3.4).

Several types of internal boundary conditions exist, only two will
be discussed here. The first is a junction where three or more river
channels meet. At a junction, the first condition is the equation of
continuity:

$$\sum_k Q_k = 0 \qquad (9.4.5)$$

when Q_k are the discharges at each of the channel boundaries, taking care
of the correct signs. The remaining boundary conditions are related to
the water level. The simplest assumption is that the water level in all
branches will be equal near the node:

$$h_1 = h_2 = h_3 = \ldots \qquad (9.4.6)$$

However, the physical situation may be such that this is not a realistic
assumption. Then, relations between water levels or energy levels must
be specified taking possible energy losses into account. Formulating
these requires a detailed inspection of flow conditions at the junction.
If n branches exist, n − 1 conditions must be specified, which together
with Equation (9.4.5), give exactly the number of n boundary conditions
required.

The second main type of internal boundary condition is a structure
(dam, weir, power station, etc.). Denoting water levels and discharges
on either side of the structure by h_1, Q_1 and h_2, Q_2 respectively, two
conditions are needed, one of which is the equation of continuity:

$$Q_1 - Q_2 = Q_w \qquad (9.4.7)$$

where Q_w is a possible discharge withdrawn from the river at the structure (for irrigation, pumped storage, etc.). The second condition spec-
ifies the operation of the structure in terms of discharge at a given
difference in head, e.g.:

$$Q = f(h_1 - h_2) \qquad (9.4.8)$$

or more generally

$$f(h_1, h_2, Q_1, Q_2) = 0 \qquad (9.4.9)$$

Again, (9.4.7) and (9.4.9) are the two boundary conditions needed for the two end points of the river branches involved.

9.4.4 Method of solution

The finite-difference method described in subsection 9.4.1 yields a system of non-linear difference equations. These are solved by a generalized Newton procedure. If the equations are written symbolically as:

$$f_i(h_1,Q_1,Q_2,h_3,\ldots,Q_k,h_k) = 0 \qquad\qquad (9.4.10)$$

$$(i = 1,\ldots, k)$$

an iteration is defined by requiring

$$f_i\left[h_i^{(k)} + \Delta h_1,\ Q_1^{(k)} + \Delta Q_1,\ldots\right] = 0 \qquad\qquad (9.4.11)$$

which, upon linearization, gives

$$\frac{\partial f_i}{\partial h_1}\,\Delta h_1 + \frac{\partial f_i}{\partial Q_1}\,\Delta Q_1 + \ldots = -\,f_i\left[h_1^{(k)},\ Q_1^{(k)},\ldots\right] \qquad (9.4.12)$$

The starting values $h_1^{(0)}$, $Q_1^{(0)}$,... of the iteration are taken to be final values of the previous time step. The system (9.4.12) is linear in terms of the corrections Δh_j, ΔQ_j. The procedure is

(i)   either executed only once in each time step, which means that in fact the linearized difference equations are solved instead of the non-linear ones,

(ii)   or repeated until a certain accuracy of the results is obtained, according to some criterion of convergence. In this case, excepting a small final error due to the iterative process, the non-linear difference equations are solved.

In order to avoid large sets of linear equations, every set of k branch equations with k + 2 unknowns is reduced by elimination to 2 equations with 4 unknowns. The remaining unknowns are discharges and water levels at the two end points of the branch.

The system of equations for the entire network now comprises:

a.   two equations for each branch

b.   the equations in the nodes, either in linearized form or not (subsection 9.4.3)

c.   external boundary conditions, either in linearized form or not.

The system is much smaller than the complete system involving all grid points, particularly if the number of nodes is small. On this level, a Newton procedure may be used again to solve the reduced system. Subsequently this solution is used to compute the values in all grid points.

9.4.5 Accuracy and stability

The stability and accuracy of the computations depend on the implicity
factor Θ, the time step Δt, the mesh width Δx and the relevant wave
period T. From a theoretical analysis graphs have been prepared to
enable a suitable choice of parameters, which guarantees both stability
and accuracy of the computations.

   Numerical errors generally have the form of a change in the velocity
of propagation (phase error) and numerical damping in addition to the
physical damping (amplitude error). A quantitative estimate of these
numerical errors can be obtained by investigating the propagation of long
waves as described by the linearized wave equations. A comprehensive
analysis of this problem is presented by Jansen (1979) and Grijsen et al.
(1976). Here only some results are summarized for dynamic waves (such
as tidal waves), where the inertia terms in the equation of motion play
an important role.

   The following definitions apply:

$$\text{damping factor } d = \frac{\text{computed amplitude}}{\text{exact (physical) amplitude}} \text{(after one wave period)}$$

$$\text{relative velocity } c_r = \frac{\text{computed velocity of propagation}}{\text{exact velocity of propagation}}$$

Both quantities should ideally be equal to one. The errors are a function of the following parameters:

Θ = implicity factor in the finite difference scheme
$n_x = L/\Delta x$ = number of grids per wave length L

$\sigma = c \dfrac{\Delta t}{\Delta x}$ = Courant number.

If a wave runs in the model during n wave periods (n may be less than
one), the numerical damping will be d^n.

   The wave length L is related to the wave period T and the celerity
$c \approx \sqrt{(gA_s/B)}$ by:

$$L = cT \tag{9.4.13}$$

For the computation of d and c_r the following formulae can be used.

$$d = |\rho|^{n_t} \tag{9.4.14}$$

$$c_r = - \frac{n_x \arg(\rho)}{2\pi\sigma} \tag{9.4.15}$$

$$n_t = n_x/\sigma = T/\Delta t \tag{9.4.16}$$

$$\rho = \frac{1 - (1 - \Theta)i\sigma \sin \xi}{1 + \Theta \, i\sigma \sin \xi} \tag{9.4.17}$$

$$\xi = 2\pi/n_x \tag{9.4.18}$$

For values of $n_x > 20$, which generally occur, c_r and d tend to be functions only of $\Delta t/T$ and Θ. In that case Fig. 9.4.4 can be used.

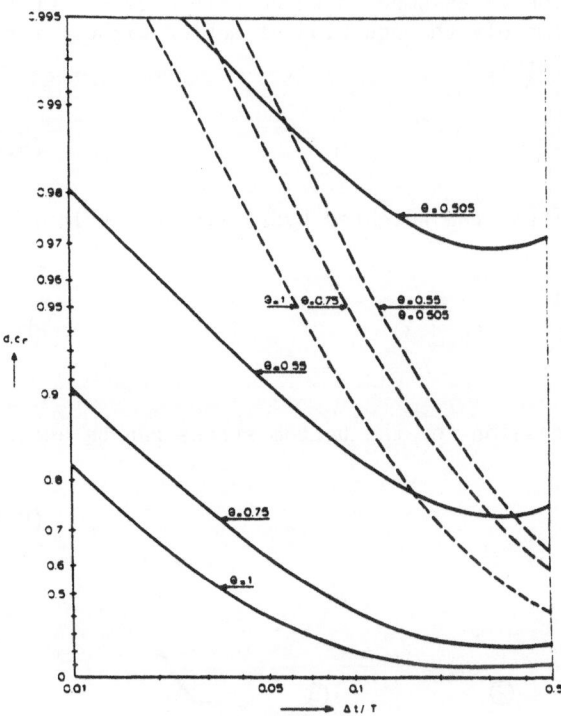

Fig. 9.4.4.   Wave damping d per wave period (drawn lines) and relative velocity of propagation c_r (dashed lines) for dynamic waves, computed with implicit method.

For $\Delta t/T < 0.1$ the value of d can easily be derived from the equation:

$$d = \exp \{-4\pi^2(\Theta - 0.5)\Delta t/T\} \tag{9.4.19}$$

Clearly the best accuracy will be obtained for values of Θ near 0.5. For $\Theta > 0.5$ the computations are unconditionally stable (see also subsection 9.3.3). Therefore often a value $\Theta = 0.55$ is chosen.

9.4.6 Non-uniform flow in a cross-section

In a normal situation, the average velocity (Q/A_s) supposedly describes the flow in a cross-section. In cases where this assumption is not warranted, a special algorithm is used to take into account the non-uniform distribution of the velocity in a composite cross-section. The cross-section is divided into a maximum of three sub-sections (Fig. 9.4.5). The slope of the water level and the water level itself are assumed to be the same for the entire cross-section. The distribution of flow in the cross-section is assumed to be described by the Chézy-equation, which is approximately the equation of motion without the inertia terms. This leads to:

$$Q_i = C_i\, A_i \left(R_i \frac{\partial h}{\partial x} \right)^{1/2} \tag{9.4.20}$$

for each sub-section i. Thus, the weighted hydraulic radius is

$$R_w = \frac{\{\sum_{i=1}^{n} C_i\, A_i\, \sqrt{R_i}\}^2}{C^2 A_s^2} \qquad (n \le 3) \tag{9.4.21}$$

with which the normal expression for the bottom stress can be used.

$$\frac{\partial h}{\partial x} = \frac{Q^2}{C^2\, R_w\, A_s^2} \tag{9.4.22}$$

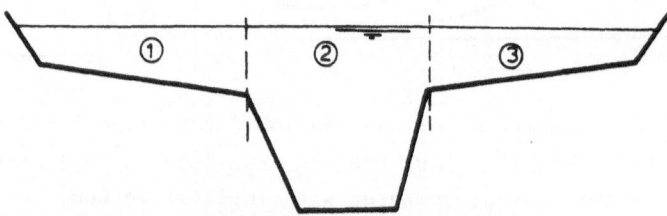

Fig. 9.4.5. Channel with flood plains

The coefficient α in the equation of motion is then computed from

$$\alpha = \frac{A_s \sum_{i=1}^{n} C_i^2\, A_i\, R_i}{\{\sum_{i=1}^{n} C_i\, A_i\, \sqrt{R_i}\}^2} \qquad (n \le 3) \tag{9.4.23}$$

In principle this procedure is only valid in case of quasi-steady
flow, where the inertia terms are small, and does not apply for tidal
flow conditions. A satisfactory solution for this problem is not yet
available. Therefore, the above procedure has also occasionally been
applied for tidal flow conditions. The results of these applications
have indicated clearly that good results can be obtained, although theoretical considerations do not completely justify this. The procedure is
of special value in cases where flood plains and rivers are combined in
one cross-section in order to simplify the structure of the model.

9.4.7 Treatment of dry channels

A way to overcome numerical problems when the water depth becomes very
small, is to define the bottom stress term only at the upstream end of
a grid section, and fully implicit. Due to this treatment, the calculation can be continued when the discharge decreases to zero. The depth
and the discharge become very small but the depth remains positive.
When the discharge increases again no special steps need to be taken.

In the case of a deep main channel and a shallow lateral channel,
the water level in the main channel may fall below the bottom level of
the lateral channel (Fig. 9.4.6). To cope with this situation the water
level in the lateral channel (at the junction) can be kept at critical
depth from the moment the water level in the main channel falls below
this depth; this already occurs if the water level in the main channel
is slightly higher than the bottom level of the lateral channel.

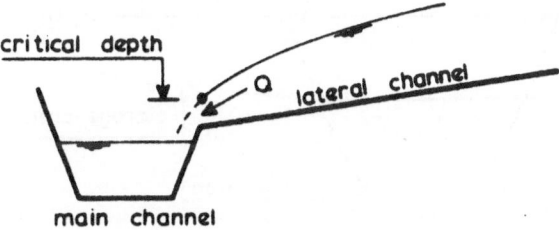

Fig. 9.4.6. Definition sketch

9.5 Practical aspects

9.5.1 Schematization

Application of these numerical methods on river systems always requires
a certain degree of schematization. This is particularly true for
reproducing the geometry of the system (cross-sections). The most
important values B, A_s, R all depend on the water level h, as well as

the results of the model (propagation, velocity, wave damping etc.)
Hence it is important to define the cross-sections with sufficient accuracy. One alternative is to express the width as a function of height,
from which the remaining data can be calculated (Fig. 9.5.1).

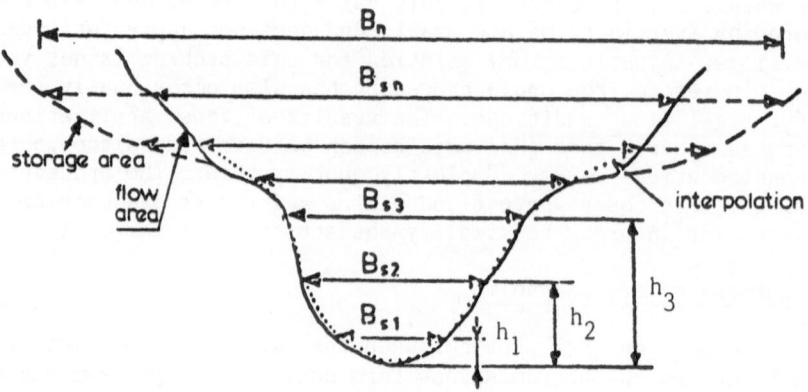

Fig. 9.5.1.   Schematized cross-section

Linear interpolations can be made between the selected values, as
indicated. For non-symmetrical profiles this information is, in fact,
insufficient. It appears, however, that making it symmetric only causes
a small error.

It is important that sufficient points are selected to prevent an
effect such as indicated in Fig. 9.5.2, where between the given levels
the storage width is determined incorrectly, with the result that the
propagation velocity of the wave is not reproduced correctly.

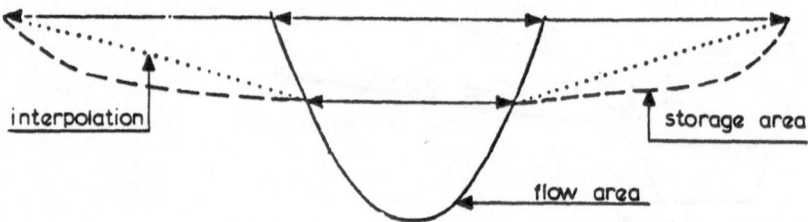

Fig. 9.5.2.   Insufficiently accurate representation of the cross-
section

Another point of schematization is the choice of the step size Δx.
It is useful to have a method which allows unequal step sizes to be used
in different parts of the network, so that the local geometry of the
river system can be correctly followed. From numerical considerations
an order of magnitude of Δx, necessary to describe the wave propagation
with sufficient accuracy, follows. A reasonable reproduction of the
local geometry (width- and depth-variations) forms a supplementary
criterion. Consequently quantitative information for a proper selection
of Δx is difficult to give. In many cases the waves are very long
(hundreds of kilometres), so that small irregularities in the cross-
section will not be very important. Eventually the results of various
schematizations can be compared with each other (sensitivity analysis).

In general, adequate reproduction of global properties of wave propagation (propagation velocity and damping) should be carefully performed.

As can be seen from the origin of the celerity c, this implies that
the slope of the rating curve must be correct, or in other words, that

$$\frac{1}{B} \frac{d}{dh} \left(C \ A_s \ R^{\frac{1}{2}} \right)$$

must be correct. For a wide and more or less rectangular cross-section,
this reads

$$\frac{3}{2} \frac{1}{B} C \ B_s \ h^{\frac{1}{2}}$$

and it is concluded that the ratio B_s/B is of particular importance for
a correct propagation velocity. In addition, a possible dependence of
the Chézy-coefficient C on the water depth and on changing riverbed
forms may have their effects and must be correctly taken into account.

Flood plains and forelands can be included in one cross-section with
the actual riverbed, provided that some relation between the discharges
of the minor river bed and the forelands exists (subsection 9.4.6). If
this relation is not evident, the network schematization allows the
inclusion of additional branches, representing the forelands. The
process of exchange of water and momentum between the different parts of
the river bed deserves careful treatment (subsection 9.6.2 and Ogink
(1985)).

9.5.2 Boundary conditions

An important problem in this type of mathematical model is the specification of boundary conditions (Section 3.4). Points of inflow generally
do not present problems, the discharge being given as a function of time.
Still, this may not be satisfactory when seen in relation to the initial
condition. This may also apply to other boundaries where a fixed condition is applied. If the initial condition is not known in detail, it is
customary to start at a condition of rest or uniform flow. The influence
of the initial situation will then be assumed to have disappeared after
some time. However, if the departure from the steady initial conditions
is not smooth, waves are generated which reflect against the 'fixed'
boundary conditions, and quite some time may elapse before they
disappear. This means that it is not advisable to let a steady situation
abruptly be followed by some transient regime. If such a change is
unavoidable, it is possible to use non-reflecting downstream boundary
conditions. This can also be used to simulate an infinitely long channel
without friction. Waves exit from the model in that channel and do not
return.

The downstream boundary conditions may be relatively simple if a
physical boundary exists in the form of a ship lock, a dam, a power
station, etc. A well-defined relation between the downstream water level
and discharge can be formulated and used as a boundary condition, i.e.
a steady-state rating curve.

If a river reach ends at any point other than such a physical boundary it will not suffice to introduce, for example, a fixed water level
as a boundary condition because this will cause a reflection of a flood
wave, thus obscuring the results. If a stage-discharge relationship is
applicable, this may be used, either in its theoretical form or experimentally determined. However, such a relationship will not always exist,
especially if backwater effects are to be expected from nearby obstructions in the river of tributaries, etc. In such cases it is necessary
to move the boundary to another location for which better information is
available. If a rating curve is used this will be applicable in principle only in steady flow. Often, but not always, this will be the case
during quite some part of the computations. In addition the deviation
from the non-steady behaviour will extend upstream only over a limited
distance related to the behaviour of backwater curves. In unsteady
conditions it may be also advisable to use Jones' formula (see Jansen,
1979) in some appropriate finite-difference form as a boundary condition,
to account for the major part of the effect of unsteadiness.

9.5.3 Model development and operation

The stages in the development and operation of a mathematical model,
which are common to any type of model, are briefly reviewed here.
    Once the engineering problem has been defined and a mathematical
model chosen, the first stage consists in gathering data. These include
geometrical conditions and design conditions. Additional data are also
required for calibration and verification. The model can then be constructed with the aid of the data collected, i.e. computer programs written
and a schematisation made of the situation to be investigated. Questions
like the choice of time and space increments (grid sizes) are settled at
this stage.
    As soon as the model produces useful results it should be calibrated.
Some steady-state flow computations are generally very useful in this
respect. Calibration is usually necessary because empirical parameters
are involved (such as the bottom roughness) and schematizations have
been made. Parameters can be adjusted to obtain good correspondence
between model results and prototype values (field measurements). Of
course the adjustment may not be extended beyond physically acceptable
values. If, for example, good results can only be obtained by using a
Chézy coefficient with twice the value initially estimated, there must
either be an explanation for this exceptionally high value, or the
schematization is not correct in some way. Cross-sections are to be
checked then, or even additional field surveys and measurements may be
necessary. The process of adjusting parameters by running the model at
different parameter values until a satisfactory result is obtained is
called calibration. It is also a very useful method to determine parameters which are difficult to measure directly. From these adjustment
runs an impression of the sensitivity of model results to changes in the
data can also be obtained. This may lead to the need for more accurate
measurements of the type of data for which the model is very sensitive
to small changes. If possible the calibration phase should also comprise
a check of the numerical accuracy, for example, by varying numerical

parameters such as the time step. Note, however, that agreement of two
runs at a different time step may imply that both are equally wrong.
The next stage, which is not always possible or simply omitted for convenience' sake, is the verification of the model. Once the coefficients
are calibrated, computations for a different set of prototype data should
be made. If a model run satisfactorily reproduces the measured prototype conditions without further adjustment of parameters, a reasonable
confidence is gained for application of the model to design conditions
which have never occurred in the prototype. Moreover an impression can
be obtained of the accuracy of the model results.

After these preliminaries the operation of the model can start. The
model should however still be used with care, especially if the conditions
for which it is applied include some kind of extrapolation. If the
model involves a physical process the mechanism of which has been formulated in mathematical terms and has been verified in various conditions,
extrapolation can be made with some confidence. This is not true if the
model results are largely determined by an empirical parameter which has
not been verified for extreme conditions.

9.5.4 Flow resistance in alluvial rivers

The flow resistance or bottom roughness, usually expressed in terms of
Manning's n or the Chézy-coefficient C, is probably the most important
parameter to be calibrated. Both in Chapter 3 and in the present chapter
it was assumed so far that the boundaries of a river channel have a
certain constant roughness. A river, however, does not only transport
water, but also sediments. Because a river flows in its own sediment,
its water therefore does not move along rigid boundaries. In the lower
reaches of many rivers, for example, sediment can be classified as sand
(particle size 0.062 to 2.0 mm) and at lower Froude numbers, the transport of sand is coupled with bed-forms of a certain regularity called
ripples and dunes.

The resistance of the river bed is then mainly determined by the
dimensions of the bed forms, i.e. height, steepness (van Rijn 1983) and
shape (Ogink 1984). The equivalent sand roughness increases when the
height and/or steepness of the bed form becomes larger; strong reductions
will occur in case the leeside slope decreases, i.e. when the flow
separation point is located downstream of the dune crest. Grain roughness itself has only a minor effect on the hydraulic resistance of a
ripple- or dune-covered bed.

For flow along rigid boundaries the Chézy-coefficient C increases
only with depth (ref: formulae of White-Colebrook and Strickler) as
Nikuradse's equivalent sand roughness remains constant. In rivers with
a sand bed the development of C with increasing depth or discharge is
more complex, since the equivalent sand roughness is no longer constant.
Starting off from a flat bed with little or no transport, where the
roughness is only determined by the grain diameter, ripples will develop
with increasing flow velocity (flow depth). With a further increase,
ripples transform into dunes, sometimes with ripples superimposed on
their surfaces; the bed resistance thereby grows. When the sand bed
consists of coarse material, the ripple phase will not occur, so in that

case dunes will develop once the critical shear stress is exceeded.
If the flow velocity further increases, the growth of the dunes will
diminish, stop and will become even negative. As a consequence the
equivalent sand roughness then decreases until ultimately flat bed conditions with transport will be reached at the transition from the lower
to the upper flow regime, see Figure 9.5.3, taken from Engelund and
Hansen (1967).

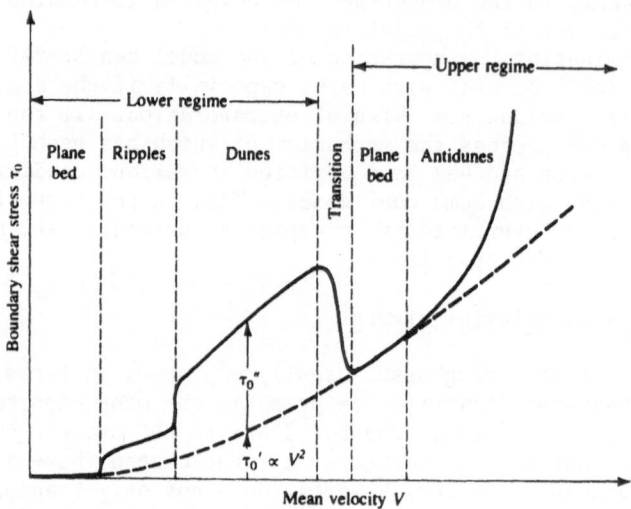

Fig. 9.5.3 Effect of bed forms on the relation between the boundary
            shear stress τ_0 and the mean velocity V. The dashed
            line shows the relation τ_0' due to skin friction alone,
            while τ_0'' represents the effect of bed forms.

Recent results (Wybenga and Klaassen 1983) of flume studies and
observations during flooding suggest that the adaptation of bed forms
lags behind the newly developing flow conditions even though the rates
of transport of water and sediment vary concurrently. A lag in development of dune height means less resistance at the moment of maximum
flooding during a sharp-peaked flood wave compared to a very long wave,
if the conditions comply with those of growing bed forms.

The bed form length, however, appears to lag behind more than the
bed form height does; so the bed forms may be steeper during a flashy
flood than under (nearby) steady flow conditions, which is a resistance
enlarging effect although of less importance.

Note that in the transitional regime the development of the resistance with time may differ from the typical lower flow regime. The
variation in resistance can be considerable: during the 1979 spring
flood in the Pannerden Canal (distributary of the Rhine) the Chézy
coefficient varied between 44.2 and 55.7 $m^{1/2} s^{-1}$.

Figure 9.5.4 was copied from Vanoni (1975). It shows dramatic changes

in the discharge-depth relationship with changes in the bed forms. Here
the particle size of the bed sediment is 0.3 mm, thus rather coarse.

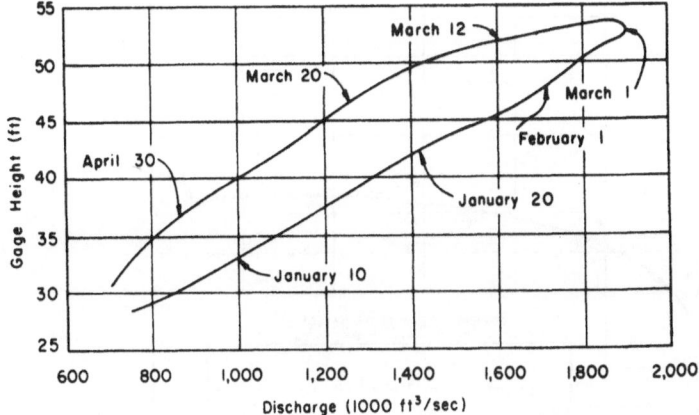

Fig. 9.5.4. Discharge-depth relationship Mississipi River at
             Tarbert Landing

Figures 9.5.5 and 9.5.6 were taken from Shen (1971). Fig. 9.5.5
shows the stage discharge relation in a river bed with various forms of
bed that were supposed to develop as a function of "stream power" $\tau_0 u$.

In this river cross section the three successive phases of ripples,
dunes and smooth bottom are all present. The median diameter of the
bed material is 0.17 mm and here the effect of increasingly high flows
is apparently erasure of ripples and dunes attended by a gradual decrease in flow resistance, as expressed in Manning's n.

In the same book Simons and Richardson (1971) note that minor changes
in one of the governing parameters (e.g. a slight temperature change)
can change the pattern of bed forms considerably. Vanoni (1975) demonstrates the effect of temperature with the flow conditions in the
Missouri River at Omaha (bed sediment size 0.2 mm). With regulated
nearly constant flow, the river was 3 m deep and 200 m wide (Figure
9.5.7).

With a temperature drop from 22 to 4°C in the autumn of 1966, the
Manning friction coefficient dropped from 0.020 to 0.016 $m^{-1/3}$ s and
consequently the water level dropped about 0.6 m.

Finally attention should be paid to seasonal changes in bed roughness due to growth and decay of vegetation in the flood plain.

Floods in the River Meuse seldom occur in summer and Figure 9.5.9
shows that the river stages during the flood in July 1980 were significantly higher than was expected from the local stage-discharge relationship at Borgharen (Fig. 9.5.8).

The local sediment is gravel and Klaassen (1981) reports that alluvial roughness of gravel rivers is well established. The increase in
water level is therefore ascribed to the seasonal increase in roughness

of the flood plain (Veraart 1983, personal communication).

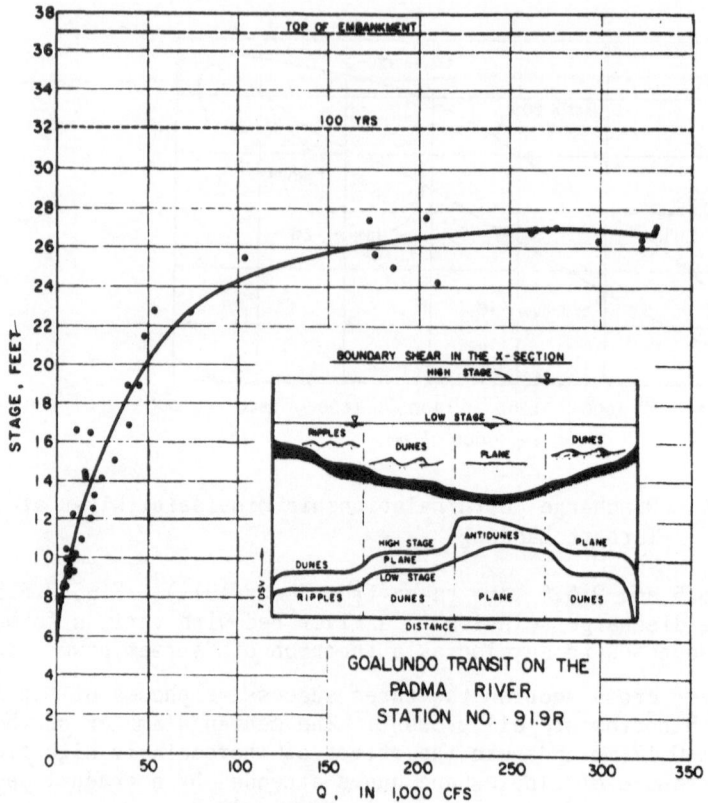

Fig. 9.5.5. Stage discharge relation for the Padma River,
Bangladesh

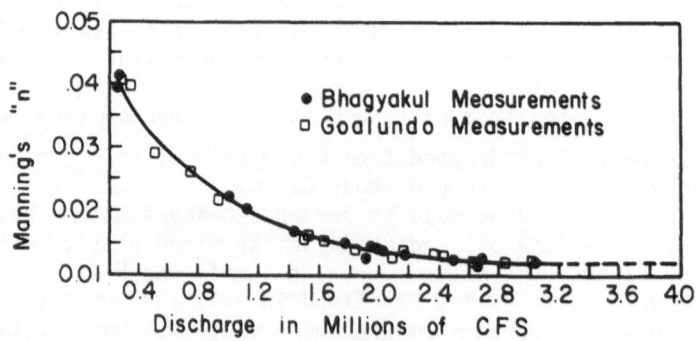

Fig. 9.5.6. Decrease in Manning's n with discharge for the Padma
River

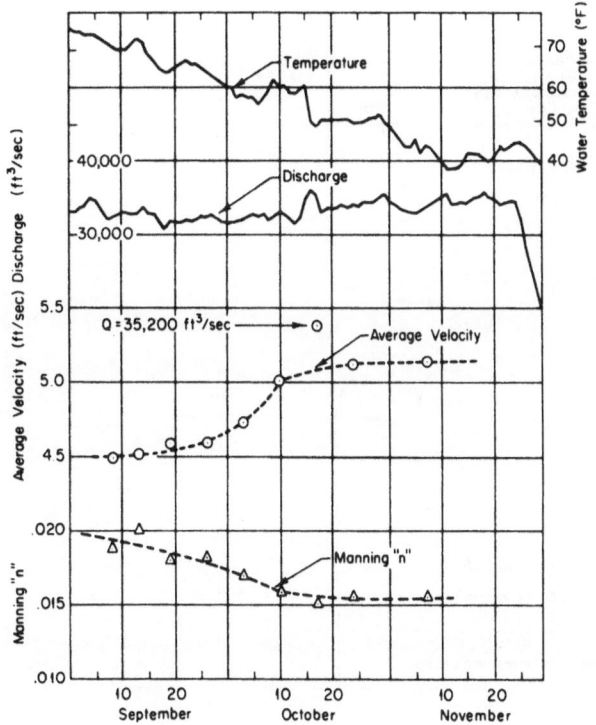

Fig. 9.5.7.   The effect of temperature with the flow conditions.

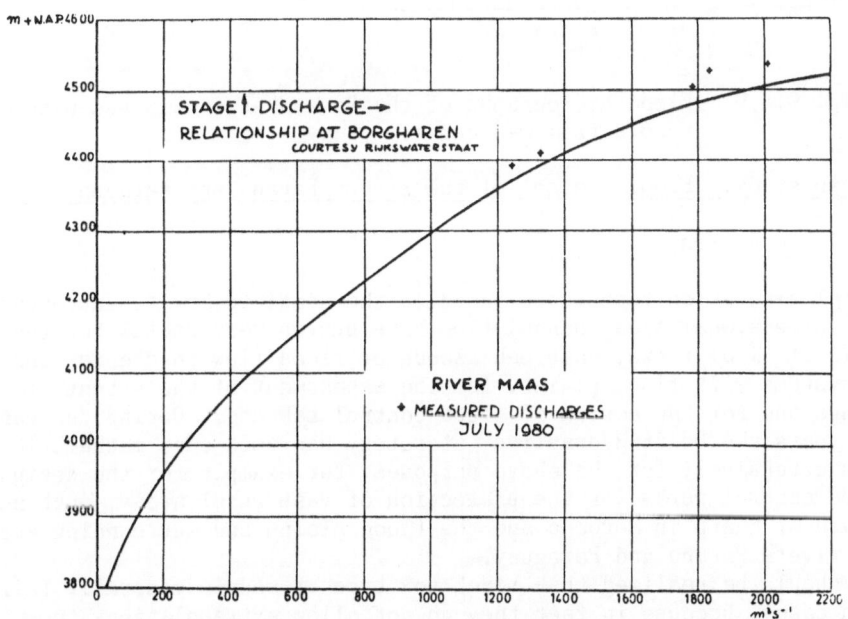

Fig. 9.5.8.   Stage-discharge relationship at Borgharen on the River
              Meuse.

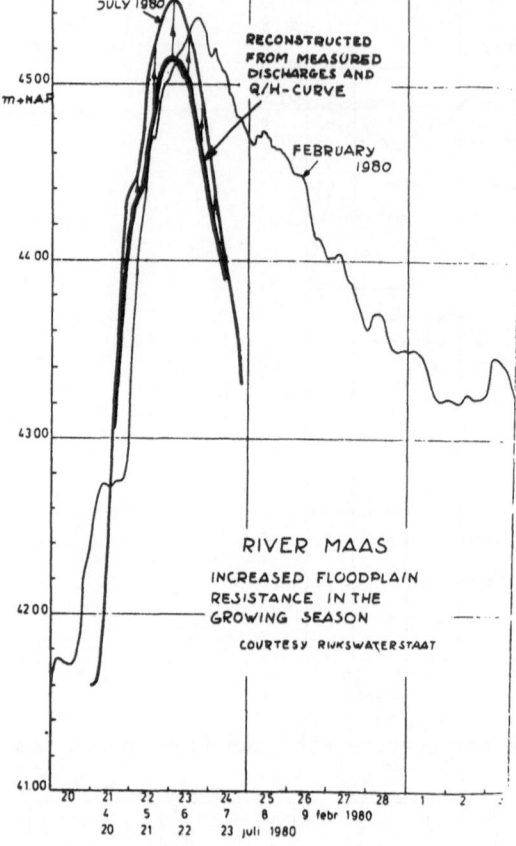

Fig. 9.5.9. Flood hydrographs of the River Meuse compared with a
            record from Q-H curve.

9.6 Case study: Flood control of the rivers Parana and Paraguay

9.6.1 Introduction

Numerical simulation techniques based on the complete St. Venant equations
for one-dimensional open channel flow have proven very useful for the
analysis of tidal flows, dambreak floods or flood flow in channel and
river systems with flood plains, for the assessment of the extent of
floodings and for the design of flood control schemes. During the past
twenty years the Delft Hydraulics Laboratory has used such mathematical
methods extensively for the above purposes, for example for the design
of flood control works for the protection of vast rural areas, such as
the Plain of Rharb in Morocco and the flood plains and surrounding areas
of the rivers Parana and Paraguay.
    It should be realized that black box type of models generally fail
in such cases, because in fact they do not allow extrapolations to
conditions they are not calibrated for, and it is often impossible to

simulate the effects of human interference in the natural system with
such models in a reliable way.

The latter study covered a 1700 km long stretch of the River Parana,
as well as part of the River Paraguay between Asuncion and its confluence
with the River Parana. Vast flood plains exist downstream of this confluence, varying in width from 15 to 60 km. These areas are regularly
inundated during prolonged floods and are of utmost importance for the
attenuation of flood waves, in particular of floods with sharp peaks
such as the "Probable Maximum Flood" and dam break waves. In order to
obtain a proper understanding of the inundation phenomena under various
flow conditions a mathematical model of the river system was developed,
including the flood plains and adjacent areas which are inundated only
during extraordinary floods.

9.6.2 The Parana-Paraguay model

Fig. 9.6.1 shows the river system and its schematization in model form.
Generally, the flood plains and the main river channels are treated as
separate parallel channels to allow for transversal slopes of the water

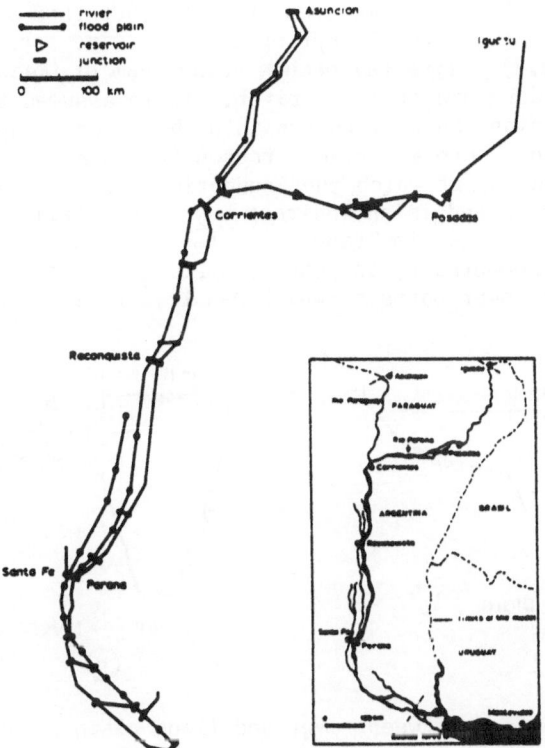

Fig. 9.6.1. Model of Parana and Paraguay rivers; situation and
schematization.

level. The distances between the 475 available cross-sections are 5 to
15 km along the rivers and 15 to 20 km in the plains. The latter have
been derived primarily from the limited information on topographic maps.
The total length of the channels in the model is about 4000 km.

In further detail, the river proper often consists of a narrow deep
channel and a wide shallow part in which the distribution of the velocity
in the cross-section is accounted for with the algorithm discussed in
subsection 9.4.6. With this method the larger part of the Upper Parana
could be simulated by one composite cross-section with two roughness
coefficients, one for the river valley and one for the river itself.

Downstream of Corrientes, the small streams running parallel to the
main river have been included in the cross-section of the flood plain,
where the above mentioned algorithm is also applied. In other places
no secondary streams exist. These flood plain channels dry up during
low river stages and then require the numerical treatment for dry channels
as outlined in subsection 9.4.7 to guarantee successful progression of
the computations.

The exchange of water between the river and plain can be simulated
by introducing short transverse channels at regular intervals. This,
however, significantly increases the complexity of the system. Therefore, an explicit method was used at the time to complete these exchange
flows using the results of the last time step, to introduce them as
lateral flows into and from the channels during the next time step. Two
situations occur (Fig. 9.6.2). Type (a) occurs downstream of Corrientes
with flood plains sloping downward from the river. It is assumed that the
water spilling into the plains can only return into the river at junctions,
where the plain is narrow or where the river crosses from one side of the
plain to the other. The levels at which the inundations start have been
derived from topographic data and in an indirect way from satellite
images, reconnaisance flights and simultaneously observed stages. The
lateral discharges q were computed by the Chézy equation through simple
shaped cross-sections, the lower parts of which had triangular forms.

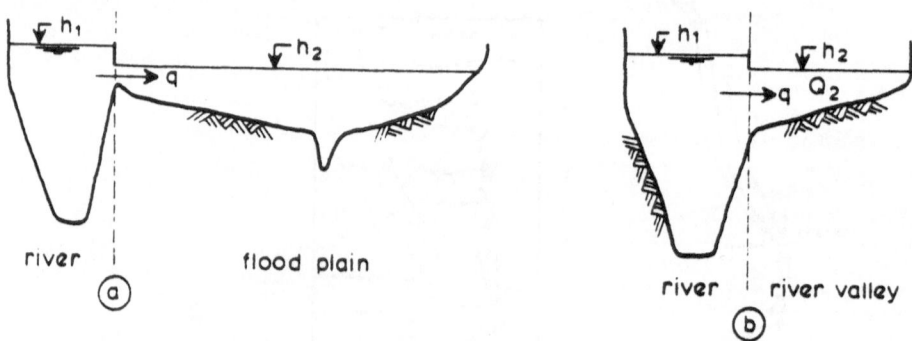

Fig. 9.6.2. Exchange flow between river and flood plain or valley.

Type (b) occurs along the river Paraguay with a river valley sloping
upward from the river. The water flowing into the valley can return
everywhere. A time lag ΔT was assumed between the water levels in the
river and the valley at the same location:

$$h_2 (t + \Delta T) = h_1(t)$$

As storage is the most important function of this valley, the exchange flow was computed by the equation of continuity of the valley.

$$q = - B \, \partial h_2/\partial t - \partial Q_2/\partial x$$

All terms are determined from the last time step completed, while $\partial h_2/\partial t$ is evaluated as $(h_1 - h_2)/\Delta T$.

As outlined above an explicit method was used to compute the flow between river and flood plain from the water levels in both channels, as computed during the last time step. This may give rise to numerical instabilities in case the model is used for the routing of peaky floods. This can be improved by treating one of the two water levels in an implicit way, which does not yet complicate the structure of the model.

9.6.3 Calibration and verification of the model

The model has been calibrated for a period of 4 years between September 1964 and August 1968 including a small flood without inundations, a moderate flood and an extreme flood with a recurrence period of about 50 years. The model was verified with data from a high flood observed in 1977. Daily recorded water levels at 52 stations and discharges from 23 catchments draining into the main river. A total of 100,000 hydrometric data were collected and processed for the calibration and verification. Floods usually last for several months with discharges ranging from 5000 to 40,000 m^3/s. A distance step of 10 km and a time step of 2 days were used but the latter had to be reduced in periods with peaky floods. The Manning coefficient of the flood plains is approximately 0.1. Some results of the computations are shown in Fig. 9.6.3.

9.6.4 Use of the model

The attenuation of normal prolonged floods by storage in the plains turned out to be rather small and was partially compensated by inflows from tributaries. The opposite occurs during extreme events such as probable maximum floods and dam break waves with high peaks and relatively small volumes. The simulation of these phenomena was an important aim of the model. Another objective was to study the effects of engineering works for flood protection, including reservoirs, lateral dikes and diversion schemes, on the inundation phenomena in the region. In this way the model provided basic information for related studies in the fields of agronomy, sociology, economy and civil engineering.

9.7 Strategy for implementation of forecasting models

A river flow simulation model as described in this chapter cannot directly be employed as part of a river flow forecasting system. Its operation requires powerful computers and expert technicians usually not

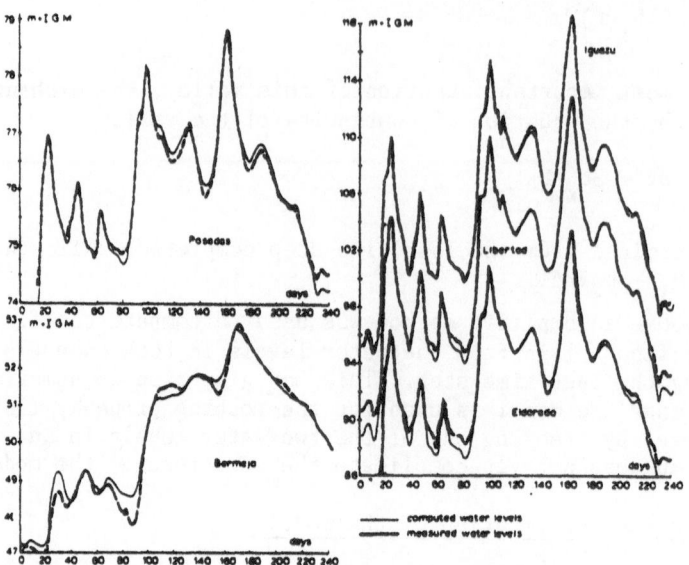

Fig. 9.6.3. Comparison of water levels for several stations

available on a continuous full-year-round basis.

In Chapter 7 of this volume Schultz already pointed out that forecasting centres prefer simple forecasting tools that fit in with the
data-collection, -transmission and -processing system and yield reliable
results even under non-optimal conditions.

For the development and implementation of important forecasting systems Cunge (1980) recommends a three-step strategy:

"(i) First, a detailed model based on the full St. Venant equations
is built, and is calibrated with all data available in the same way as a
normal flood simulation model. Based on the full equations and sound
physical hypotheses, the model allows not only for the simulation of
exceptional, unrecorded events beyond the range of calibration, but also
for the inclusion of river system modifications such as dams, dykes, etc.,
without losing essential predictive capacity.
(ii) A simplified model is conceived for use in the forecasting itself.
For this purpose, several possible methods are studied; each tested,
using the complete calibrated model in step (i) to simulate flood events,
and their respective applicabilities and forecasting capabilities assessed. Once the simplified method to be used is chosen, its coefficients
are calibrated by repeated use of the full model. Then the real-life
system (flood forecasting on a small computer using a simplified method,
recording of input data, transmission, lack of data at some input stations, random transmission incidents, etc.) is simulated in order to
check its applicability and improve its efficiency.
(iii) The simplified model is implemented on the smaller computer at the
forecasting centre.
Use of such a strategy allows one to benefit from the efforts of the

specialized and qualified staff who built and operated the full model
using a powerful computer. This original model is then maintained in
operating condition – an action which allows for quick revision of the
simplified model. The revision is either periodic, based upon the forecasting experience, in each year, or infrequently, when the river is
modified significantly. Such a strategy makes the forecasting systems
a durable investment, useful for studies of future developments and their
consequences. These advantages accrue from the fact that it is based on
a full-equation model."

SYMBOLS

h	water depth	L
o	order of magnitude	1
c	celerity, velocity of wave propagation	LT^{-1}
h_j^n	finite difference approximation of $(h\Delta x, n\Delta t)$	L
Θ	weighting factor (implicit models)	1
α (see 9.3)	modification factor	1
σ	Courant number	1
A	total wetted area	L^2
A_s	flow area	L^2
Q	discharge	L^3T^{-1}
B	total channel width at the water surface	L
B_s	width of flow area at the water surface	L
τ_0	bottom shear stress	$ML^{-1}T^{-2}$
τ_s	wind shear stress	$ML^{-1}T^{-2}$
$u_{1,2}$	longitudinal velocity components of the lateral outflows q_1, q_2	LT^{-1}
α (see 9.4)	coefficient concerning velocity distribution	$L^{\frac{1}{2}}T^{-1}$
C	Chézy coefficient	
n	Manning coefficient	$L^{-1/3}T$
R	hydraulic radius	L
O	wetted perimeter	L
q	lateral flow	L^2T^{-1}
d	damping factor	1
c_r	relative velocity	1
n_x	number of grid points per wave length L	1
L	wave length	L
n_t	number of time steps in period T	1
ρ	amplification factor	1
R_w	weighted hydraulic radius	L

REFERENCES

Cunge, J.A., Holly, F.M. and Verwey, A. (1980): Practical Aspects of Computational River Hydraulics, Pitman Publ. Ltd., London.

Engelund, F. and Hansen, E. (1967): A Monograph on Sediment Transport in Alluvial Streams, Copenhagen: Hydraulics Laboratory of the Technical University of Denmark.

Grijsen, J.G. and Meyer, Th.J.Q.P. (1979): 'On the Modelling of Flood Flow in Large River Systems with Flood Plains', Proc. XVIII IAHR Congress, Cagliari, Italy, also Publication 227, Delft Hydraulics Laboratory.

Grijsen, J.G. and Vreugdenhil, C.B. (1976): 'Numerical Representation of Flood Waves in Rivers', Int. Symp. on Unsteady Flow in Open Channels, Newcastle upon Tyne, also Delft Hydraulics Laboratory Publication No. 165, 1976.

Jansen, P.Ph. (ed.) (1979): Principles of River Engineering, Pitman Publ. Ltd., London.

Klaassen, G.J. (1981): 'Morphologische Verschijnselen Grensmaas' (Morphologic Phenomena: River Meuse bordering on Belgium and the Netherlands, Delft Hydraulics Laboratory Rep. R863.

Liggett, J.A. and Cunge, J.A. (1975): Chapter 4 of Unsteady Flow in Open Channels, edited by K. Mahomoad and V. Yevjevich, Water Resources Publication. Numerical Methods of Solution of the Unsteady Flow Equations.

Ogink, H.J.M. (1984): 'Hydraulic Roughness of Bed Forms', Delft Hydraulics Laboratory, Report M2017/M1314.

Ogink, H.J.M. (1985): 'On the Effective Viscosity Coefficient in 2-D Depth-Averaged Flow Models', to be published in Proc. XXI IAHR Congress, Melbourne, Australia.

Preismann, A. (1960): 'Propagation des Intumescences dans les Canaux et Rivières' ('Flood Wave Propagation in Channels and Rivers), Proc. A.F.C.A.L. p. 433.

Rijn, L.C. van (1983): 'Sediment Transport Phenomena: 1. Sediment Transportation in Heavy Sediment Laden Flow; 2. Prediction of Bed Forms and Alluvial Roughness; 3. Equivalent Roughness of Alluvial Bed', Delft Hydraulics Laboratory, Publication No. 311.

Shen, H.W. (ed.) (1971): River Mechanics Vol. I and Vol. II, Ft. Collins, Colorado USA 80521.

Simons, D.B. and Richardson, E.V. (1971): 'Flow in Alluvial Sand Channels' in River Mechanics (ed. H.W. Shen).

Vanoni, V.A. (1975): 'River Dynamics', Advances in Applied Mechanics, Vol. 15, Acad. Press Inc.

Wybenga, J.H.A. and Klaassen, G.J. (1983): 'Changes in Bed Form Dimensions under Unsteady Flow Conditions in a Straight Flume', Spec. Publ. Int. Ass. of Sedimentology 6, 35-48.

10. THE FORECASTING AND WARNING SYSTEM OF THE 'RIJKSWATERSTAAT' FOR THE RIVER RHINE

J.G. de Ronde

Rijkswaterstaat Dir. W & W
Hooftskade 1
2526 KA Den Haag
The Netherlands

10.1 Introduction

The sea and rivers would flood the largest part of the Netherlands, if
it were not protected by hundreds of kilometers of dikes. Dikes which
must be carefully maintained and monitored. Especially during storms
and river floods, the 'Dike-guard' performs detailed inspection and when
necessary makes emergency repairs. Clearly, good warning systems giving
accurate forecasts far enough in advance are important. In the following
section, the organization and tasks of 'Rijkswaterstaat', the government
agency which operates these warning systems, are discussed. In section
10.3 the riverflood warning system will be dealt with in more detail.
In sections 10.4 and 10.5 the "old" empirical and the statistical forecasting models currently applied are discussed. In the final section
long term forecasting of low flow on the River Rhine is discussed.

10.2 General description of Rijkswaterstaat and its warning services

In the Netherlands, 'Rijkswaterstaat', a department of the Ministry of
Transport and Public Works, is responsible for the management, maintenance and administration of the major rivers and canals, the estuaries,
the coastal waters and the Dutch part of the North Sea. This department
has a similar responsibility for the main roads. Most related engineering works, such as bridges, locks, weirs and harbours, are built and
maintained under the supervision of Rijkswaterstaat. However, only part
of all dikes are state property, most dikes are owned by waterboards.
  The operation of different warning systems is an important task of
Rijkswaterstaat. The "Berichtendienst" (Information Service), part of
the Directorate for Water Management and Hydraulic Research (one of the
17 directorates of Rijkswaterstaat) implements these systems. Five
different warning or information services are used by the Information
Service:
I.   "S.V.S.D." Stormsurge Warning Service.
      Warning notices are broadcasted when very high water levels due to
      stormsurges are expected along the coast and estuaries.
II.   Riverflood Warning Service.

D.A. Kraijenhoff and J.R. Moll (eds.), River Flow Modelling and Forecasting, 273-286.
© 1986 by D. Reidel Publishing Company.

Alerts are issued when very high water levels are expected on the
Rivers Rhine or Meuse.
III.   Daily River Information Service.
       Information on actual and forecasted water levels, discharges,
       locks and weirs etc. is broadcast daily. The information also
       covers the Rhine and Meuse in Switzerland, Germany and Belgium.
       It is of particular interest for shipping.
IV.   Ice Information Service.
       If ice occupation occurs along the coast or on the rivers and
       canals in the Netherlands, information bulletins on ice conditions
       and shipping possibilities are issued.
V.   Information Service on Water supply.
       During periods of drought and of very low river discharge, infor-
       mation is issued concerning the allocation of the available water
       being discharged by the Rhine and Meuse and supplied from the main
       reservoirs, the IJssel lake and the Marker lake.

10.3   Organization of the riverflood warning system

The Daily River Information Service and the Riverflood Warning Service
make use of daily forecasts of water levels of the Rhine at Lobith (at
the Dutch-German border). The lead times of these forecasts are up to
four days. If a water level of N.A.P. + 14.00 m (the N.A.P. reference-
level corresponds closely to mean sea level) is reached and it is expected to exceed N.A.P. + 15.00 m (corresponding to a discharge of
$7500 \ m^3 \ s^{-1}$ and a return period T of 3 years), the Riverflood Warning
Service is set in motion. Warning telegrams containing the necessary
information are then issued to the:
-   Rijkswaterstaat services
-   Provincial services and
-   8 Water boards.
       If the situation becomes more threatening, the Minister of Traffic
and Public Works can proclaim the 'B.R.C.' (Emergency Supervision).
B.R.C. implies a special assignment, with legal powers, for the regulation officers of 'Rijkswaterstaat', who are then completely in charge
of emergency operations. These officers can, for example, seize any
material they deem necessary for the maintenance of the dikes. These
powers exceed considerably those of the Water boards. The B.R.C.
was established when the Water boards controlled smaller areas (and
smaller dike lengths) and were less capable of adequately maintaining
the dikes in real emergency. The latest B.R.C. proclamation occurred in
1948, when a level of N.A.P. + 16.18 m was reached at Lobith (T = 20
years). In 1970, the level of N.A.P. + 16.13 m occurred; however, no
B.R.C. was announced owing to the fact that the organization of the
Water boards had, by then, been improved considerably. These authorities
were judged capable of managing an emergency situation.

10.4   The empirical forecasting model

In the 1950's, an empirical model for making forecasts of the water
levels at Lobith for 1, 2, 3 and 4 days in advance was developed by

Wemelsfelder (1953). This model is based on water levels observed on the Rhine and its tributaries (for the gauging stations used, see Fig. 10.4.1). This is done by transforming these levels into discharges, combining the tributary discharges with the main river discharge, and also taking flow-time into account. Corrections are also made for non-stationary flow and for systematic errors.

The flow-times are taken into account as follows: to make a forecast n days in advance, discharges at gauging stations with a flow-time to Lobith of approximately n days are used. For example, to forecast 1 day in advance, the gauging stations at Cologne (on the main river) and Wetter (on the Ruhr tributary) with flow-times to Lobith of respectively $1\frac{1}{4}$ and $1^{1}/_{6}$ day are used. Time corrections are required; a forecast at Lobith for 8 a.m. tomorrow is based on discharges at Cologne at 2 p.m. today.

To account for the influence of the smaller tributaries, empirical relations between the discharge of a tributary (e.g. the Ruhr at Wetter) and the discharges of the smaller tributaries in the neighbourhood (in the case of Wetter the tributaries between Cologne and Lobith) are used.

A correction for non-stationary flow is made by subtracting a fraction of the discharge increment of the preceding day (from Maxau up to Cologne one half is taken, and between Cologne and Lobith a third part). Conversely, a corresponding amount is added when the discharge is decreasing. Additional empirical corrections are made, either a constant flow rate or a percentage of the discharge. Final corrections are based on errors observed in preceding forecasts. These corrections are weighted moving averages: one half of the error of the preceding day, plus a quarter of the error of 2 days before, plus an eighth of the error of 3 days before, and so on.

In formula, the discharge at Lobith 1 day in advance, if t = 10 for today, is:

$$
\begin{aligned}
\text{Lobith}_{11} = \; & C_{10} + W_{10} \\
& - 0.5\,(C_{10} - C_9) - 0.5\,(W_{10} - W_9) \\
& - 0.08\,C_{10} \\
& - 0.5\,E_{10} - 0.25\,E_9 - 0.125\,E_8 \ldots
\end{aligned}
\qquad (10.4.1)
$$

with: C_{10} = discharge at Cologne on day 10

W_{10} = discharge at Wetter on day 10

E_{10} = error in forecasting for day 10

= L_{10} (observed) − L_{10} (forecast)

A forecast 2 days in advance would then be:

$$
\begin{aligned}
\text{Lobith}_{12} = \; & C_{11} + W_{11} \\
& - 0.5\,(C_{11} - C_{10}) - 0.5\,(W_{11} - W_{10})
\end{aligned}
\qquad (10.4.2)
$$

etc.

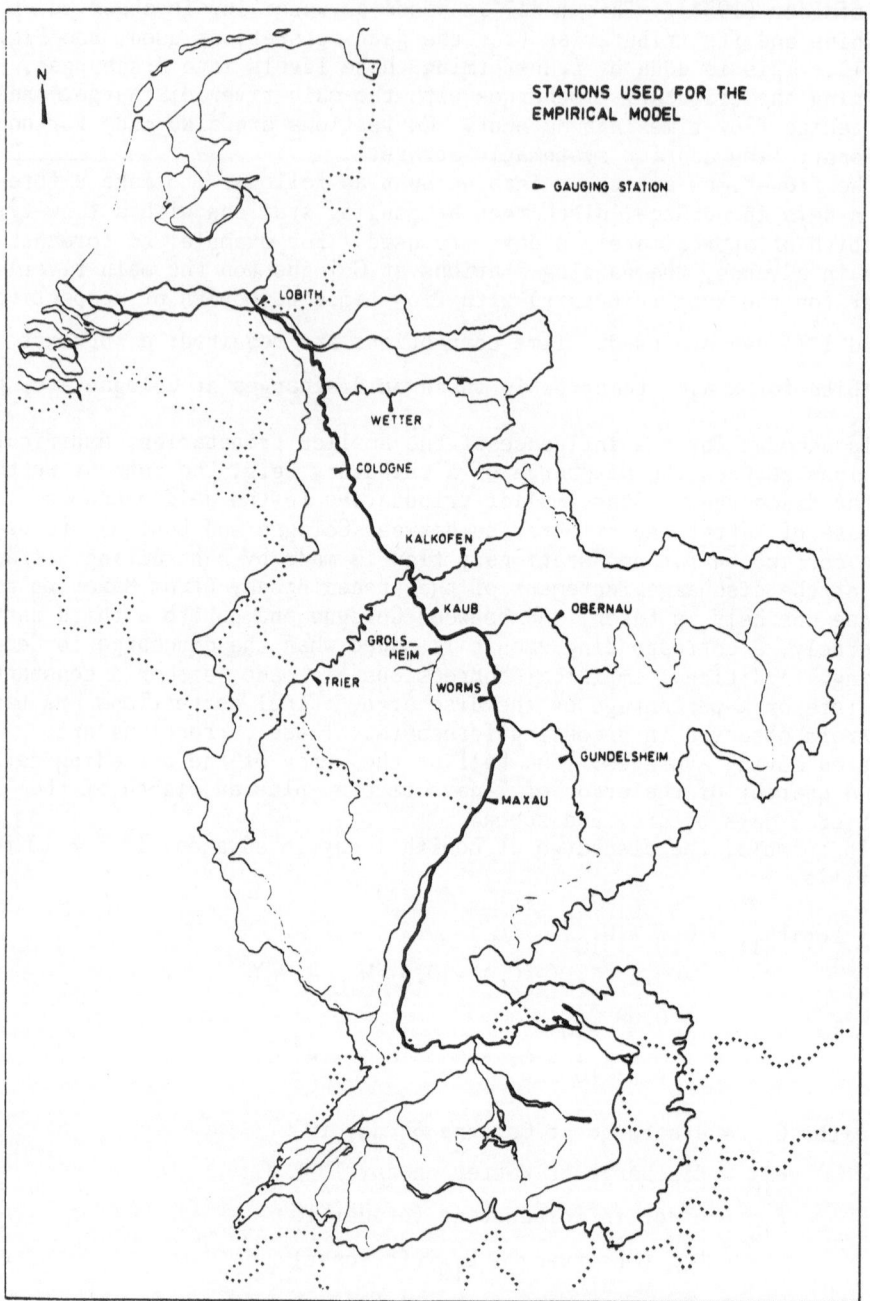

Fig. 10.4.1.   Stations used for the empirical model.

with: C_{11} derived from discharges at Kaub, Kalkofen and Trier in a
manner analogous to the one presented in Eq. 10.4.1.
W_{11} to be estimated, taking into consideration the weather
forecast, according to the personal view of the person in
charge.

The model's forecasts were not very accurate and had systematic
errors. This was presumably partly due to the fact that the model had
never been updated with revised discharge curves and adapted to changed
relations. Updating the model would have required much effort and time,
whereas it was doubtful whether this effort would significantly improve
the results. It was therefore decided to employ a multiple-linear regression model of considerably larger scope.

10.5 The multiple linear regression model

Daily water levels and precipitation data, which were both operationally
available were initially chosen as input for the linear regression model.
From a hydrological viewpoint, use of discharges seems to be more appropriate, but in actual practice, other problems arise. Investigations
of this subject were made at a later stage; the outcome will be discussed.
The water levels at Lobith 1, 2, 3 and 4 days in advance are output of
the model.
    Although a linear model is assumed, the relationship between the
water levels themselves, and especially the relationship between water
levels and rainfall, are actually non-linear. Therefore three different
models are used for the three discharge (c.q. water level) categories:

$$\text{I} \quad Q < 2500 \ m^3 s^{-1} \quad H < 10.70 \ m \ (\text{Lobith N.A.P.} +)$$
$$\text{II} \quad Q > 2500 \ m^3 s^{-1} \quad H > 10.70 \ m$$
$$\text{III} \quad Q > 4500 \ m^3 s^{-1} \quad H > 13.00 \ m$$

For comparison: the average river discharge at Lobith is 2200 $m^3 s^{-1}$.
A discharge of 4500 $m^3 s^{-1}$ is exceeded by 2 floodpeaks during a year on
the average. The third model pertaining to discharges exceeding 4500 $m^3 s^{-1}$
is of special importance for accurate forecasting of extreme water levels
during very large floods. The other two models are intended for daily
use.
    An investigation into the optimal combination of stations and time-
intervals demonstrated that 5-day records of 10 river-gauging stations
(including Lobith) and of 8 precipitation stations should be preferred.
    These stations are shown in Fig. 10.5.1.
    The equation for the forecast, one day in advance for example, is
as follows:

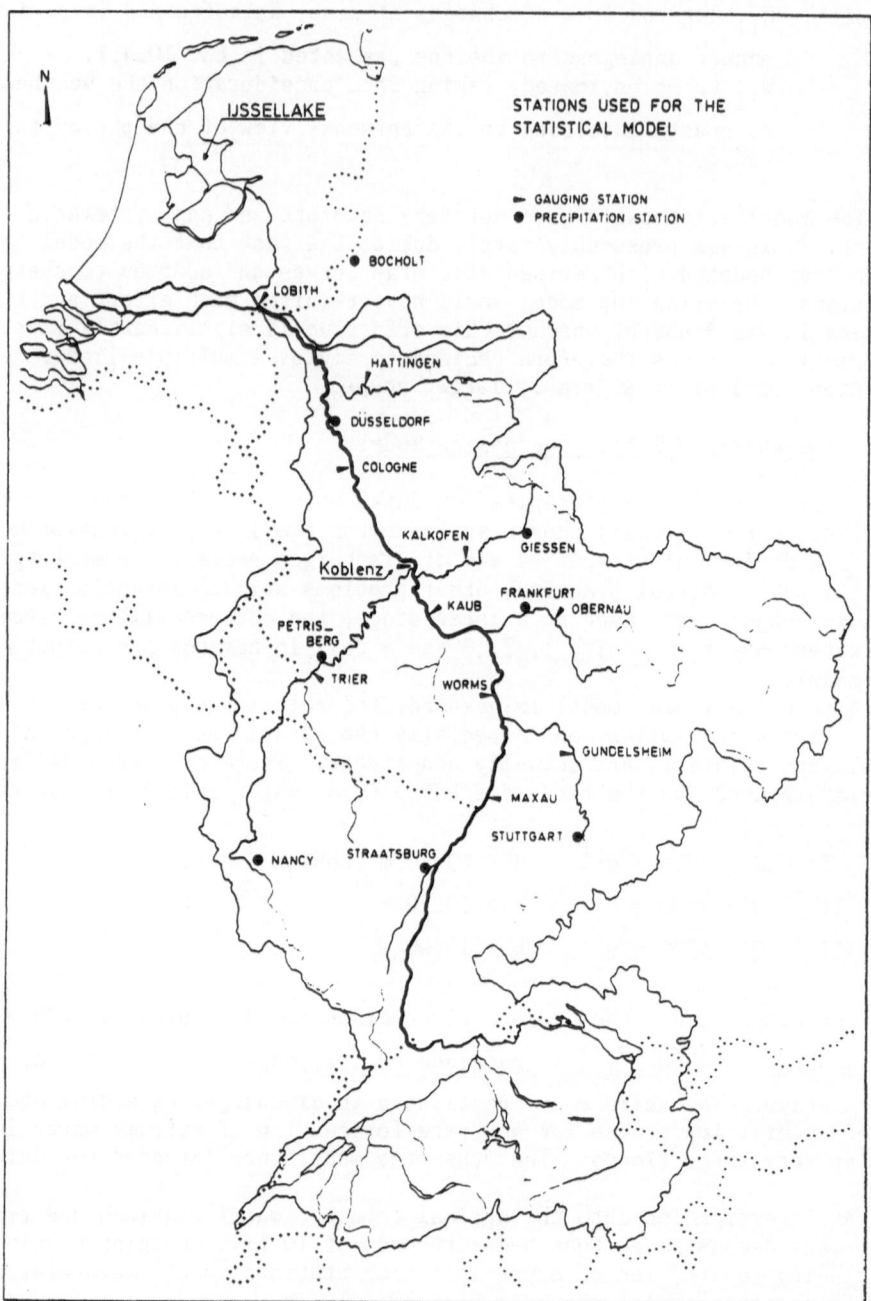

Fig. 10.5.1.   Stations used for the regression model of the Rhine

$$L_t = A_0 +$$
$$A_1 \times L_{t-1} + A_2 \times L_{t-2} + \ldots + A_5 \times L_{t-5} +$$
$$A_6 \times C_{t-1} + A_7 \times C_{t-2} + \ldots + A_{10} \times C_{t-5} + \ldots$$
$$\ldots + A_{90} \times Stu_{t-5} \qquad (10.5.1)$$

where A_0 = constant

L_t = river level at Lobith at time t

C_t = river level at Cologne at time t

Stu_t = precipitation at Stuttgart at time t

The parameters A_0 to A_{90} are obtained by means of the least squares method. Inputs for the third model, used to forecast very high discharges, consist of data from only 6 stations (two of which are rainfall stations) over 3 time-intervals. It therefore has 19 parameters. The parameters of the three models are calculated, using data of the period 1977-1980.

The model for extreme water levels/discharges was calibrated on only 100 observations; the model for high water levels/discharges on 500. A sensitivity analysis was performed on the number of stations to be used for the forecast and the intercorrelation of the model parameters was taken into account.

As mentioned previously, a study was made to find out whether the use of discharges instead of water levels would improve the outcome. To compare results the forecasted discharges must be converted into water levels. Some of the results obtained with the models are presented in Table I.

As regards the low-flow models, no use is made of discharges. Under such conditions, the water level at some stations is elevated by weirs so that the relationship between stage and discharge is disturbed.

From these results, it is clear that using water levels or discharges as input does not affect the results much.

For forecasts 1 to 2 days in advance, the use of discharges usually improves results, whereas the approach employing water levels is better for forecasts 3 to 4 days ahead.

The forecasts 2 days ahead of time for 4 different models are presented in Fig. 10.5.2. As an example, this Figure illustrates clearly that this is a case where the model using discharges (Q + P) to forecast floodpeaks is better than the model based on water levels (H + P). In the latter model, the forecasts for the peak-flows are often too high. On the basis only of the top levels attained, the 2 models for H > 13.00 m perform best with water levels as well as with discharges.

In Fig. 10.5.3 forecasts for 1, 2, 3 and 4 days in advance are presented for the same period, derived by means of the model for discharges corresponding to H > 10.70 m. Clearly, forecasts 3 days in advance become considerably less accurate; large errors are associated with forecasts made 4 days in advance, especially as regards the rising limb of a floodwave.

Table I.   Standard deviations and maximum discrepancies of the results
           of the models starting from water levels (H) and discharges (Q)
           in cm.

Model used	Period[1]	H/Q[2]	Number of days ahead in time							
			1		2		3		4	
			S	max	S	max	S	max	S	max
$Q < 2500$ m^{3}s^{-1}	1977...80	H	3.3	21	7.4	74	13.1	113	20.3	133
$H < 10.70$ m	1982	H	3.5	–25	7.0	–33	12.4	59	19.4	149
$Q > 2500$ m^{3}s^{-1}	1977...80	H	3.7	–14	11.0	51	26.5	193	46.1	210
$H > 10.70$ m		Q	3.1	12	10.2	49	28.3	197	50.4	240
	1982	H	5.3	–21	16.9	–68	36.3	157	60.4	267
		Q	5.2	–27	15.8	64	36.9	146	64.3	241
$Q > 4500$ m^{3}s^{-1}	1977...80	H	4.0	–10	12.0	49	21.8	55	34.9	115
$H > 13.00$ m		Q	3.7	11	11.9	45	23.2	70	36.6	138
	1982	H	4.6	13	16.0	63	35.1	79	56.7	113
		Q	4.0	14	16.7	71	33.8	94	56.2	110

(1)   1977–80   period used for calibration
      1982   period used for verification

(2)   H   model using water levels as input
      Q   model using discharges as input

      In practice, the output of the model is the point of departure for
preparing a forecast. Prior to issuing a forecast, this output is
adjusted on the basis of the hydrological situation in the region, anticipated rainfall, and, most important, on the basis of the experience of
the personnel involved. If there is uncertainty on which model should
be used – the one for extreme or for high discharges – both are used
and the results compared. When heavy rainfall is forecasted, Rijkswaterstaat relies less on the model, and more on the experience of its
Forecasting Service personnel. In case of a breakdown at a gauging
station, estimated data are used in the model. During the periods of
very high discharge, German forecasts are used, if available. From
Switzerland, forecasts of the water level at Rheinfelden are received
daily. The objective is to develop a set of models that completely
cover the main river: the Swiss model up to Rheinfelden, the German
model between Rheinfelden and Cologne, and the Dutch model downstream
from Cologne.

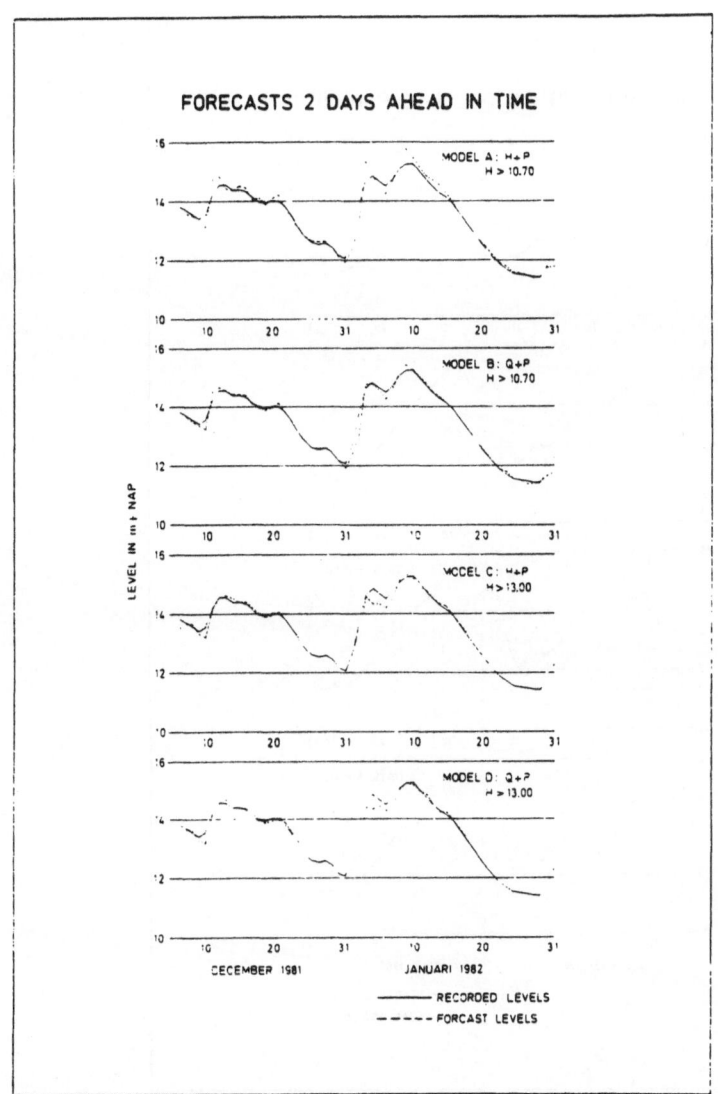

Fig. 10.5.2. Forecasts 2 days ahead in time.

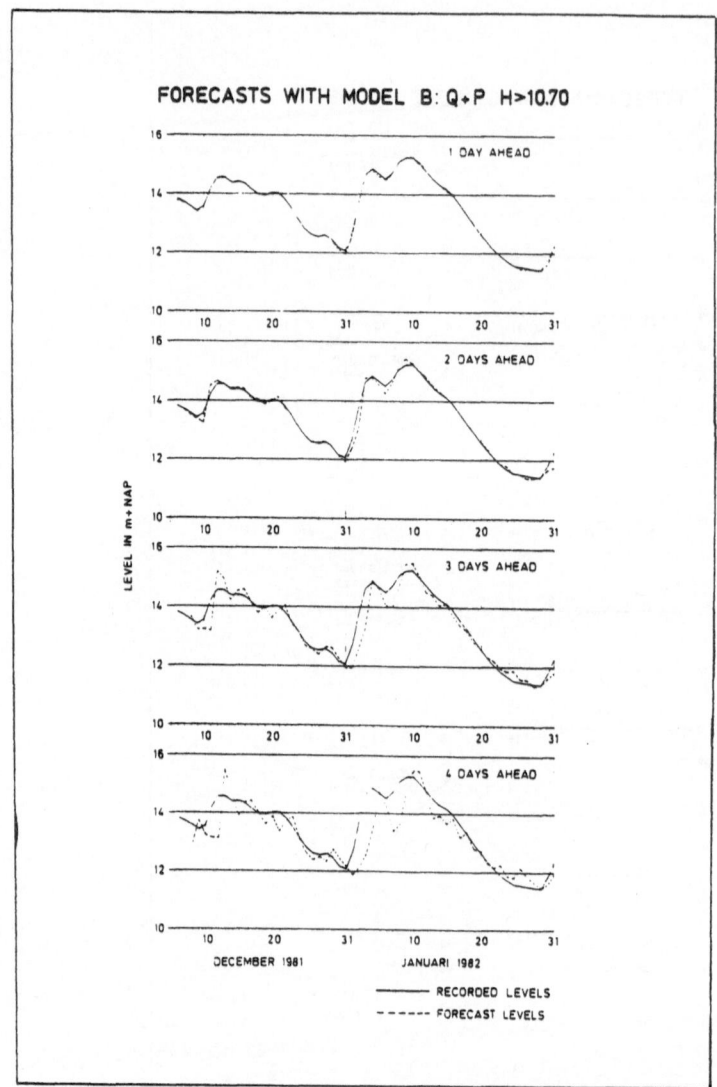

Fig. 10.5.3.   Forecasts with model B: Q + P H > 10.70

10.6 Low flow forecasting

Forecasts of low flow are also of economic interest, especially for water supply, shipping and water management of the Yssel Lake (Fig. 10.5.1). For this reason Rijkswaterstaat developed a method for sub-dividing the Rhine flow Q_L at Lobith into a component of "direct runoff" Q_s and a component of "base flow" Q_b.

This concept was introduced by Wemelsfelder (1963) and elaborated by

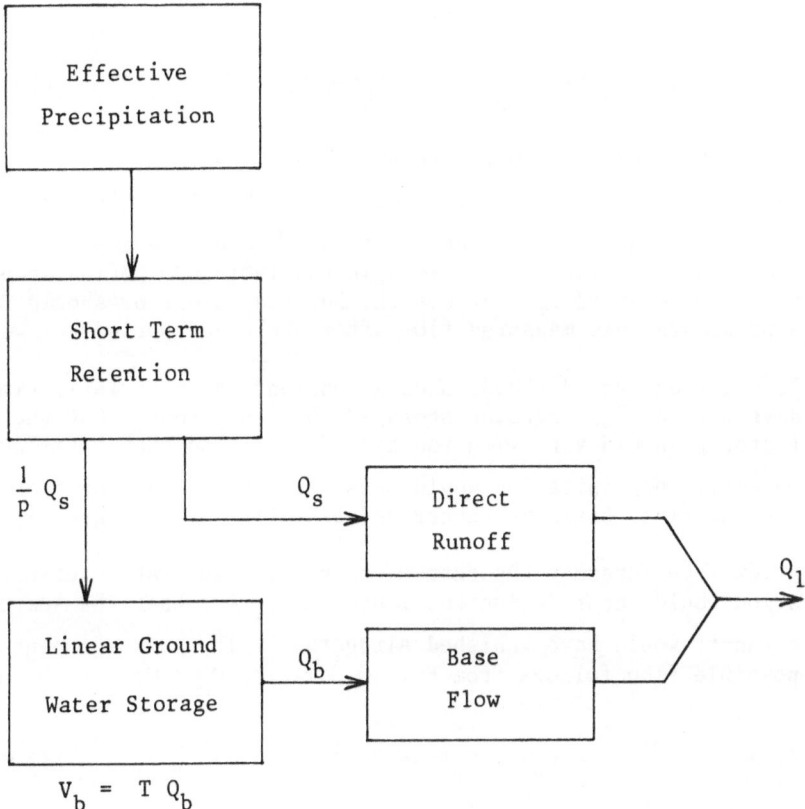

Fig. 10.6.1.

van der Made (1982). The model consists of an unspecified detention
storage that transforms the hyetograph of effective precipitation into
a hydrograph of runoff. This hydrograph is subsequently divided into:
-   A component of direct runoff Q_s that passes Lobith without further
    delay and attenuation.
-   A proportional component $1/p\, Q_s$ that flows into a linear reservoir

    with a long characteristic time T, representing "groundwater storage".
    The outflow Q_b from this "groundwater storage" is called the "base

flow" at Lobith and it can be expressed in a recursive model (this
volume, Young, Eq. 6.2.3 and 6.2.4):

$$Q_b(t + \Delta t) = Q_b(t)e^{-\Delta t/T} + \frac{1}{p}\, Q_s(t)(1 - e^{-\Delta t/T}) \qquad (10.6.1)$$

Series expansion of $e^{-\Delta t/T}$ for $\Delta t \ll T$ yields:

$$Q_b(t + \Delta t) = Q_b(t)(1 - \Delta t/T) + \frac{1}{p}\, Q_s(t)\Delta t/T \qquad (10.6.2)$$

After substitution of $Q_s(t) = Q_L(t) - Q_b(t)$ this equation reads:

$$Q_b(t + \Delta t) = Q_b(t)\left[1 - \frac{p+1}{p}\frac{\Delta t}{T}\right] + Q_L(t)\frac{\Delta t}{pT} \qquad (10.6.3)$$

It can now be used for the recursive calculation of base flow Q_b as influenced by the measured river flow Q_L at Lobith. The criterion for optimising the two parameters p and T was that, excepting cases of a frozen river, the calculated base flows in the 1901-1980 period should never exceed the measured flow at Lobith, but the base flow should closely approximate this measured flow after prolonged periods of drought.

Wemelsfelder and van der Made used a constant characteristic time T = 150 days for the "groundwater storage", but they found that the divider factor p should vary with the base flow Q_b, so that a smaller part of effective precipitation would pass through the "groundwater storage" as the river basin is wetter and base flow is consequently higher.

For a low-flow forecast the extreme assumption is that no effective precipitation would occur during the lead-time t_v and that the component of direct runoff would have vanished altogether. The corresponding minimum possible flow follows from Eq. 10.6.1 with $Q_s = 0$:

$$Q_L(t + t_v) = Q_b(t + t_v) = Q_b(t)e^{-t_v/T} \qquad (10.6.4)$$

Obviously this lowest possible flow does not equal the most probable river flow at the end of the lead-time t_v. Van der Made used all calculated base flow rates for the period 1901-1975 and derived straight-line correlations between the 75 values of $Q_b(t_0)$ at a certain date, say April 1st, and the corresponding $Q_b(t_0 + t_v)$ values. For instance, for t_v = 5 months after April 1st the expected base-flow on September 1st was found to be:

$$Q_b(\text{September 1st}) = 613 + 0.404\ Q_b(\text{April 1st})(m^3s^{-1}), \qquad (10.6.5)$$

with a standard deviation of 145 m^3s^{-1}.

Such standard deviations enabled van der Made to forecast base-flow values with the required confidence interval.

In the flow hydrograph of Lobith on Figure 10.6.2 the calculated base flow is shown. The figure also shows "the lowest possible base flow" according to Eq. 10.6.4 and the "expected base flow" after April 1st, including its September 1st value according to Eq. 10.6.5.

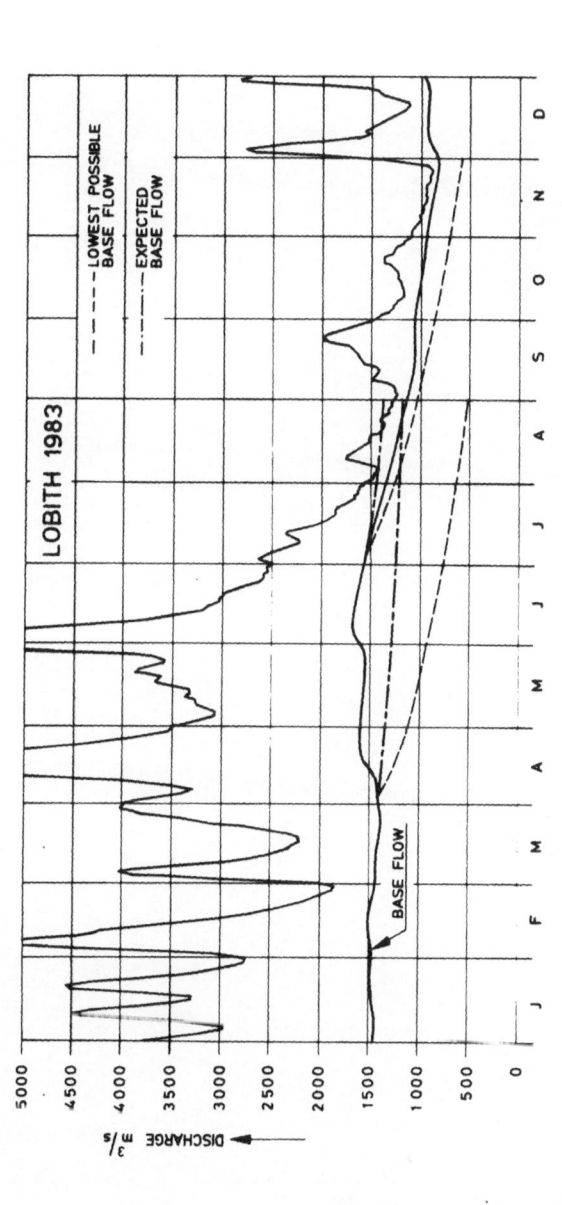

Fig. 10.6.2.   FORECASTS BASE FLOW

REFERENCES

Made, J.W. van der 1982. Commission Internationale de l'Hydrologie du
  Basin du Rhin (CHR/KHR) - Report no. II - 1 'Analyse Quantitative des
  Débits'.
Wemelsfelder, P.J. 1963 'The persistency of River Discharges and Ground-
  water Storage', IAHS publication no. 63, Berkeley.

11. SHORT RANGE FLOOD FORECASTING ON THE RIVER RHINE

J.R. Moll

'De Voorst' Laboratory
P.O. Box 152
8300 AD Emmeloord
The Netherlands

11.1 Introduction

In this chapter, a real-time flow forecasting method is presented and applied to the Rhine floods which occurred in the winter of 1970 (CHR 1978). This method is based on extending a deterministic hydrological flood-routing model with a stochastic component thereby describing the information contained in the forecast errors of the model. Flood forecasting is emphasized in the application, but forecasting of low flows can be accomplished using the method. The rather short forecast lead time used can be increased as demonstrated.

11.2 Flow forecasting

The most important step in any forecasting procedure is establishing a criterion for evaluation of the procedure. The designer of a forecasting method may be very satisfied with a minimum-variance property of the constructed forecaster; his employer is only concerned that forecasters do not underestimate the relevant variable (flow, water level).
    To quote the Dutch Rijkswaterstaat: 'If we forecast 10 cm too low, we are wrong, if we forecast 50 cm too high, we did a good job' (press statement 14 April 1983). Possible criteria for forecasting procedures are:
- minimum variance of forecast errors
- no autocorrelation in forecast errors
- minimization of maximum errors
- minimum error in peak timing
    The method presented in this chapter will be evaluated on several criteria.
    Having established the criteria, a forecasting problem can be subdivided into three problems:
(a) estimation of the actual state of the basin
(b) forecasting of basin inputs during the selected lead time
(c) description of the water-movement during the lead time: the main forecasting procedure.
These three problems require some explanation.

D.A. Kraijenhoff and J.R. Moll (eds.), River Flow Modelling and Forecasting, 287-297.
© 1986 by D. Reidel Publishing Company.

11.2.1 State of the basin

A list of values of hydrological variables, gauged at time t, will, in
most cases, not suffice to describe the actual state of a basin at time
t. The history of these variables, as measured at the gauging stations,
provides additional information: the actual form of a flood-wave in a
river cannot be assessed from the values at time t, but must be derived
from the series of antecedent values. This is illustrated in Fig.
11.2.1. It is impossible to derive the exact form of the flood-wave in
the river section at $t = t_1$ (indicated with stripes in Fig. 11.2.1) from
the measurements $y(0,t_1)$ and $y(x_1,t_1)$ only.

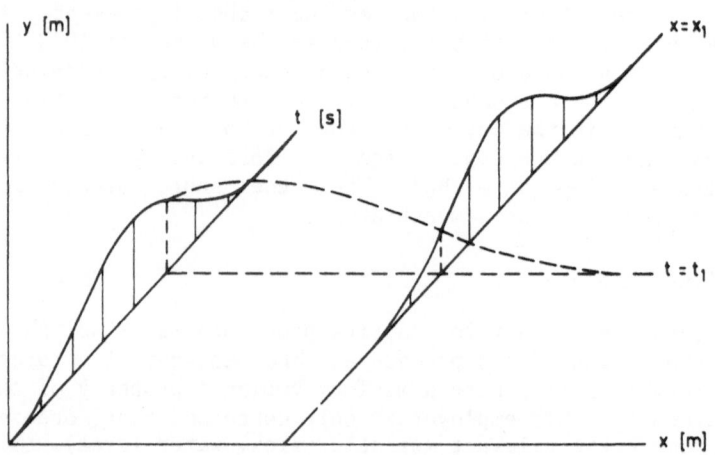

Fig. 11.2.1. The waterdepth $y(x,t)$ during the passage of a flood
          wave.

Problems in state-estimation arise from incomplete measurements in
the basin, moreover the data may include a significant measurement-noise
component. An optimal state-estimation can be obtained by using a
Kalman filter (Chapter 6).

11.2.2 Forecasting inputs

For an accurate forecast the information to be derived from the state of
the basin may, however, be incomplete, because new inputs, occurring
during the lead-time after the issue of the forecast, may also affect
the events at the point of interest. If we only consider the wave motion
in the channel system, some of these interfering inputs, like tributary
and lateral inflows, can also be forecasted in some cases. Rainfall-
runoff models (Chapter 2), quantitative precipitation forecast models,
or time-series models (Chapter 6), can be used to generate these inputs.
However, if unpredictable inputs during the lead-time may have significant effects, they do limit our forecasting potential.

11.2.3 Main forecasting

In the case under study, water-movement is described in terms of the
propagation of floods in a river. The basic equations of motion (St.
Venant) for long waves in a prismatic channel with a rectangular cross-
section and no lateral inflow are:

$$\frac{\partial Q}{\partial x} + b\frac{\partial y}{\partial t} = 0 \tag{11.2.1}$$

$$\frac{\partial y}{\partial x} + \frac{v}{g}\frac{\partial v}{\partial x} + \frac{1}{g}\frac{\partial v}{\partial t} + S_f - S_0 = 0 \tag{11.2.2}$$

where Q is the discharge, b is the channel width, v is the average velocity, y is the waterdepth, S_0 is the bottom slope and S_f is the friction
slope.

Our approach involves dropping the second and third term on the left
side of Eq. (11.2.2) and applying the Chézy-friction law for a wide
rectangular channel:

$$S_f = \frac{Q^2}{A^2 C^2 y} \quad \text{or} \quad Q = Cby^{3/2}\left[S_0 - \frac{\partial y}{\partial x}\right]^{1/2} \tag{11.2.3}$$

where C is the Chézy-friction coefficient.

The so-called convection-diffusion analogy can be obtained by diff-
erentiating Eq. (11.2.1) with respect to x and Eq. (11.2.3) with respect
to t and y and eliminating y and its derivatives:

$$\frac{\partial Q}{\partial t} + c\frac{\partial Q}{\partial x} - D\frac{\partial^2 Q}{\partial x^2} = 0 \tag{11.2.4}$$

with

$$c = \frac{1}{b}\frac{\partial Q}{\partial y} \tag{11.2.5}$$

and

$$D = \frac{Q}{2B\left[S_0 - \frac{\partial y}{\partial x}\right]} \tag{11.2.6}$$

The linear convection diffusion model, Eq. (11.2.4) with constant
values for c and D, was proposed by Schönfeld (1948) as a method for
flood routing. Forecasts obtained with this model tend to have strongly
autocorrelated forecast errors, as will be illustrated in this chapter.
An additional feature can therefore be included, namely recursive updating of the parameters of the forecasting function using a Kalman
filter.

11.3   A deterministic hydrological model for the river Rhine

For flow forecasts at Lobith, the river reach illustrated in Fig. 11.3.1 is of major interest. This reach is divided into sections in such a way that a gauging station is located on the upstream and downstream boundary of every section. A flood is routed through a section using the convection-diffusion model with constant parameters:

$$\frac{\partial Q}{\partial t} + c \frac{\partial Q}{\partial x} - D \frac{\partial^2 Q}{\partial x^2} = 0 \qquad\qquad (11.3.1)$$

The simplifications made by deriving this model, form a source of errors in our forecasts, the most important being that the parameters c and D are assumed to be constants during a flood event, which contradicts Eqs. (11.2.5) and (11.2.6).

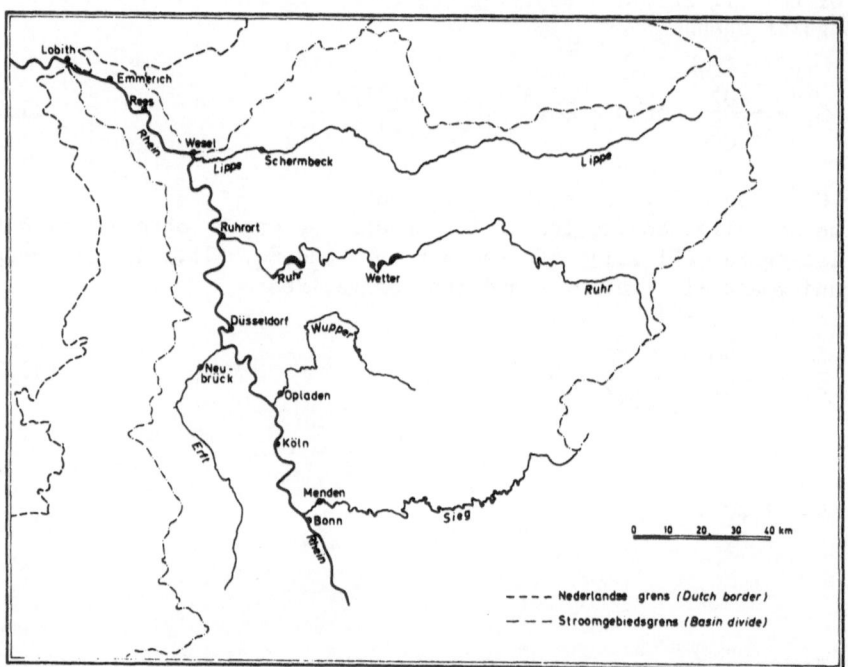

Fig. 11.3.1.   Basin area and gauging stations.

The partial differential equation (11.3.1) is solved numerically, using finite-difference approximations. A Stone and Brian scheme of second order accuracy with weighting factor θ = 0.5 was used in this study, cf. Jansen (1979).

A non-reflective downstream boundary condition is obtained by extending the numerical grid in the downstream direction and using an explicit formulation, (see Fig. 11.3.2).

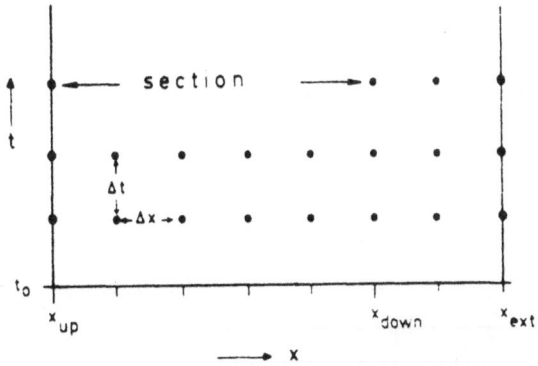

Fig. 11.3.2. Numerical grid and boundary conditions.

The forecasted downstream value for a particular section serves as the upstream boundary condition for the next section. The upstream boundary condition for the first section is constructed using an ARIMA time series model for the upstream gauging station discharges (see Chapter 6).

11.3.1 Application to the Rhine

Continuous water level recordings at eight stations along the Rhine and average daily discharge values at five stations on the main tributary rivers during the period 1 January-31 March 1970 served as basic data for this study (see Fig. 11.3.1). Results for the section Wesel-Lobith will be presented. Fig. 11.3.3 shows the Lobith measurements. Water levels were sampled at intervals of 6 hours and converted into flows using the calibrated stage-discharge relationship and the Jones formula, cf. Henderson (1966).

$$Q = Q_0 \left[1 + \frac{1}{S_0 \cdot c} \frac{\partial y}{\partial t} \right]^{1/2} \tag{11.3.2}$$

where Q_0 is the discharge according to the stage/discharge curve.

Performance of the model (11.3.1) has been evaluated for the section Wesel-Lobith. The length of this section is 48.2 km. The tributary basin area (see Fig. 11.3.1) is relatively small, 600 km^2, in comparison with the total basin area upstream of Lobith: 160800 km^2, therefore absence of lateral inflow can be assumed. An ARIMA (1,2,0) model was fitted to the Wesel inputs, cf. Box and Jenkins (1976):

$$(1 - \phi) \nabla^2 Q_t = \varepsilon_t \tag{11.3.3}$$

where ϕ is a model parameter, ε_t a residual and ∇ is defined by

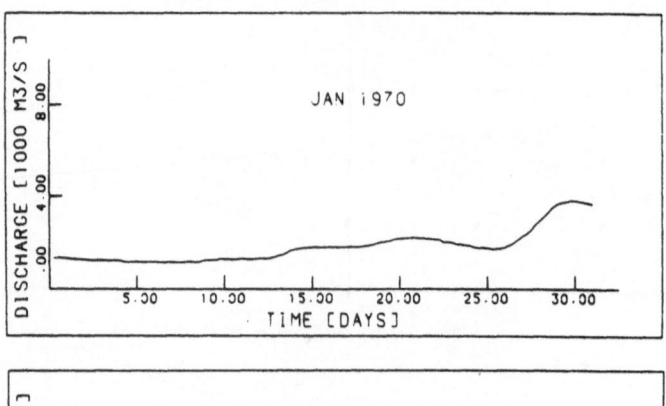

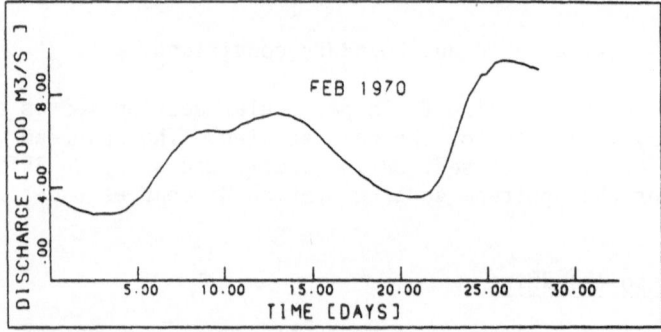

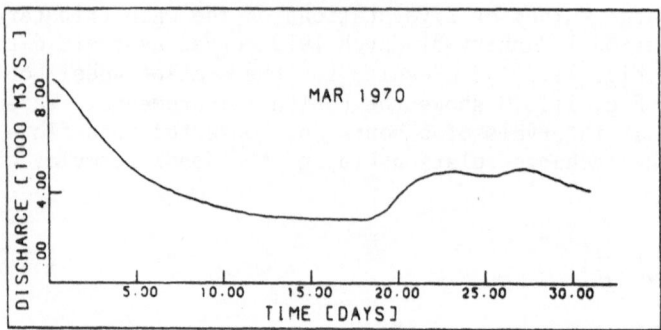

Fig. 11.3.3.   Lobith discharge measurements January–March 1970.

$$\nabla Q_t = Q_t - Q_{t-1} \tag{11.3.4}$$

Equation (11.3.3) can also be written as

$$Q_t = (2 + \phi)Q_{t-1} + (-1 - 2\phi)Q_{t-2} + \phi Q_{t-3} + \varepsilon_t \tag{11.3.5}$$

The parameters c and D were calibrated on the basis of the January data:
the resulting values were maintained throughout the forecasting period,
1 February–31 March 1970. Results for the entire period, using a lead
time of 6 hours, are summarized in Table I and illustrated in Fig. 11.3.4.

Table I. Summary of forecast results of model (11.3.1).

Analysis of forecasts			Analysis of forecast errors		
error at largest peak Q	[%]	3.5	mean	$[m^3s^{-1}]$	-144
time delay largest peak	[hr]	12	standard dev.	$[m^3s^{-1}]$	184
largest error Q	$[m^3s^{-1}]$	603	autocorrelation	[6 hr]	.97
largest error Q	[%]	9.5			
largest error y	[m]	0.30			

The large errors and the significant autocorrelation demonstrate that improvements in the model are required.

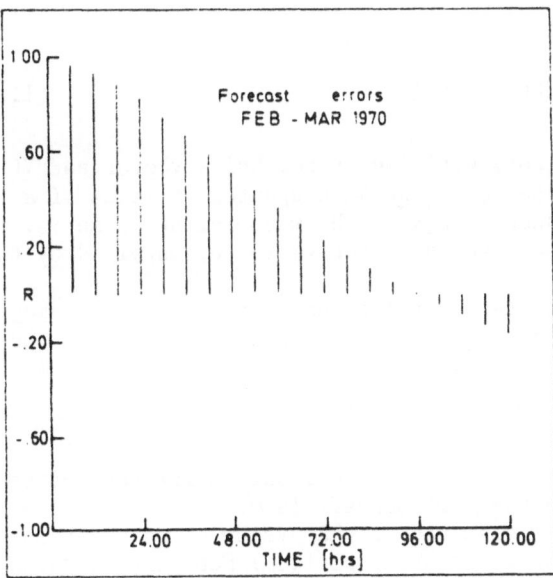

Fig. 11.3.4. Autocorrelogram of forecast errors.

The flow at the gridpoints of Fig. 11.3.2 was therefore considered a state-vector, describing the actual state of the basin (problem a) and a Kalman filter was used to make an optimum estimation of the state. The results of these improvements were insignificant.

Considering the short lead time used, the results are not influenced by the basin input forecasts by equation (11.3.5) (problem b).

Much attention was given to updating the forecasting function (problem c). Possible courses of action are: (i) applying the variable parameter diffusion method (NERC 1975), (ii) an Extended Kalman Filter approach to estimate both the state variable Q (x,t) and the parameters c and D (Jazwinsky 1970), (iii) extension of the model with a noise model (see Chapter 6 or Gutknecht and Kirnbauer 1976, Moll 1984). Our approach uses the last alternative mentioned above.

11.4 A stochastic real-time forecasting model

In this approach, the forecast from model (11.3.1) is considered a preliminary forecast: $Q_1(t)$. This forecast is then used as input to a second, regression-type model (cf. Miquel 1981):

$$Q_2(t) = x_1(t) + Q_1(t).x_2(t) \qquad (11.4.1)$$

where $Q_2(t)$ is the final forecast. When new data are available, the vector $\underline{X}(t) = \left[x_1(t), x_2(t)\right]^T$ is updated, the initial value being $\underline{X}(0) = [0,1]^T$. The model can be expressed in a state-space form.

$$\underline{X}(t) = \underline{X}(t - 1) + \underline{W}(t) \qquad (11.4.2)$$

$$Q_2(t) = \left[1, Q_1(t)\right]^T.\underline{X}(t) + \underline{V}(t) \qquad (11.4.3)$$

where $\underline{X}(t)$ is the state-vector, $\underline{W}(t)$ the system noise vector and $\underline{V}(t)$ the measurement noise. The state-vector $\underline{X}(t)$ is updated by means of a Kalman filter. The covariance matrix of the measurement noise is, in this case, a scalar: the measurement error variance, supplied by the gauging station.

The covariance matrix of the system noise is assumed to be diagonal, thus the two diagonal elements form the only unknown factors in the specification.

11.4.1 Application

For a lead time of 12 hours, the second model was calibrated for the section Wesel-Lobith on the basis of January 1970. The parameters calibrated were the two diagonal elements of the system noise covariance matrix. The results for the forecasting period 1 February–31 March 1970 are summarized in Table II and illustrated in Figures 11.4.1 and 11.4.2.

Table II. Summary of results of model (11.4.1).

Analysis of forecasts			Analysis of forecast errors		
error at largest peak Q	[%]	1	mean	$[m^3s^{-1}]$	$- 4$
time delay largest peak	[hr]	0	standard dev.	$[m^3s^{-1}]$	53
largest error Q	$[m^3s^{-1}]$	263	autocorrelation	[6 hr]	0.27
largest error Q	[%]	3			
largest error y	[m]	0.13			

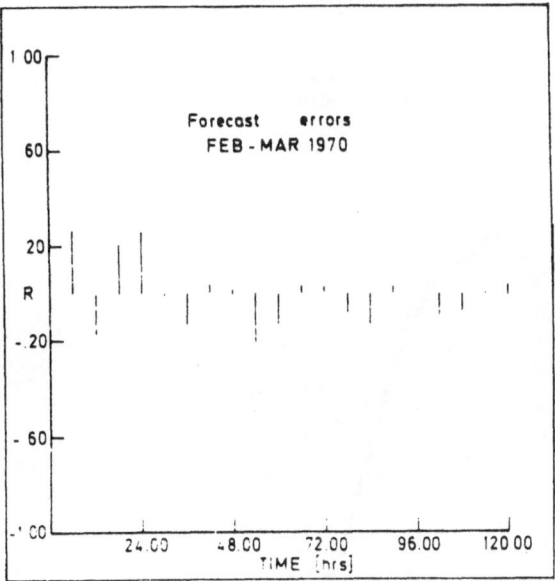

Fig. 11.4.1. Autocorrelogram of forecast errors

11.5 Conclusions

Extension of a deterministic hydrological forecasting model (11.3.1) with a stochastic component (11.4.1) gives good results.

Of the three aspects of a forecasting procedure considered in this chapter, the forecast quality was most sensitive to the third, the updating of the forecasting function.

In the case of the Rhine, forecast lead time can be extended by including river sections upstream of Wesel in the calculation, using the same procedures. For extended lead times, tributary flows and forecasted tributary flows become more important.

SYMBOLS

Q	discharge	L^3T^{-1}
y	waterdepth	L
b	width of channel	L
v	average velocity of water	LT^{-1}
S_0	bottom slope	1
S_f	friction slope	1
Q_0	discharge according to stage-discharge curve	L^3T^{-1}
g	acceleration due to gravity	LT^{-2}
A	wetted area	L^2
C	Chézy coefficient	$L^{\frac{1}{2}}T^{-1}$
c	wave celerity	LT^{-1}

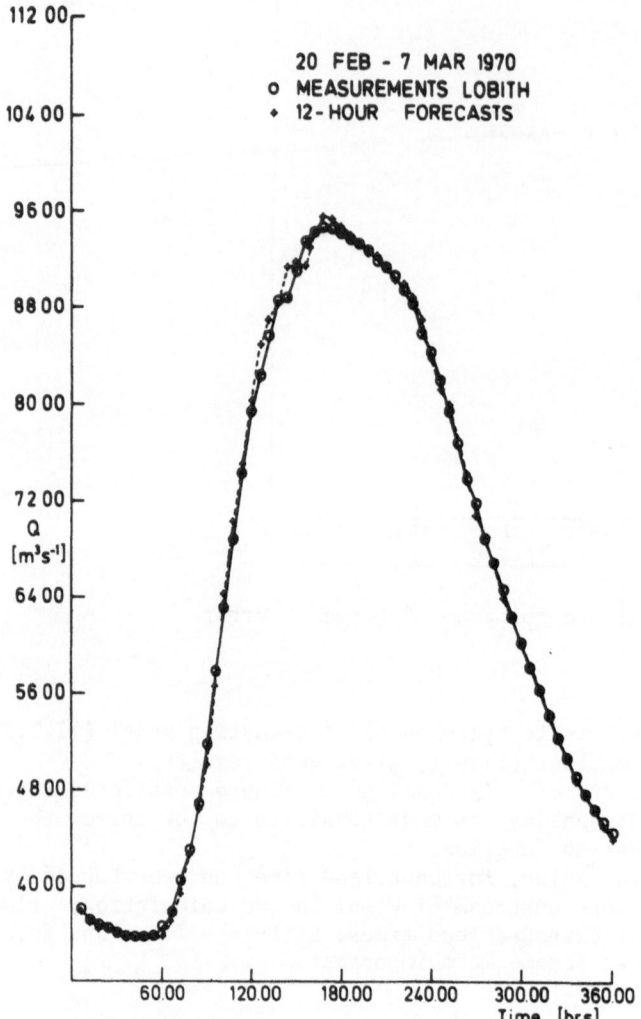

Fig. 11.4.2.   Forecasts and measurements at Lobith during passage
               of the peak.

D	Diffusion coefficient	L^2T^{-1}
∇	difference operator	1
$\underline{X}(t)$	state vector	1
$\underline{W}(t)$	system noise vector	1
$\underline{V}(t)$	measurement noise vector	1

REFERENCES

Box, G.E.P. and Jenkins, G.M. 1976 Time Series Analysis: Forecasting and
  Control, Holden-Day, San Francisco, USA.

Commission Internationale de l'Hydrologie du Bassin du Rhin 1978, Le Bassin du Rhin, Government Publishing Office, The Hague, The Netherlands

Gutknecht, D. and Kirnbauer, R. 1976 Deutsche Gewässerkundliche Mitteilungen 20 Heft 6, p. 146, 'Arima-Modelle zur Verbesserung der Ergebnisse eines Wellenablaufmodells' ('Arima-models for the Improvement of the Results of a Flood Routing Model').

Henderson, F.M. 1966 Open Channel Flow, MacMillan, New York, U.S.A.

Jansen, P.Ph. (ed.) 1979 Principles of River Engineering, Pitman Publ., London, Great Britain.

Jazwinsky, A.H. 1970 Stochastic Processes and Filtering Theory, Acad. Press, New York, USA.

Miquel, J. 1981 XIXe Congrès de l'AIRH, New Delhi, 'Filtre de Kalman et annonce de crue' ('Kalman Filtering and Flood Forecasting').

Moll, J.R. 1984 'Afvoervoorspellingen voor de Rijn bij Lobith' ('Rhine flow forecasting at Lobith') Nota 67. Laboratory of Hydraulics and Catchment Hydrology, Wageningen University.

NERC 1975 Flood Studies Report Volume 3, Natural Environment Research Council, London, Great Britain.

Schönfeld, J.C. 1948 De Ingenieur, 60, No. 4, pB1, 'Voorplanting en verzwakking van hoogwatergolven op een rivier' ('Propagation and subsidence of flood waves on a river').

12. DESIGN AND OPERATION OF FORECASTING OPERATIONAL REAL-TIME HYDROLOGICAL SYSTEMS (FORTH)

J. Němec

WMO
Case Postale No 5
1211 Geneva 20
Switzerland

12.1 INTRODUCTION

12.1.1 Definition of a FORTH (Forecasting Real-Time Hydrological) system

Hydrological forecasting is the prior estimate of future states of hydrological phenomena in real time. It comprises technical activities connected with hydrological and non-hydrological subjects, such as network design, data processing, hydrological analysis and synthesis (modelling), remote-sensing techniques, telecommunications, operational use of computers, etc. In view of this, the subject of hydrological forecasting should not be viewed as one particular hydrological technique, but as an economic activity using many technological developments, both hydrological and non-hydrological.

12.1.1.1 Flood forecasting

It is safe to say that flood forecasting is often compared with hydrological forecasting in general. But the emphasis of flood forecasting has changed in recent times. When the writer attended university some 40 years ago, flood forecasting was considered as a poor relation of actual flood prevention by structural measures (dams, dikes and levees), then considered the only effective disaster-mitigation measure. This philosophy prevailed for many decades until new evidence indicated that the point is "not to keep the water away from the people, but people away from the water". Firstly, it is impossible in many countries not to use the parts that the river regularly floods – the flood plain. In Asia and other parts of the world, floods are not only a curse, but also a blessing. Proper flood plain management, which foresees flood forecasting, can reduce the curse while retaining the blessing. Furthermore, flood forecasting as a means of flood damage reduction, has another, more subtle, advantage over structural methods of flood control. Sugawara (1974) has pointed out that flood control reservoirs are effective for small- and medium-sized floods but are of little value for the control of large, very infrequent events. He further notes that a population which depends on methods of controlling any type of disaster shields itself

D.A. Kraijenhoff and J.R. Moll (eds.), River Flow Modelling and Forecasting, 299–327.
© 1986 by D. Reidel Publishing Company.

from the more frequent events and so, having no chance to learn how to
contend with any disaster, suffers even more from the uncontrollable
large events. It seems that every method of disaster prevention has this
unfortunate characteristic - it increases the damage from large disasters.
A flood-oriented FORTH system serving well-established disaster-prevention
operations should, in most cases, prove more efficient in mitigating the
effects of major floods than would structural measures.

12.1.1.2 Classification of forecasts

Hydrological forecasts can be classified mainly by three, mutually
interdependent, characteristics:
(a) The forecasted variable;
(b) The forecast purpose;
(c) The lead time also known as forecasting or forewarning period.
    According to the forecasting period WMO recognizes four types of
forecasts (WMO, 1983):

Short hydrological forecast:	Forecast of the future value of an element of the regime of a water body for a period ending up to two days from the issue of the forecast.
Medium-term (extended) hydro-logical forecast:	Forecast of the future value of an element of the regime of a water body for a period ending between two and ten days from the issue of the forecast.
Long-term hydrological forecast:	Forecast of the future value of an element of the regime of a water body for a period extending beyond ten days from the issue of the forecast.
Seasonal hydrological forecast:	Forecast of the future value of an element of the regime of a water body for a season (usually covering a period of several months or more).
Hydrological warning:	Emergency information on an expected hydrological phenomenon which is considered to be dangerous.

12.1.1.3 Use of the WMO Hydrological Operational Multipurpose Sub-programme (HOMS) in FORTH system design

It will be recalled that the WMO Hydrological Operational Multipurpose
Subprogramme (HOMS) is intended to promote the transfer of hydrological
technology between Members of WMO for use in their water-resource projects. It does this by making the technology available to users in the
form of components. These components are of various kinds, for instance,

manual and computerized techniques for data collection, processing and analysis; commonly-used hydrological models; manuals describing field or office procedures; instruments specifications.

The functioning of the system is described by the HOMS Reference Manual published by WMO which includes summary descriptions of all the components at present available through HOMS and advice on selecting and using appropriate components in water-resource projects and in hydrological forecasting.

The subject matter is laid out according to the major activities of an operational hydrological service, and is presented under the following headings: policy, planning and organization; network design; instruments and equipment; remote sensing; methods of observation; data transmission; data storage, retrieval and dissemination; primary data processing; secondary data processing; hydrological models for forecasting; analysis of data for planning, design and operation of water-resource systems; mathematical and statistical computations.

Larger sections are further subdivided to give a comprehensive classification of the subject matter covered. This classification scheme, together with a degree-of-complexity indicator for each component, forms the basis of a component numbering system. Complexity is graded from the simplest manual calculation methods to the most complex, such as large computer-based rainfall/runoff models or sophisticated automatic instrumentation.

The transfer of components has already taken place in 600 cases and many hydrological services have received components to be included in forecasting systems. In some cases expert assistance is offered together with components. A list of some HOMS components pertaining to hydrological forecasting is included in Annex 1.

12.1.2 Relationship of meteorological and hydrological forecasts and relevant services

Prior knowledge of the meteorological conditions increases the scope and efficacy of hydrological forecasting, lengthens the validity of such forecasts and increases their accuracy and reliability. Quantitative precipitation forecasts (QPF) and other meteorological forecasts (temperature, wind, and snow conditions) constitute an important and essential input to the present and future procedures and methods of hydrological forecasting. Considering the present trend in making use of rainfall/runoff models to simulate the catchment response to precipitation inputs, any improvements in timeliness of hydrological forecasting for any but the very large rivers will hinge chiefly on progress made in rainfall forecasting. The input to these models, usually the rainfall quantity during a given time period, is normally obtained from the actual observations and for most rivers the flood forecast from the model based on this input is but a few hours ahead of the actual event. As a consequence, the alert of flood-plain residents and subsequent remedial measures for their protection are under a severe time constraint in their execution. The alternative of taking remedial action under the assumption of flooding on the basis of only qualitative rainfall forecasts may result in large expenses in the long run, since many of these actions

will prove unjustified "a posteriori". And with far more serious consequences, this alternative may bring about disbelief on the part of the
community after forecasters "cry wolf" for many events that subsequently
do not materialize, and ultimately may cause disastrous damage when the
"real" but unbelieved event actually occurs. Quantitative precipitation
forecasts also represent the first and most important input to any operational on-line forecasting system and to the decision mechanism of
computerized short-term control strategies for the management of multipurpose reservoir systems and river-regulation. There is also an increasing need for more precise and quantified forecasting of intense rainfall
liable to cause flooding on small catchments (so called "flash floods")
and urban storm-water drainage systems.

However, despite the compelling need, no reliable method of quanti-
tatively forecasting precipitation sufficiently in advance has yet been
developed. More recently the development of numerical weather prediction,
based on a large-scale multi-level representation of the atmosphere, has
considerably increased the potential for the description and prediction
of the smaller-scale meteorological phenomena. Nevertheless, the production of such meteorological forecasts is still a matter of research and
development and its use for hydrological forecasting is not operationally
reliable. Detailed discussion of these problems is included in a more
comprehensive publication of the author (Němec, 1985).

12.2 COMPONENTS OF A FORTH SYSTEM

12.2.1 General description

12.2.1.1 A concrete FORTH system depends on many conditions which, in
addition to those of a technical nature and the natural environment
(basin and river), include social and administrative structures of the
specific country. For this reason a general description of the design
and operation of the components of a FORTH system can include only those
sub-systems which are indispensable for the system in general. These
sub-systems are:
- Historical and real time data collection
- Data transmission
- Data processing and storage (files of the data base or data bank)
- Forecasting procedure (modelling) in development and operational mode
- Forecast dissemination services
- Forecast evaluation and updating.

The different sub-systems and their interdependence are illustrated
by the flow-chart in Figure 12.2.1. The selection of any particular
sub-system is dependent not only on the conditions mentioned above but
also on the other sub-systems, which in many cases may already exist.
This is particularly the case with the data-collection sub-systems,
because the historical data will have to be collected from an existing
network, generally some time before the actual FORTH system is put into
operation. Similarly, an institutional set-up (for example legal status
of a government agency responsible for forecasting) normally exists
particularly in countries subject to regularly occurring, hydrology-
related hazards (floods and droughts) and the introduction of new sub-

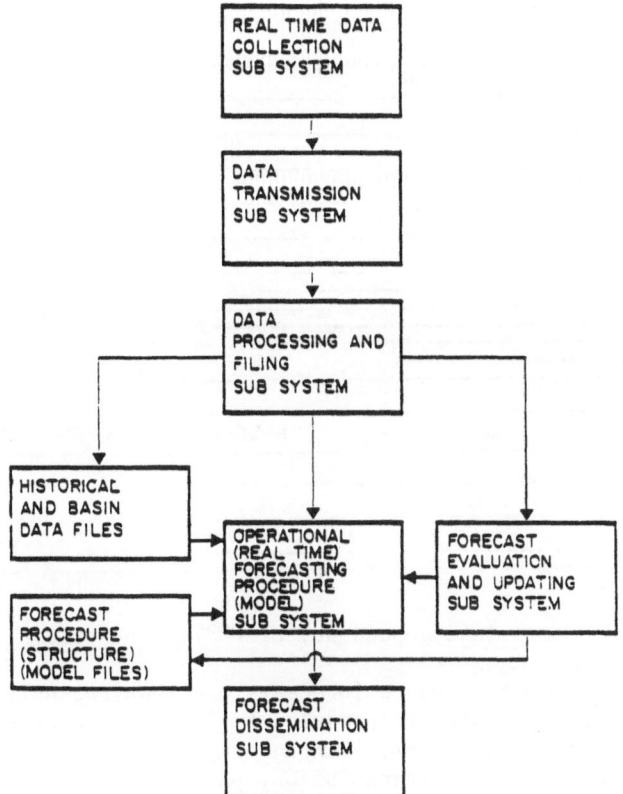

Fig. 12.2.1.   FORTH subsystems

systems of FORTH has to take into consideration the existing set-up. The different sub-systems and their action in the FORTH system as time-space interdependent is illustrated in Figure 12.2.2.

12.2.1.2 On the time scale of Figure 2 and on the parallel scale of basin areas of forecast, elementary basins, average times of concentration are indicated, calculated and averaged from hydrographs of many rivers in basins of Europe, USA, Latin America and Asia, corresponding to the time necessary from the beginning of the runoff-producing event to the culmination of a flood wave (maximum flow).

From Figure 12.2.2, it is evident that past data collection, transmission and processing FORTH sub-systems will be most important in elementary basins up to 1000 km^2, and only in larger ones, where forecasts are based on composition of hydrographs from elementary basins of up to 1000-3000 km^2 by different techniques, mainly routing. This will be necessary in particular for basins in which forecasts are required in different flow-points (forecasting points) across the basin. In cases

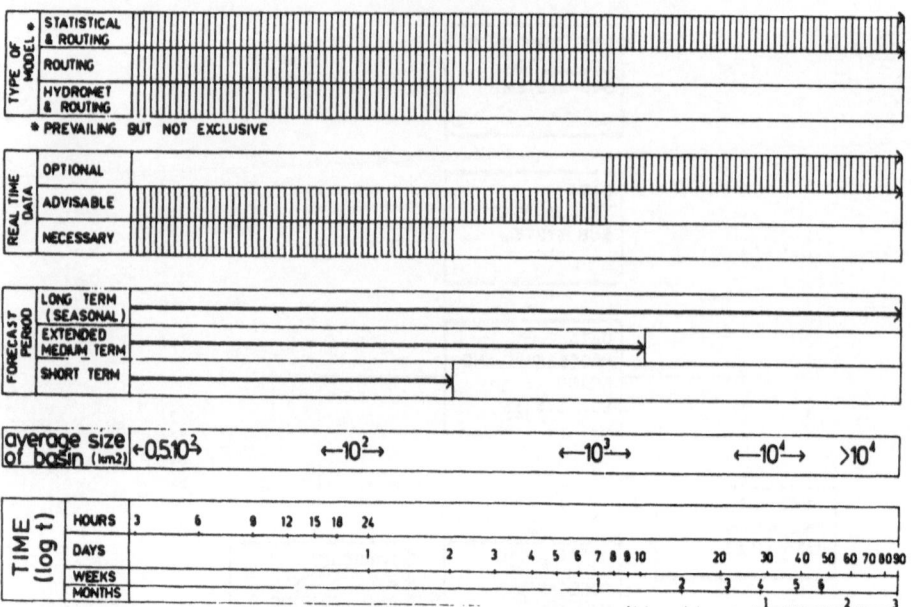

Fig. 12.2.2.   Time-space interdependency of FORTH sub-systems.

of basins larger than 10,000 km^2, where forecasts are required only on
the lower reaches of the basin and are produced by lumped parameter
models, the data collection sub-system, although important, does not
necessarily require real-time data transmission, particularly if the
forecast is produced by routing of flood waves from upstream to downstream.

12.2.2   Historical and real-time data collection sub-systems

12.2.2.1   Data needed for FORTH

The data used in hydrological forecasts can be divided into two groups:
the first includes those required for developing the forecasting method,
and the second includes the information needed to operate the forecast
system.

The first group, encompassing the conventional time series as well as
historical hydrological information necessary for testing and evaluating
the trial forecast procedure, also includes those constant basin and
river characteristics, for example sub-catchment areas, woodland, soil,
channel dimensions and slopes, needed for pre-computation of the processes occurring over the basin and in the channels. The second group
includes the hydrometeorological data specified by the forecasting
scheme to characterize the state of the catchment immediately before the
issue of the forecast transmitted in real-time. It also includes a
measurement of the forecast element itself which may be used to monitor
the forecast performance or update the forecast model.

The reliability of a forecast procedure is directly related to the amount of data available for the development of the procedure, and to the consistency, i.e., no change in relative values of the records with time. Care must be taken to ensure that there is no bias between the data used to develop the forecast procedure and the data used for operational forecasting. For this reason, consistency in the records is as important as quantity of data.

12.2.2.2   The networks of stations from which both above types of data are to be provided may not be the same. In general the development of the forecasting procedure (model) requires historical data covering a longer time and more space than those needed for actual issuance of the forecast. Although network design is a hydrological technique of its own, the general principle for a FORTH network may be summarized in two simple rules:
(a) Use for the development of the system, use <u>as much</u> of the data available <u>as possible</u>, providing it is homogeneous, consistent and relatively reliable (the reliability being often difficult to ascertain, particularly for floods and low flows).
(b) For the operation of the system, establish <u>as few</u> real-time transmitting stations <u>as possible</u>, the criterion being satisfaction of accuracy needed for the purpose of the forecast.
More details on network design practices may be found in Němec (1985).

12.2.2.3   The sensors installed in the stations collecting data in real-time are in general the same as those in all other hydrological networks. Nevertheless the interface with the telecommunication equipment may, in some cases, require a particular type of sensor. Thus tipping-bucket raingauges are more common than floats (siphon) types. While the nature of the river is decisive in the selection of the stage recorder type, both bubble gauges and float-type guages are being used. The new pressure transducers which are reaching the market appear quite promising as real-time stage recorder sensors. Several producers have their preferred sensors for instruments in package real-time telephone or radio data transmission sub-systems.

12.2.2.4   <u>Remote sensing</u>

The need for a detailed and rapid spatial coverage in real-time data collection gives particular importance to remote sensing of data. At this point it is important to emphasize the difference between remote sensing of data and their telemetering. In the former the sensor is always removed from the physical element to be sensed (rain, water level, snow cover etc.). In the latter data from a sensor installed <u>in situ</u> are transmitted over a distance sometimes even without recording data at the location of the sensor (a solution which, incidentally, as a rule, is not recommended). According to WMO (1983), remote sensing from ground-based radar, satellite, and aeroplane, offers many distinct advantages for certain classes of forecasts, primarily because of the possibility of directly observing areally extensive variables which are otherwise only amenable to point sampling, and providing observations over inaccessible terrain or over the sea. It can be used to provide direct inputs

to forecasting procedures in the following areas:
(a) Areal rainfall, both qualitative and quantitative indications;
(b) Areal extent of flood plain inundation;
(c) Cloud image indicating tropospheric wind as input to meteorological and hence hydrological forecast models;
(d) Tropical cyclone or hurricane movement;
(e) Area of snow cover and its water equivalent;
(f) Soil moisture in the uppermost layers of the soil;
(g) Water quality, in particular, turbidity.

Most of these techniques are not yet operational. The only ones used in operational forecasting systems at present are determination of the area of snow cover from satellites and use of radar. The main uses of radar in hydrological forecasting are in assessing areal rainfall and in forecasting heavy rainfall and hence the resulting flood discharge, particularly on small catchments. Detailed descriptions of these two operational remote sensing methods and their evaluation are in other publications (Němec, 1985).

12.2.3 Data transmission sub-system

12.2.3.1 The communication system transmits data from the field monitoring station back to the forecast centre. It may report back at regular intervals or may warn when some specified event has occurred. In existing and future FORTH systems, particularly in developing countries, the following real-time data transmission sub-systems do and will exist:
-   Manual;
-   Semi-automated (mixture of man-operated and automatic station);
-   Fully automated.

12.2.3.2 The real-time data transmission sub-system can use either wire (telephone) or wireless (radio) links. Most common are radio links using either earth surface network of stations only, or a link of these via a satellite or meteorite scattering paths. Two types of satellite are employed for data collection; polar-orbiting and geostationary. Meteor-scatter or, more commonly, meteor-burst communications rely on the ionization in meteor trails to re-radiate or reflect radio waves for distances up to 2,000 km. More information on the different sub-systems is in Němec (1985).

12.2.4 Data processing, filing and retrieving sub-system of FORTH

12.2.4.1 Data processing

Part of this sub-system is normally carried out in the forecasting centre, but modern systems foresee some preprocessing at the collection site prior to its transmission. In fact, the design of FORTH at present in preparation in several countries of the world foresees a considerable part of the processing taking place before transmission (due to rapid advances of microprocessors), and thus achieving economy in data transmission.

The interface between the data collection, transmission and processing

is the coding of the data, which depends on the processing needs. If
several users require the data, in particular in international basins,
it has also to satisfy some nationally or even internationally-agreed
principles of coding, for transmission and processing.

To provide easily applicable hydrological codes, WMO, on the recommendation of its Commission for Hydrology, established two code forms,
one for coding and transmission (and possible processing) of hydrological
data, called HYDRA, the second for transmission of forecasts, called
HYFOR. They are included in the WMO Manual on Codes (WMO, 1974). More
information on these are in Němec (1985).

12.2.4.2 Transfer and processing of historical data

The processing of historical data is the subject of many standard hydrological procedures. Their general description is in the WMO Guide (1983),
Němec (1972) and in national publications; individual procedures are
referenced in HOMS Reference Manual (WMO, 1981) and available through
HOMS. To create files of such data for a FORTH system, however, their
transfer from other sources is necessary. The general rules for such
transfer are given in Němec (1985).

12.2.4.3 Creation of FORTH files

The data processing, filing and retrieval sub-system of any FORTH system
depends on many conditions: those of the hydrological nature of the
basin, type of forecasting procedure, type of available hardware which
can range from simple desk calculators, passing through microcomputers
(Apple, Tandy, Cromemco, etc.), ending with medium (mini) and main-frame
computers. As a rule this sub-system absorbs the computing operation of
the model development and model execution (operational forecasting)
program, as well as updating and often even data bank creation programs.
The files created can be of many types, the most common are:
-   Station files (time series);
-   Catchment parameter files;
-   Model parameter files; and
-   Program execution files.

12.2.4.4 Selection of a computer for FORTH

Depending on the sub-system described above and on the forecasting procedure selected, if the computer to be used for the FORTH operation is not
already available, the question arises as to what type of computer is to
be chosen. The selection depends on many circumstances, some of which
are entirely independent of the technical design and character of the
FORTH system. The selection is discussed in Němec (1985).

12.2.5 Forecasting procedures (models)

12.2.5.1 Two modes of existence of forecasting procedures

Every forecasting procedure (model) exists in two modes (computational forms):

(a) development mode (calibration); and

(b) operational mode (actual forecasting).

The complexity of the first mode may not be relatively equivalent to the second one, in other words, a model which has very complicated and large requirements on computation time and facilities (computer size) in the development (calibration) mode may be operated for the actual issuance of the forecast with very simple mathematical formulation and minimal computational facilities, bordering on manual use of a pocket calculator. Reciprocally, a relatively simple, mathematically formulated model needing less time and relatively smaller computational facilities in development (calibration) may be more difficult to operate for the issuance of an actual forecast, and may require larger computational facilities, in some cases requiring at least a micro or mini-electronic computer.

12.2.5.2 Classification of models for hydrological forecasting purposes

It is necessary to stress that the classification of models for hydrological forecasting purposes is not necessarily identical to the classification of models in general or for other purposes such as computation of design data, extrapolation or records, research purposes, etc. Some models while based on stochastic principles or using exclusively statistical approaches may function as deterministic models, when used for real-time forecasting. In this connexion, it should be stressed that any model used for real-time forecast is, by definition of the latter, a deterministic model. The classification proposed by the author is as follows:

A.   Purely deterministic forecasting procedures (models)

A.1   Hydrometric data-based (involving only streamflow processes)

A.1.a   Correlations of stages and/or volumes (discharges)

A.1.b   System's approach to streamflow (hydrologic routing)

A.1.c   Hydrologic routing using
        (i)   Dynamic wave
        (ii)   Diffusion analogy
        (iii) Kinematic wave

A.2   Hydrometeorological and hydrometric data-based models (involving rainfall/runoff streamflow processes)

A.2.a   Correlations using physical variables and parameters or indexes (such as API); and

A.2.b   System's approach to the basic basin response to rainfall

A.2.c   Distributed parameters approaches (hydrological and hydraulic)

A.2.d   Conceptual moisture accounting using
        (i)   soil moisture indexes
        (ii)   implicit moisture accounting
        (iii) explicit moisture accounting.

B.   Hybrid-stochastic-deterministic forecasting procedures (models)

B.1   Using only time series stochastic parameterization.

B.2   Using system's approach to the basic basin response and time series stochastic parameterization.

It should be noted that often some models of the categories A.1 and A.2 are combined in one single forecasting procedure and that lately categories A and B are being combined in the so-called "self-tuning algorithms", namely models of category A are updated in operational mode (see 12.2.5.1 above) by procedures of category B.

All above procedures (models) are described in Němec (1985) and several of them are indicated elsewhere in this book.

12.3 SELECTION OF FORECASTING PROCEDURES

12.3.1 The lonely forecaster in a crowd of models

From the above classification of a crowd of models, it appears that in the design of a FORTH system the choice of a model will not be an easy task, particularly if the designer is left alone, unless only one model is for some reason available to the designer. Indeed this is often the case and the reason is mainly administrative. However it can be said off hand that such a system will be far from optimal. To ensure a minimum of optimization in the design and operation, after a preliminary selection, at least two or three models of different type should be tested on the basin. Details of such a selection can be found elsewhere (Němec, 1985). To help the lonely forecaster in this crowd of models, the WMO conducted from 1970 to 1975 a project on intercomparison of deterministic models used in forecasting.

The aim of the project was:
(a) To proceed with an evaluation and intercomparison of operational conceptual hydrological models, which use electronic computers to provide short-term forecasts of streamflow;
(b) To provide information and guidance on the use of such models in various forecasting situations, with regard to specific conditions and accuracy requirements.

It should be emphasized that the intention of the project was not to determine which model fits best in all circumstances.

12.3.1.1 The project involved the testing of ten operational conceptual hydrological models submitted by seven countries on six standard river catchment data sets from climatologically and geographically varied conditions in six countries. Each data set consisted of two distinct periods: a calibration period (six years), and a verification period (two years) which immediately followed one another. For each data set, the model owners were supplied with the necessary concurrent observed input data (precipitation, evaporation and other meteorological data) and observed output data (streamflow) for the six-year calibration period and only observed input data for the two-year verification period. The observed output data fro the two-year verification period were retained by the WMO Secretariat. The ten models as well as the catchments for the six standard data sets are listed in Table I.

12.3.1.2 Standard data sets tested on participating models

For each data set the model owners used the concurrent observed input

Table I. Standard data sets tested on participating models

Model name (and abbreviation)	Name and address of model developer	Standard river catchment data sets tested in the project					
		Bird Creek U.S.A. 2344 km³	Bikin River U.S.S.R. 13100 km³	Wollombi Brook Australia 1580 km³	Kizu River Japan 1445 km³	Sanaga River Cameroun 131500 km³	Nam Mune River Thailand 104000 km³
Bureau of Meteorology Model (CBM)	Bureau of Meteorology, P.O. Box 1289 K, Melbourne. Vic. 3001, Australia	•	•	•	•		
Girard I Model	ORSTOM, 19, rue Eugène Carrière, 75018 Paris, France	•	•				
Serial Storage Type Model (Tank I)	National Research Centre for Disaster Prevention, 1 Ginza Higasi 6, Chuo-ku, Tokyo, Japan	•	•	•	•	•	•
Serial Storage Type Model (Tank II)	Ditto	•	•	•			
The Flood Forecasting Model (IMH2-SSVP)	Institute of Meteorology and Hydrology Şos. Bucureşti-Ploieşti 97, Bucarest, Romania	•	•				
Streamflow Synthesis and Reservoir Regulation Model (SSARR)	Corps of Engineers, Portland, Oregon, U.S.A.	•	•	•	•	•	•
National Weather Service Hydrologic Model (NWSH)	National Weather Service, Silver Spring, Maryland, U.S.A.	•		•	•	•	
Sacramento River Forecast Center Hydrologic Model (SRFCH)	National Weather Service River Forecast Center, Sacramento, California U.S.A.	•	•	•	•	•	
Rainfall Runoff Model of the Hydrometeorological Centre of the U.S.S.R. (HMC)	Hydrometeorological Centre of the U.S.S.R., Bolshevistkaya 13, Moscow 123 376, U.S.S.R.	•	•			•	
Constrained Linear System Model (CLS)	Hydraulic Institute of Pavia University and IBM Pisa Scientific Centre, Italy	o	•		•	•	o

Notes: o = Test results received in the WMO Secretariat after the conference (1974), that is after the modellers had had the opportunity to acquaint themselves with the data on the output during the two-year verification period of the data sets. These two test results will hereafter be referred to as "delayed test results".

The Kostroma river data set (U.S.S.R.) was tested on the Snowmelt-runoff Model of the Hydrometeorological Centre of the U.S.S.R.

and output data for the six-year calibration period to calibrate and develop the parameters of their models and employed the additional two years of observed input data in the verification period to produce a simulated discharge (computed output). The simulated discharges produced by the tested models for both the calibration and verification periods in each data set were then centrally evaluated and compared by WMO using several graphical and numerical verification criteria agreed upon by all modellers. On the basis of this evaluation and intercomparison, several conclusions and recommendations were made concerning the performance and use of the models in various forecasting situations.

12.3.1.3 Recommendations on verification criteria

(a) As one of the many factors involved in model selection, it would be advantageous if the verification and intercomparison of models in general could be carried out in accordance with at least some generally accepted verification criteria. The numerical verification criteria for such general use should, as far as possible, be selected from among those used in the WMO project.

(b) With respect to the graphical verification criteria, the double mass curves of observed versus simulated flows and in particular the flow duration curves should be available for all verification and intercomparison procedures as they convey a maximim of information.

(c) As an alternative to double mass plots a curve of the sums of the residual errors versus time can be used. The sum of residual errors is expressed as: $\sum(y_c - y_o)$ where y_c and y_o are the computed and observed discharges respectively. This curve, plotted on the same graphs as the observed and simulated hydrographs, assists in the analysis, particularly with respect to the model's ability to keep account of volumes.

The graphical and numerical verification criteria used in the evaluation of the simulated discharges produced by the tested models are given in Table II.

12.3.1.4 General conclusions and recommendations

The selection of a model for a specific forecasting situation should be guided by the following criteria:

(a) General and specific purpose and benefits of the forecast (e.g., continuous hydrograph of discharges, floods, water quality, water resources management, etc.):

(b) Climatic and physiographic characteristics of the basin;

(c) Length of record of the various types of input data;

(d) Quality of data field, both in time and space;

(e) The availability and size of computers, for both development and operation of the model, as well as the possible use of the model by relatively non-expert hydrological forecasting personnel;

(f) The possible need for transposing model parameters from smaller catchments to larger catchments, usually downstream where sufficient data for development are not available and the possible application of models to larger river systems with important human interference (man-made structures);

TABLE II. The Graphical and numerical verification criteria used in the project

Graphical criteria	Numerical criteria	Flow variables for which numerical criteria were indicated
1. Linear scale plots of simulated and observed hydrographs	1. Coefficient of variation of residual of errors given by: $$Y = \left[\frac{\sum (y_c - y_o)^2}{n} \right]^{1/2} \Big/ \bar{y}_o$$	1. Mean daily discharge (in $m^3\ s^{-1}$)
2. Double mass plots of simulated versus observed monthly discharge volumes	2. Ratio of relative error to the mean given by $$R = \frac{\sum (y_c - y_o)}{n \bar{y}_o}$$	2. Maximum daily discharge ($m^3\ s^{-1}$) for each month in which the flow equals or exceeds the mean flow for the whole period of calibration and verification
3. Flow duration curves of simulated and observed daily discharges	3. Ratio of absolute error to the mean given by $$A = \frac{\sum |y_c - y_o|}{n \bar{y}_o}$$	3. Monthly volumes of flows expressed in centimetres of depth over the catchment area
4. Scatter diagrams of simulated versus observed monthly maximum daily discharges (peak flows)	4. Arithmetic mean given by $D = \dfrac{\sum y_{o,c}}{n}$	4. Mean daily discharge for low flow days defined as those days during which the flow is above zero and is below the flow not exceeded during a period of 130 days during the verification period.
	5. Phasing coefficient (PH) for the monthly peak flows given by the number of times that the simulated peak is shifted in time from the corresponding observed peak by at least one day	
	6. Coefficient of persistence (PE) given by $\dfrac{\sum\limits_{i=1}^{k} B_i^2}{V}$	

In the above equations:

y_o = observed discharge

y_c = computed discharge

n = total number of observations

$\bar{y}_o = \dfrac{\sum y_o}{n}$

k = number of positive and negative runs

$V = \sum\limits_{i=1}^{n} (y_o - y_c)^2$ for the n items

B = the individual areas of each segment

(g) The ability of the model to be conveniently updated on the basis of current hydrometeorological conditions.

12.3.2 Mechanism of moisture accounting in the basin

12.3.2.1 From the classification of forecasting models above, it appears that those using as input measured or forecasted rainfall precipitation necessarily have to deal with the concepts of "net rainfall" also called "effective rainfall (or precipitation)" producing "direct runoff" as opposed to "base flow" and "basin recharge", the latter, together with the evapotranspiration from the entire basin representing "losses". Thes terms are to some extent a convention resulting from the analysis of the hydrograph of a runoff (flood) event and they are being challenged on several accounts, although their physical basis is indubitable. The main point challenged (Nash, 1970) is the lack of the strictly physical identification of the processes such as "direct runoff" or "base flow" in an actual basin. Yet any simple forecasting rainfall-runoff model, unless it uses input/output correlation, is based on the "net rainfall" concept.

12.3.2.2 The physical mechanism of the derivation of the "rainfall excess" on a micro-scale of a runoff plot of a few square meters or a small agricultural catchment (field) can be represented by exact mathematical expressions, often differential equations of flow in porous media used by soil scientist and/or agricultural hydrologists. Such models can be supplied with the necessary data on the soils only on this small scale. In basins on which real-time hydrological forecasting of runoff from rainfall has a practical importance (see Figure 12.2.2) such micro-scale approach is difficult if not impossible. If the unit hydrograph forecasting procedure described in Chapter 2.5 is used, the derivation of the "rainfall excess" and "direct runoff" and "base flow" can be done by a variety of methods described in general hydrology textbooks or manuals (Linsley et al., 1982, Shaw E., 1983, WMO, 1983, Němec, 1972). They can be divided into the following approaches, listed in the order of decreasing complexity (or detail), and need of observational data (Gray et al., 1970):
(a) Infiltration curves
(b) Infiltration indices
(c) Soil moisture indices.

12.2.2.3 The infiltration curves can be derived by direct measurement by field infiltrometers, by analysis of observed hydrographs, in which a standard infiltration ("f-capacity") curve is derived or by analysis of mass curves of the different elements of the runoff process, where the same concept of "f-capacity" is used.
The simple infiltration index method is this of the ϕ index, which is the average rainfall intensity over and above that when the mass of rainfall equals the mass of runoff. In this method no allowance is made for infiltration or depression storage during rainless periods. A somewhat more complex derivation is required by the "W-index", which is

the average infiltration rate during the time when rainfall intensity
exceeds an average infiltration rate. The most detailed and complex
infiltration index method is the "f-ave" approach, which computes the
average infiltration rate during a period of continuous water availability for infiltration.

The soil moisture indices are based on the assumption that the higher
the soil moisture the lower the infiltration and the higher the rainfall
excess. It should be noted that while this approach appears extremely
simplified, indeed in some cases contrary to the physical evidence when
a very intense rain falls on a very dry soil, caked and virtually impervious, on the average and in the majority of cases, particularly for
larger basins, this approach gives good practical results. The simplest
soil moisture index is the "antecedent precipitation index (API)" derived
from spatially averaged rain observations for a certain period antecedent
to the rainfall-runoff event. This method, introduced in the USA flood
forecasting practice by the Weather Service many years ago, was first
used with a co-axial graphical correlation derivation procedure (Linsley
et al., 1975) and later computerized (Sittner et al., 1969, Němec, 1985).
Although today replaced in the USA by conceptual models, it is widely
used in many parts of the world in practical flood forecasting activities
in larger basins. A similar "soil moisture index" (SMI), although
derived by different methods, is used by the conceptual model of the US
Corps of Engineers, the "SSARR (Streamflow Synthesis and Reservoir Regulation)" model (US Army Corps of Engineers, 1975). This index represents,
however, a bridge to the more realistic and also more complicated method
of representation of the moisture movement processes in the basin and
soil, the conceptual moisture accounting (category of models A.2.d in
the classification in para 12.2.5.2 above).

12.3.4.4 The conceptual models represent a continuous movement of the
water from rainfall and/or snowmelt in the basin through the soil by a
simplified series of conceptual storages commonly called "upper (soil)
zone", "lower (soil) zone" or even more "zones". Such models thus need
not separate a priori either "base flow" or "direct runoff", in every
stage of the movement of the moisture on the surface of or in the soil.
The models are necessarily continuous, in order to avoid initialization
of the parameters at each runoff event. This type of models is described
and discussed in some detail in general by O'Donnell in Chapter 2.6,
and in case studies by Fleming in Chapter 13. The latter is mainly
concerned with the "HYDROCOMP SIMULATION PROGRAM" based on the so-called
"Stanford" conceptual model. A systematic review of such models used in
FORTH systems can be found in Němec (1985).

12.3.3 Sensitivity of a FORTH system to the interdependence of its
        components and their relative significance

The correct design of a FORTH system requires an evaluation of the significance of all the sub-systems (components) and should not be biased
toward one of them, which is very often the case. Indeed, many hydrologists and as a result managers of FORTH systems consider the forecasting
procedure of "the model" as the most important component and often pay

little attention to the interrelation of this component with others, in particular the real-time data collection sub-system.

The two sub-systems mentioned above are often considered separately, at least as far as their influence on the final result of the forecast is concerned, although undoubtedly everybody agrees that they have to be technically compatible and properly interfaced. What is often less evident is that the data collection sub-system may be crucial for the accuracy and cost-effectiveness of the FORTH system and that its design or selection is at least as important as that of the hydrological model. A more detailed analysis of the interdependence of these sub-systems is in Němec (1985).

A sensitivity analysis can be made for other components (sub-systems) of a FORTH system. Nevertheless, the data collection on one side and the hydrological model on the other are, in the opinion of this author, the most sensitive sub-systems with respect to their relative interdependence. They should never be considered separately and a decision on their selection should not be taken without at least a summary cost/benefit or cost/effectiveness analysis.

Similar analysis of interdependence of FORTH sub-systems, in the opinion of this author, may provide a design and management of a FORTH system as an entity without technically unwarranted bias. However, the author recognizes that a large number of non-technical considerations may influence the design and are often unavoidable. Thus, as in many similar situations, a technician provides his evaluation and then hopes for the best.

12.4 FORECAST UPDATING AND EVALUATION (WMO, 1983)

12.4.1 Forecast update techniques

If an observation is made of the forecast output, y_1, then the opportunity exists to adjust subsequent forecasts in the light of the known forecast error $e_i = y_1 - y_i$ where y_i is the forecast estimate. Most adjustments are the result of the subjective judgement of the forecaster but various mathematical techniques have been developed which allow the process to be formalized. The underlying principles of the formal approach are described below. Several other parts of this book are concerned with these techniques.

At its simplest, the adjustment may take the form of an addition of the current error e_i to the new forecast $y_{i+1, i+2}, \ldots$. To avoid discontinuities, the adjustment is usually blended into the computed hydrograph over several time periods. A more complicated procedure is to subject the error series e_1, $e_2, \ldots, e_i$ to a time series analysis to extract possible trends or periodicities which can then be extrapolated to estimate the new error e_{i+1} to add to the new forecast y_{i+1}.

The observed values y_1, $y_2, \ldots, y_i$ can be used to redefine the state variables of the forecast model. This is termed "recursive estimation" and, if the forecast model can be cast in a sufficiently simple form,

provides a basis for a formal and optimal strategy for adjusting model
output.

The proper choice of adjustment procedure depends on several factors
such as user requirements, amount and quality of the available data, the
equipment used for data collection, transmission and processing of data
and the qualifications and experience of the personnel.

12.4.2 Evaluation of hydrological forecasts

The main objectives of evaluation are:
(a) To evaluate the accuracy and effectiveness of each forecasting method;
(b) To determine the degree of success of an individual operational fore-
    cast.

The first objective is to evaluate the usefulness and reliability of
a newly-developed method before it is applied operationally. The second
objective is to determine the effectiveness of the method in actual
practice, when it is used in conjunction with operational data. The
statistical evaluation of the accuracy and effectiveness of each forecasting method constitutes the important final stage in the development of
forecasting methods. Operational forecasts should be checked routinely
for possible revision of procedures.

12.4.3 Principles underlying the evaluation of forecasting methods

The method used in forecasting any hydrological phenomena can be regarded
as effective and its practical use justified only when the forecast error
of a given probability is less than the equally probable deviation of the
forecast variable from its mean. If this is not the case, then the
method has no advantage over the use of hydrological or climatic long-
term means. It therefore follows that the statistical evaluation of the
effectiveness of the forecasting method should be based on comparisons
of forecast errors with deviations of the forecast variable from its mean
value. The simplest and at the same time most common method of evaluation entails the use of the coefficient of determination d_y:

$$d_y = 1 - \frac{S^2}{\sigma_y^2}$$

where S is the standard error of the forecast, and σ_y is the standard
deviation of the forecast variable.

The standard error of the forecast is defined as:

$$S = \sqrt{\frac{\Sigma(\hat{y} - y)^2}{n - m}}$$

where y is the true (observed) value of the variable, $\hat{y}$ is the forecast
value, n is the number of observations, and m is the number of degrees
of freedom in the relationship employed in the forecast method.

12.4.4 Evaluation of the degree of success of operational forecasts

Here the main problem is the calculation of errors in each individual forecast. This is important in order to compile a sample of statistical data for checking previous evaluations of the accuracy and effectiveness of the forecasting method. This evaluation of effectiveness is virtually indispensable when the method is developed on the basis of a short series of observations. Strictly speaking, statistical determination of the effectiveness of an individual forecast is impossible, as the forecasting method itself can indicate only the confidence interval within which the variable is expected to appear with a certain degree of probability. Even if the forecast is issued in the form of a confidence interval, not covering the entire range of the possible values of the variable, we are still unable to solve conclusively the problem of its verification. Indeed, if we anticipate with probability p that the variable y will be contained within the interval between y_1 and y_2, we also know that the probability that it may fall outside the range of this interval is q = 1 - p.

The expression generally used for computation of tolerable error in the individual forecast is:

$$\sigma \text{ tolerable} = \pm0.674\sigma$$

or

$$\sigma\Delta \text{ tolerable} = \pm0.674\sigma$$

Experience shows that the forecasting procedure is acceptable for practical application if 80 per cent of all forecasts have errors smaller than $\sigma\Delta$ tolerable.

The evaluation and verification criteria are discussed in more detail in Němec (1985).

12.4.5 Forms in which forecasts are issued

The correct presentation of hydrological forecasts is of great importance for their use and practical evaluation of their effectiveness. Being stochastic by its nature, it is possible to include an indication of the probability of errors of various magnitudes. This may be expressed technically in three ways:
(a) In the form of the mean value of the forecast variable, with an indication of the error of a given probability;
(b) In the form of a confidence interval within which the forecast variable is anticipated with a given degree of probability;
(c) In the form of a cumulative probability distribution function, from which the probability of exceedance of various values of the forecast variable can be found.

12.5 BENEFIT AND COST ANALYSIS OF HYDROLOGICAL FORECASTS

12.5.1 The importance of setting up a forecasting system derives from
the following statistics: it has been reckoned that in the hundred years
between 1870 and 1970 tropical hurricanes, and the flooding which accom-
panies them, caused an average $1,500 million worth of damage and more
than 5,000 deaths per year. Out of the world total the countries of
Asia and the Far East suffered losses to the sum of $950 million a year,
the countries of the Caribbean and the United States about $400 million,
and the countries on the south-western shores of the Indian Ocean about
$46 million per year. Again in Asia and the Far East alone, it is
believed that river floods damaged or destroyed annually about 10 million
acres of land and crops and affected the lives and well being of over
17 million persons.

12.5.2 Despite these facts not all authorities of the concerned countries
are aware of the potential benefits of a flood forecasting service,
particularly in view of the necessary investments. It is therefore often
necessary to find ways of quantifying the benefits of the forecasting
system in order to prove to the governing or financial bodies the utility
of such services.

12.5.3 The determination of a realistic benefit-cost ratio for a particular application of hydrological forecasting is often a difficult and
time-consuming task. Concepts of hydraulics, hydrology and economics,
have to be combined to develop comprehensive basin-analysis procedures.
The essence of such procedures may be stated simply as:
(a) Evaluate the economic consequences of man's presence in the river
    basin without any forecasting service;
(b) Evaluate the same economic consequences of man's presence when a
    forecast service is provided;
(c) On the basis of steps (a) and (b), evaluate the benefits versus costs
    of a forecasting service, taking into account the reliability of the
    hydrological analysis.
    Difficulties in evaluating intangible benefits and costs, especially
those associated with the potential loss of life, continue to prevent
their inclusion. However, in case studies undertaken to determine the
benefit-cost ratio for flood forecasting, the results have shown an
overwhelming margin of benefit over cost. A discussion of other problems
of cost/benefit analysis can be found in Němec (1985).

12.5.4 While benefit/cost analysis may be an important source of information in the decision-making process in developed countries, this is
not and should not be the case in developing countries, where the intan-
gible benefits accruing from the prevention of loss of human life and
human suffering are of paramount importance and cannot be quantified to
obtain a meaningful benefit-cost ratio. For developing countries the
concept of cost-effectiveness ratio, particularly for floods and
droughts, should instead by the main criterion used for evaluating the
design of the forecasting system.

12.6 EXAMPLES OF ESTABLISHED FORTH SYSTEMS

The author, during his service with WMO, has been entrusted in the past 15 years with technical supervision of the establishment of a large number of FORTH systems in developing countries of Asia, Latin America and Africa. In Asia, major systems are being established in Bangladesh, Burma, China, India, Indonesia, Republic of Korea, Malaysia, Pakistan, and the Philippines. In Latin America, the largest system is being established on the River Amazon in Brazil. In Africa, an older system was established on the River Niger in Guinea, Mali, Ivory Coast, Benin, Burkina Faso, Cameroon, Niger, and Nigeria. As an example, the case of Pakistan is included in Annex 2 to this paper. Further examples and the discussion of their design are included in Němec (1985).

ACKNOWLEDGEMENT

The author wishes to express his gratitude to the Secretary-General of the World Meteorological Organization for his authorization to publish this text and use materials of WMO. The views expressed are those of the author and should not be considered or construed as the official views of the Organization, unless specifically indicated as such.

SYMBOLS

E	standard relative error	1
A	basin area	L^2
C_v	coefficient of spatial variation of rainfall	1
n	number of raingauges	1
X	point precipitation for given duration	L
P	areal precipitation for given duration	L
Z-R	relation precipitation intensity/reflecting energy	1
G-R	raingauge/radar sampling difference	1
σ	standard deviation	1
S	standard error of forecasts	1
$\hat{y}$	forecasted value of y	1
y	observed value	1
d_Δ, d_y	determination coefficient	1
Δ_i	change in the forecast variable during period Δt	1
	corresponding mean change	1

REFERENCES

Gray, D.M., Editor-in-Chief, (1970): Principles of Hydrology Handbook, Canadian National Commission for the IHP, Ottawa.

Linsley, R.K., Kohler, M.A. and Paulhus, J.L.H. (1975, 1982): Hydrology for Engineers, Second and Third Edition, McGraw-Hill, New York.

Nash, J.E. and Sutcliffe, J.V. (1970: 'River flow forecasting through conceptual models', Journ. of Hydrology No. 10, pp. 282-290.

Němec, J. (1972): Engineering Hydrology, McGraw-Hill, London.

Němec, J. (1985): <u>Design and Management of Hydrological Forecasting Systems</u>, D. Reidel, Dordrecht.

Shaw, E.M. (1983): <u>Hydrology in Practice</u>, Van Nostrand Reinhold, UK.

Sittner, W.T., Schauss, C.E. and Monro, J.C. (1969): 'Continuous Hydrograph Synthesis with an API-type Hydrologic Model', <u>Water Resources Research</u>, <u>5</u>, pp. 1007-1022.

Sittner, W.T. (1982): 'Flood forecasting – an introduction'. <u>Proceedings of Regional Training Seminar on Flood Forecasting</u>, Bangkok, WMO, Geneva.

Sugawara, M. (1974): 'On natural disasters – some thoughts of a Japanese', unpublished manuscript.

US Army Corps of Engineers (1975): <u>Program Description and User Manual for SSARR</u>, US Army Engineering Division, Portland, Oregon, Revised.

WMO (1974): <u>Manual on Codes, Vol I – International codes</u>, WMO-No. 306, WMO, Geneva.

WMO (1975): 'Intercomparison of conceptual models used in operational hydrological forecasting', Operational Hydrology Report No. 7, WMO-No. 429, WMO, Geneva.

WMO (1981): <u>HOMS Reference Manual</u>, internal publication of WMO, Geneva.

WMO (1983): <u>Guide to Hydrological Practices</u>, WMO-No. 168, 4th ed., WMO, Geneva.

ANNEX 1

LIST OF HOMS COMPONENTS RELATED TO HYDROLOGICAL FORECASTING SELECTED FROM WMO HYDROLOGICAL OPERATIONAL MULTIPURPOSE SUBPROGRAMME REFERENCE MANUAL (WMO, 1984)

SECTION C, INSTRUMENTS AND EQUIPMENT

C30   PRECIPITATION, RECORDING AND TELEMETERING GAUGES

C30.2.01   Rainfall alarm

C33   PRECIPITATION, MEASUREMENT BY RADAR

C33.3.01   Radar raingauge system

SECTION D, REMOTE SENSING

D00.0.07   Snow cover from multi-channel satellite data

SECTION E, METHODS OF OBSERVATIONS

E25   METEOROLOGICAL OBSERVATIONS FOR HYDROLOGY

E25.3.01   The hydrological radar

SECTION F, DATA TRANSMISSION

F00.1.01   Automatic collection and transmission of hydrological observations

F00.1.06   Flood warning device
F00.2.01   Telephone service water level gauge
F00.2.02   Water level transmitter, Telemark
F00.2.03   Telexdat: automatic data collection unit
F00.2.04   Hydra III: regional telemetry unit
F00.2.06   Flood warning system
F00.2.07   Automatic data acquisition system
F00.2.08   Data acquisition and data transmission via satellite (the
           Argos system)
F00.2.10   Hydrological data transmission system
F00.2.11   Data transmission via geostationary satellites
F00.2.12   Data transmission via polar orbiting satellites
F00.2.13   Telemetering station land radio link
F00.3.01   Data collection and transmission system (OTT Allgomatic)
F00.3.02   River information system
F00.3.03   Level and rain digital telemetering system
F00.3.04   Automatic network for collecting real-time hydrological data

SECTION J, HYDROLOGICAL MODELS FOR FORECASTING AND DESIGN

J05   RAINFALL-RUNOFF MODELS FOR GENERAL USE

J05.1.01   Tank model
J05.1.02   Runoff calculation by the storage function method
J05.1.03   Seasonal forecast of inflow to a lake
J05.2.01   Runoff model for cultivated soils
J05.2.02   Streamflow synthesis and reservoir regulation (SSARR)
J05.3.01   Manual calibration program (NWSRFS-MCP)
J05.3.02   Sacramento soil moisture accounting model (NWSRFS-SAC-SMA)
J05.3.03   Sacramento model modified for use in the upper Nile basin
           project

J12   RAINFALL-RUNOFF MODELS FOR FORECASTING

J12.1.01   Flood forecasting by graphical correlation
J12.1.02   Deterministic rainfall-runoff model for basins of 100 to
           1000 sq. km.
J12.1.04   Flood peak forecasting by a grapho-analytic technique
J12.1.05   A method to forecast the spring flood volume
J12.1.06   Model for operation (MODRIE)
J12.2.01   Water supply streamflow and storage forecasts
J12.2.04   A conceptual watershed model for flood forecasting
J12.2.05   Forecasting inflows to a lake
J12.2.08   Conceptual watershed model (the HBV model)
J12.2.09   Model to forecast rainfall floods
J12.2.10   Model for the calculation of snow-melt and rainfall runoff
J12.2.11   Method for short-term forecasts of discharges in mountain
           rivers
J12.2.12   Short-term forecasts of spring inflow to reservoirs on plain-
           land rivers
J12.3.01   Real time streamflow forecasting model (MISP)

J12.3.02   Application of hydrological models for river forecasting

J50   STOCHASTIC MODELS, GENERAL

J50.2.04   CLSX (Constrained linear system extended) model

J72   STREAMFLOW ROUTING, SINGLE CHANNEL

J72.1.01   Channel routing Muskingum–Cunge method
J72.2.01   Lag and K routing (NWSRFS–LAG/K)
J72.3.01   Channel routing – implicit solution of full equations

J74   STREAMFLOW ROUTING, MULTIPLE CHANNEL

J74.2.01   Dynamic wave operational model (DWOPER)
J74.2.03   Method of unsteady flow calculation in braided river beds

HOMS sequences related to hydrological forecasting

007   Dynamic flow forecasting
009   Rainfall/runoff simulation using a soil moisture accounting
            model
010   ALERT flood warning system
011   Satellite data collection/relay systems

ANNEX 2

EXAMPLE OF A FORTH SYSTEM

1.   Country: Pakistan

2.   River basin

2.1 Basin name: Indus River Basin

2.2 Total area: 970,000 km^2 of which 469,000 km^2 in Pakistan

2.3 Index map: A map showing the location of the basin and the main
     rivers is given in Figure 1.

3.   FORTH system:

3.1 Network: A map of the basin with the hydrometeorological network
     is given in Figure 2.

3.2 General information:

(a) Purpose of the system

Pakistan is almost totally economically dependent upon its water resources as represented by the Indus River Basin and its five rivers: Indus,

Jhelum, Chenaab, Ravi and Sutlej. However, these rivers have also produced devastating floods which, since Independence in 1947, have been of record or near record magnitude. The system is intended to:

(i)   optimize planning, operation, regulation and management of the river system (Mangla, Chesme and Tarbela reservoirs and linked irrigation systems);

(ii)   minimize the damages to the vital hydraulic structures and irrigation works;

(iii)   provide safety for people and livestock against floods; and

(iv)   aid in planning and taking measures to prevent damage to property.

(b) History of the system

Prior to 1972, there were many stream gauging and climatological stations being operated by different agencies. However, after the catastrophic floods in August 1972, it was concluded that the system then in existence needed complete reappraisal. This conclusion resulted in World Meteorological Organization (WMO) representatives visiting Pakistan in March 1974 to assess the situation. Then in 1975, the WMO-UNDP joint co-operative projects (PAK/74/027 and PAK/75/WMO/FIT, "Improvement of the River Forecasting and Flood Warning System for the Indus River Basin in Pakistan") were initiated. The three components of these projects were: Component I - expansion and modernization of the basic Hydrological Data Collection and Processing System; Component II - establishment of a radar at Sialkot and a meteorological satellite readout station at Lahore; Component III - development of river flow and flood-forecasting techniques. Originally a snowmelt hydrology component was included, but was dropped early on with the intention of taking it up again later. The three components were completed in October 1980, after which the system was taken over by the Government of Pakistan. The system was test-run during the 1980 flood season.

(c) Resources of the forecasting system

(i)   Organization
        The two projects' activities were carried out as a single project.
        WMO was designated as the executing agency with two principal
        counterpart agencies: Pakistan Meteorological Department (PMD)
        and the Water and Power Development Authority (WAPDA). A tech-
        nical committee was formed in 1976 to co-ordinate the various
        agencies. They have continued to execute this responsibility
        after the completion of the project. The responsibilities of
        maintaining and inspecting the system were taken over by the
        Government of Pakistan in 1980.

(ii)   Computer facilities
        Two "Nova Computer - 3/12's" with 64K CPU were installed, one in
        Lahore and one in Sialkot. Also, an Automatic Satellite Picture
        Transmission ("Alden APT-3B type") was installed in 1978 in
        Lahore. There is also a 9825A Hewlett-Packard calculator.

(iii)   Staffing
        The 14 international employees (three team leaders, arriving in

turns, four experts and five consultants) had 21 national counter-
parts who included 11 meteorologists (hydrological forecasters),
five engineers, plus support staff.

(iv)   Capital cost
        The total budget contributed by the Dutch Government and UNDP was
        US$2,503,297. In addition, the Government of Pakistan contributed
        Rs.3,813,000. (NB: 10Rs. = US$1).

3.3 Data collection system

This is a continuously running, self-reporting system based on VHF radio
links. There is one weather radar for the upper catchment rivers monitoring hourly rainfall amounts.
    The data collection system includes 34 stations with 2 types of sen-
sors: river level gauges with 1 cm reporting increment and tipping
raingauges of 1 mm reporting increment.
    A total of 41 hydrometeorological stations have telecommunication
facilities.
    The telecommunication system has 36 stations with VHF automatic data
transmission, five stations with VHF voice communications. Eight stations
have no telecommunication facilities.
    The Automatic Picture Transmission satellite receiver relies on two
satellites (TIROS-N and NOAA-6) which pass twice in 24 hours. Three
passes picked up by Lahore cover the entire area.

3.4 Data-transmission system layout

The layout of the data transmission system is given in Figure 2.

3.5 Data-processing system

Data are processed on the Nova 3/12 (96) computer.

3.6 Receipt and use of meteorological forecasts

The weather radar at Sialkot covers the upper catchments of the Chenab,
Ravi, Jhelum and Sutlej Rivers. Hourly rainfall amounts were to be
supplied for 17 sub-catchments. However, for technical reasons the
radar is not used for operational forecasting.

3.7 Flood-forecasting model

Real-time data are inputs to the computer-based rainfall-runoff and
routing models. The rainfall-runoff model is used for the catchment
areas above the rim stations and the routing model is used to calculate
the movement of flood waves from the rim stations to the sea.
    Both Constrained Linear System (CLS) and Antecedent Precipitation
Index Continuous (APIC) rainfall-runoff models have been calibrated for
forecasting the discharge at rim stations for the Indus, Jhelum, Chenab,
Ravi and Sutlej Rivers. The lead time of the forecast is one day.
    The routing model divides the Indus River into 30 river reaches

related to the tributaries or link canals, etc., outflows in the form of canals, etc. The flood forecasts can be plotted for individual reaches. A lead time of 10-15 days is available in forecasting the flows based on the observed or forecasted flows at the rim stations. Thereafter these can be updated on a daily basis. Two types of flood routing models have been calibrated, based on lumped (non-linear cascade) or distributed (dynamic wave) parameters.

3.8 Warning system

With the help of models and of the telemetering system flood forecasting is performed for the "rim" stations with approximately 24 hours of lead time and for 30 sites of the Indus River System with 12 hours to 12 days lead time. The low incidence rate of high floods since 1980 (only one high flood on the Ravi River) has not permitted a full evaluation of the accuracy of the forecasts.

3.9 Problems with the system

There is a problem with the non-standardized equipment and methods being utilized for discharge and sediment transport measurements. This reduces the accuracy of the forecasts.

4. References

Water and Power Development Authority/Pakistan, Improvement of River Forecasting and Flood Warning System for the Indus River Basin in Pakistan, Brochure of the Hydrology and Investigation Directorate, WAPDA, Lahore, Pakistan, 1980.
WMO/UNDP, Improvement of the River Forecasting and Flood Warning for the Indus River Basin in Pakistan: Project Findings and Recommendations, PAK/74/027 and PAK/75/WMO TF, WMO/UNDP Geneva, 1980.

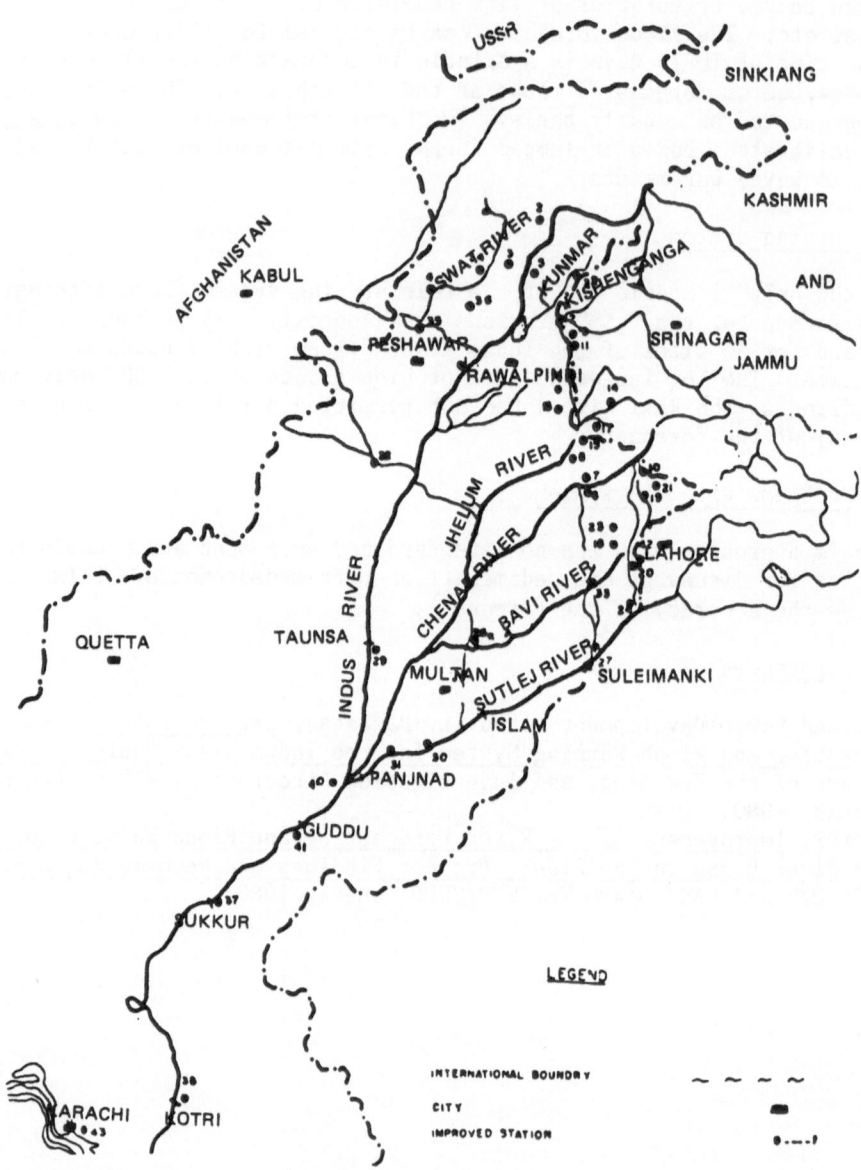

Annex 2, Figure 1. Location map of improved hydromet network

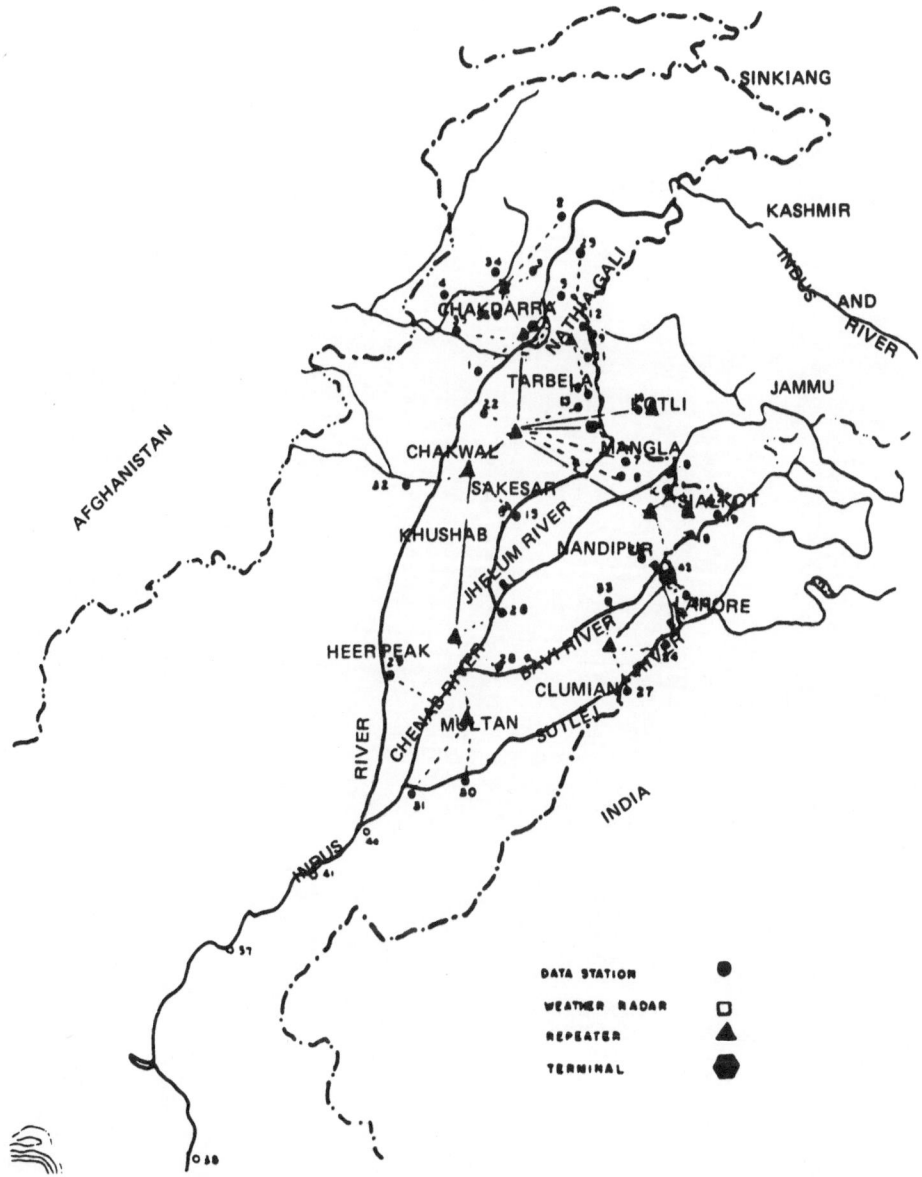

Annex 2, Figure 2.   VHF telecom network.

13. CASE STUDIES ON REAL-TIME RIVER FLOW FORECASTING

G. Fleming

University of Strathclyde
Dept. of Civil Engineering
John Anderson Building
107 Rottenrow
Glasgow G4 ONG, U.K.

13.1 Introduction

Three case studies are presented giving examples of flood peak fore-casting, drought sequence forecasting and forecasting for river basin development. Each case study will be presented as a series of steps as shown in Figure 13.1.1 (Fleming 1979). Two fundamental points should be

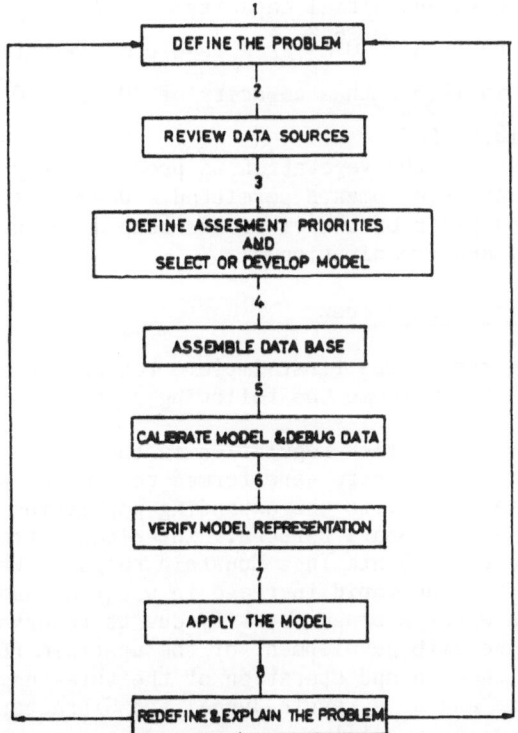

Fig. 13.1.1. Steps in applying a hydrological model (after Fleming 1979).

D.A. Kraijenhoff and J.R. Moll (eds.), River Flow Modelling and Forecasting, 329–366.
© 1986 by D. Reidel Publishing Company.

considered when developing a system for flow forecasting. Firstly in
hydrological flow forecasting there cannot be an exact solution to the
problem, but rather a step-wise development of a better quantitative
understanding of the processes involved. Secondly the problem itself is
not steady state. It evolves with time, changing as more understanding
is gained in trying to assess it.

13.2   The Santa Ynez river, California, USA

The Santa Ynez River is situated in North Central California at Lat 34D
40M and Long 120D 00M. It drains a 2317 km^2 catchment as shown in Fig.
13.2.1 with an elevation range of 2116 m to zero at the Pacific Ocean.
Slopes are relatively steep in the upper part of the basin becoming flat
downstream of Cachuma Reservoir. Three reservoirs have been formed
within the basin; in 1920 Gibraltar reservoir was initially formed, then
in 1957 the capacity was increased to $21.28 \times 10^6 m^3$. Siltation between
1957 and 1970 reduced this capacity by 44% to $11.84 \times 10^6 m^3$. Juncal
reservoir was formed in 1931 with an initial capacity of $8.73 \times 10^6 m^3$
which by 1970 reduced by 17% to $7.28 \times 10^6 m^3$ as a result of siltation.
Cachuma reservoir was formed in 1956 with a capacity of $292.5 \times 10^6 m^3$
and a surface water area of $13.61 \; km^2$.

Upstream of Cachuma Reservoir, the vegetation is predominantly
wilderness scrub forest with no developments permitted. Downstream of
this point the wide flat flood plain of the river has been developed
considerably for agriculture, and urbanisation.

13.2.1   The forecasting problem, Santa Ynez

A real-time flood forecasting system was first implemented on the Santa
Ynez in 1970 (Hydrocomp 1970) to overcome the following problem, which
had developed at that time.

The problem was basically one of flow regulation in the entire
Santa Ynez River. The upstream reservoirs were formed to provide water
storage for regulation of water supply to the expanding population on
the Pacific coastline in and around Santa Barbara. Inter-basin transfer
was made via rock tunnels under the Santa Ynez Mountain range on the
Southern boundary of the basin. The rapid increase in water demand
together with the decrease in water storage capacity due to reservoir
sedimentation led to the progressive development of the upstream reservoirs. However in each case, design and operation of the three dams was
based on water supply criteria and as a result Juncal and Gibraltar dams
were built with fixed uncontrollable spillway.

This provided regulatory control capability for diversions and
floods but the dam was operated as a statutory water supply scheme with
no attempt made to regulate floods. At flood times spillway gates were
operated to release incoming flows when the reservoir was full.

However, downstream development of the fertile flood plain underwent
similar rapid expansion as the Santa Barbara coastal strip with the

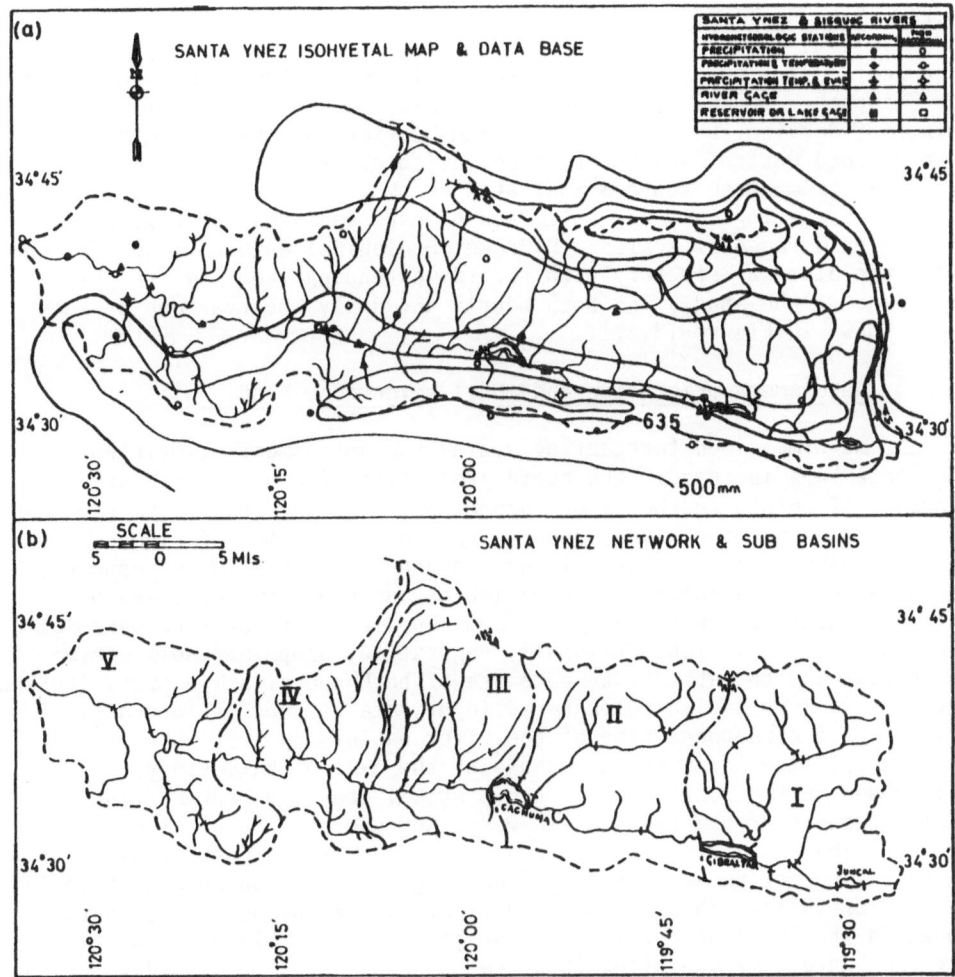

Fig. 13.2.1.   The Santa Ynez catchment

result that flood problems developed downstream of Cachuma when the river overtopped bank full conditions. By 1969 extensive use of the flood plain had taken place with the effect that a major flood in that year resulted in severe property damage and loss of life. As a direct result of this flood, a change in policy was introduced, requiring the regulation of Cachuma reservoir for the dual purpose of flood control and water conservation.

The problem could then be summarised as the need to: maximise water supply from the three reservoirs by maintaining them optimally full; minimise flood magnitudes downstream of Cachuma by maintaining this dam optimally empty; and provide civil defence operators with advanced warning of the risk of flooding. Such a problem necessitated a real-time flow forecasting system.

13.2.2 Data sources, Santa Ynez

Data sources for the Santa Ynez River basin were numerous and included
the US Weather Bureau, US Bureau of Reclamation, US Geological Survey,
US Forest Service, US Soil Conservation Service, and the Santa Barbara
County Flood Control and Water Conservation district. Review of data
available from these sources indicated a good coverage of hydrographic,
vegetation, geological, and soil data on maps and areal photographs
plotted on a range of scales. Hydrometeorological data coverage for the
area were also satisfactory with records extending from 1940 to 1970
(the date of the study) for both hourly and daily observations at locations shown in Figure 13.2.1.

13.2.3 Assessment priorities and model selection, Santa Ynez

For a real-time flood forecasting system the assessment priorities in
the Santa Ynez included: the rapid prediction of hourly flow rates at
a number of points in the river channel network; the ability to represent
flow routing in three linear reservoirs and in a number of river channel
reaches which exhibited both linear and non-linear hydraulic response;
the updating of catchment state variables and the representation of
variable rainfall-runoff response from the different sub-catchments of
the basin due to variable rainfall, vegetation, slope and soil characteristics; and the ability to make use of both quantitative precipitation
forecasts and real-time telemetered input data to provide forecasts in
advance of the response time of the river basin.

General flow forecasting methods at the time of this study (1970)
were based on the monitoring of flows at key stations representing the
early response of the basin which were then used in conjunction with
flow routing and correlation methods to predict flows downstream. Yet
another alternative was the use of co-axial correlation methods (Linsley
and Kohler 1951). These methods suffer limitations, primarily in the
speed in which repeated forecasts can be prepared. Since response time
for the Santa Ynez was often less than 24 hours it was decided to use a
computer-based forecasting system utilising an integrated data management
system, a land surface rainfall-runoff response model and a river flow
routing model capable of representing a river channel network comprising
a combination of trapezoidal open channels and reservoirs. Such an
approach would satisfy the basic assessment priorities.

A number of deterministic models existed at that time (1970) which
would provide either the components of land surface rainfall-runoff
modelling and the channel routing processes separately or an integration
of these two components into one system. The methods available for
rainfall-runoff modelling are presented in Chapter 2 by O'Donnell and
for flood routing in Chapter 3 by Dooge.

For the Santa Ynez River it was decided to employ an integrated
rainfall-runoff and channel routing model. Two models were considered
appropriate at the time (1970). The Hydrocomp Simulation programme
HSP, (Hydrocomp 1969); and the SSARR model (Rockwood 1964). The HSP
model was used in the study since it was considered more appropriate to
river basins of the size of the Santa Ynez. The organisation of the HSP

model structure is shown on Figure 13.2.2. The land surface component
of this combined model is a modification of the Stanford Watershed Model
(Crawford and Linsley 1966) which is described by O'Donnell in Chapter 2.

13.2.4 Data base assembly

All available data for the Santa Ynez River Basin were obtained from the
sources listed in subsection 13.2.2. Maps were prepared of the sub-
catchment boundaries and the primary river channel network as shown on
Figure 13.2.1(b). The location of all gauging stations was plotted and
maps prepared of the distribution of rainfall both by isohyetal method
(Fig. 13.2.1(a)) and the Thiessen method. Vegetation, soil type
topographic and geological maps were prepared and overlaid to define
subcatchment area of similar physical characteristics. From a combination of physical conditions subareas referred to as segments were delineated. Five segments were selected as shown on Figure 13.2.1(b).

A preliminary reconaissance of the river basin was undertaken to
confirm the suitability of the segment division and to provide information on the subdivision of the channel network into a set of river
channel reaches which would be most representative of the system. A
final channel network was then chosen consisting of 37 reaches, three of
which were reservoirs. The average length of a reach was 11.77 km. All
available data on rainfall, streamflow and evaporation were obtained
on hourly and daily time series. This was converted to computer compatible formats and fed into the computer where it was stored within the
data management system of the model. Analysis of all meteorological
time series was then undertaken to test the records for missing data,
errors and inconsistency. On the basis of this analysis a number of
primary rainfall stations were listed for use in representing input to
each segment of the sub-divided catchment.

The method of flow routing used by the model was based on the kinematic approach. For application of this method a survey was undertaken
of the river channel cross sections along the longitudinal profile of
the river to determine the average cross sectional characteristics for
each reach. An equivalent cross section was taken to represent the
average dimensions of each reach as shown in Figure 13.2.3. This information together with the longitudinal slope, length and contributary
land surface area was assembled for each reach. With three reservoirs
in the channel network, data were required on their physical characteristics together with diversions and discharge releases from them.

The physical characteristics of each dam included the maximum,
minimum and spillway elevations, the capacity-elevation-spillway discharge relationship, the area of the water surface at full-pool, and the
rule curves in operation. Most of the information on the physical data
for each reservoir was obtained from the reports on the design of each
dam. In the case of Juncal Dam no data existed on the spillway elevation-
discharge relationship. The spillway was therefore considered equivalent
to a weir 60 m long, and the standard weir formula Equation 13.2.1 was
used to compute the spillway discharge rating curve.

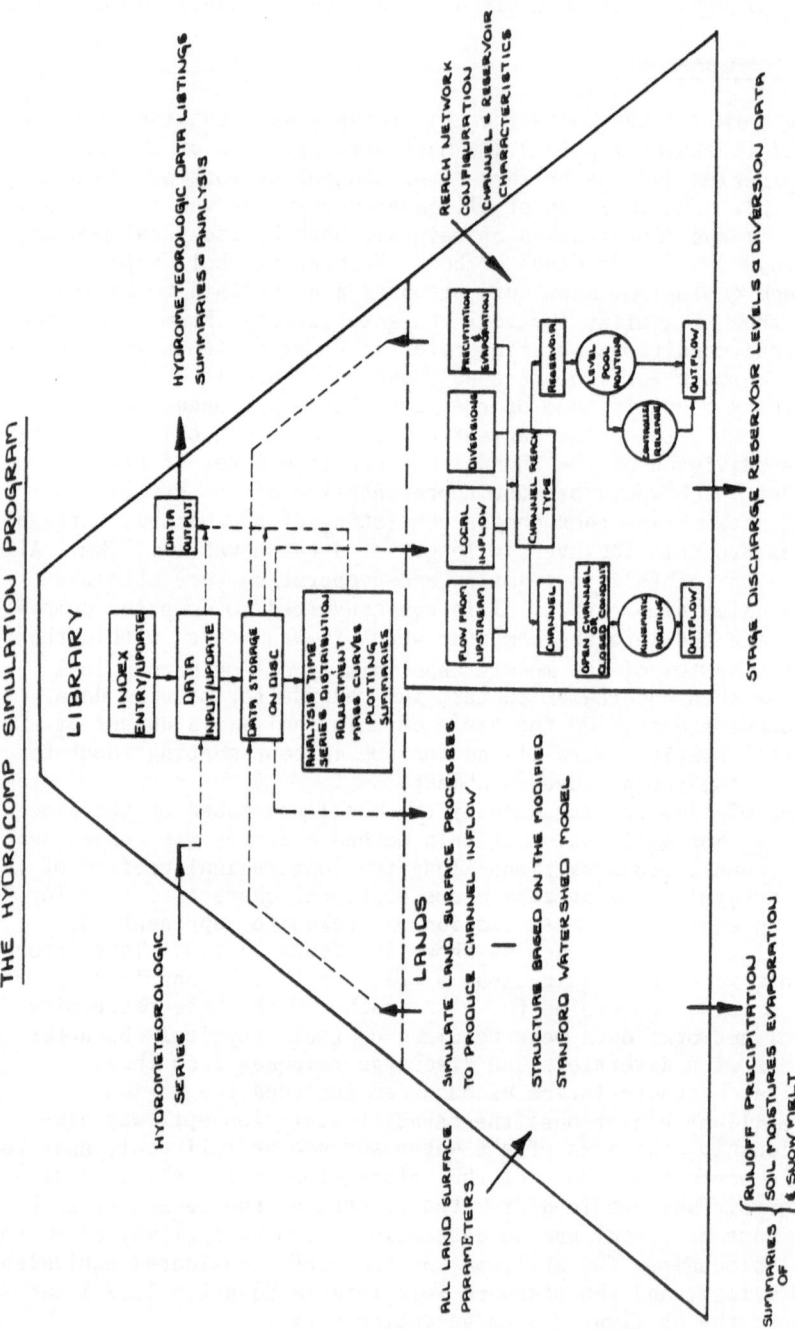

Fig. 13.2.2.   General layout of the Hydrocomp Simulation Program (after Fleming 1975)

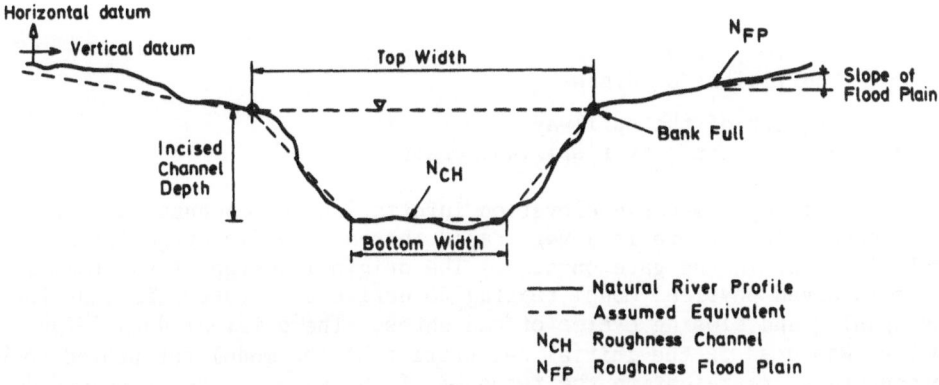

Fig. 13.2.3.   Channel reach cross section data

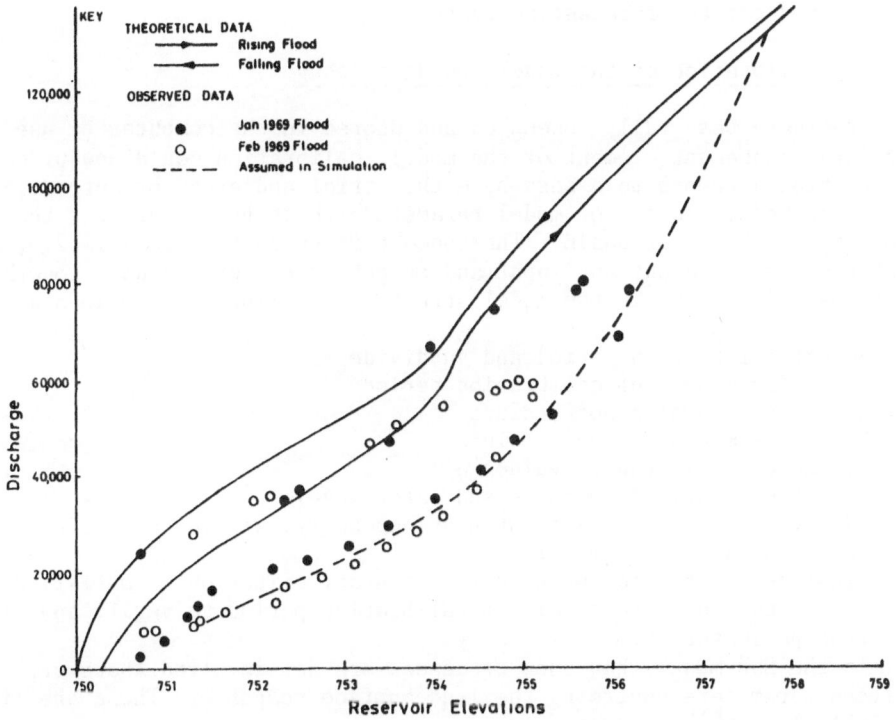

Fig. 13.2.4.   Cachuma dam discharge elevation data

$$Q = C_d b H^{3/2} \qquad\qquad\qquad (13.2.1)$$

Q = Spillway flow
C_d = Coefficient of discharge

b = Breadth of the spillway
H = Head of water over spillway crest

The spillway discharge-elevation information for Cachuma was more complicated since there is a very rapid increase in discharge for a small increase in the gate opening. The original design of the radial gates involved physical model testing to derive the hysteresis loop for the opening and closing cycles of the gates. The original data (Figure 13.2.4) was used in the initial calibration of the model but proved to be inaccurate in representing the response of the gates. Use of recorded discharge data for the 1969 flood, together with concurrent records of reservoir elevation allowed a re-examination of the rating curve which then indicated that the gates responded much more slowly on the rising limb than the original design data specified. The updated rating curve was adopted for the forecasting system.

13.2.5 Calibration of the model and data debug

With the data base fully assembled and stored in the computer by use of the data management element of the model, calibration could now proceed. Calibration attempts to assess by either trial and error or automatic search methods, the set of model parameters which best represent the response of the river basin. The concept is shown on Figure 13.2.5, and must take into account the input and output errors of the data together with possible errors in the model structure. Calibration follows a sequence of steps:
1   Select calibration period and subdivide
2   Using first half of calibration period
     a.   Check annual runoff values
     b.   Check monthly runoff values
     c.   Check daily runoff values
     d.   Check hourly flow rates and water levels
3   Adjust parameters at each step as necessary, within the criteria recommended for the model
4   Finalise calibration when chosen accuracy criterion is satisfied
5   Rerun model on second half of calibration period to verify and fine tune parameters where necessary.
     In the HSP model when snow processes are not involved there are fifteen parameters governing the land surface response. These are listed in Table I.
     Only four parameters are adjusted when calibrating lands as indicated on Table I. During the calibration of the land surface response inconsistency between the data input and output is often identified by the model and input and output errors highlighted which would otherwise be missed in routine data checks. This is most evident in the first stage of calibration. Errors are also detected in diversion data and spillway

rating curves as shown on Figure 13.2.4 where the original spillway relation for Cachuma was found to be in error.

Table I.   HSP Land Surface Parameters

Parameter	Description
K1	Ratio of Gauge rainfall to segment rainfall
A	% impervious or directly connected area
EXPM	Interception storage (mm/unit area)
UZSS *	Upper zone soil store
LZSS *	Lower zone soil store
K3	% area of vegetation drawing from lower zone
K24L	Loss of water to deep groundwater
K24EL	Loss of water from swamp land
ILFILTRATION *	Nominal infiltration rate (mm/hour)
INTERFLOW *	Interflow parameter
L	Length of overland flow (meters)
SS	Slope of overland flow (m/m)
NN	Roughness of overland flow
IRC	Interflow recession ratio
KK24	Groundwater recession ratio

*   Parameters adjusted during calibration

The most commonly used accuracy criterion for annual, monthly and daily runoff volumes is shown in Equation 13.2.2 and is termed the minimum value of the sum of the squares of the differences between observed and simulated volumes.   (Dawdy and O'Donnell 1965).

$$F_{Min}^2 = \sum_{i=1}^{n} (q_o - q_s)^2 \qquad\qquad (13.2.2)$$

where F_{Min}^2 = criterion of accuracy

q_o = observed flow

q_s = simulated flow

n = number of records used

When calibrating the hourly response of channel flow the main calibration parameters are the roughness representing the channel and flood plain.   All other data are essentially physically fixed.   At this stage three accuracy criteria are used (Fleming 1975) to fit the peak flow rate, the flow volume and the time to peak.

For the Santa Ynez system the calibration period was chosen as twenty years from 1948 to 1968.   This was split into two ten year periods and the calibration procedure followed.   For each river gauging station

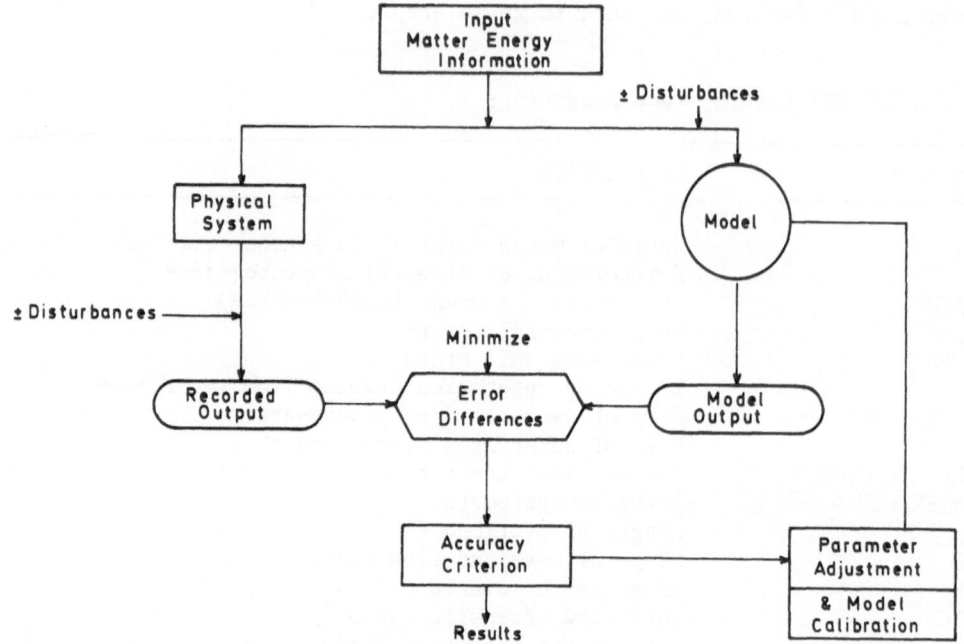

Fig. 13.2.5. Mathematical model concept (after Fleming 1975)

comparisons were made between recorded and simulated data and model
parameters for the land surface processes adjusted until a best fit was
achieved. Figure 13.2.6 shows typical results of daily flow simulation
compared to recorded response.

13.2.6 Verification of the model

The first stage of verification takes place at initial calibration.
Second stage verification is to test the flow forecasting system on an
event outside the range of flows encountered in the calibration period.
For the Santa Ynez River the 1969 flood events provided an opportunity
for second stage verification.

The maximum flood peak recorded during the 1948–1968 calibration
period was 424 m^3/sec. In 1969 two flood peaks occurred with values of
approximately 2264 m^3/sec and 1698 m^3/sec. The model was rerun using
the 1969 input data and a comparison was made between recorded and
simulated data as shown on Figure 13.2.7. With the good agreement
obtained the model can then be considered for application to the Santa
Ynez River for real-time flow forecasting.

13.2.7 Application of the model in real time

Once calibration and verification was achieved the model was loaded into

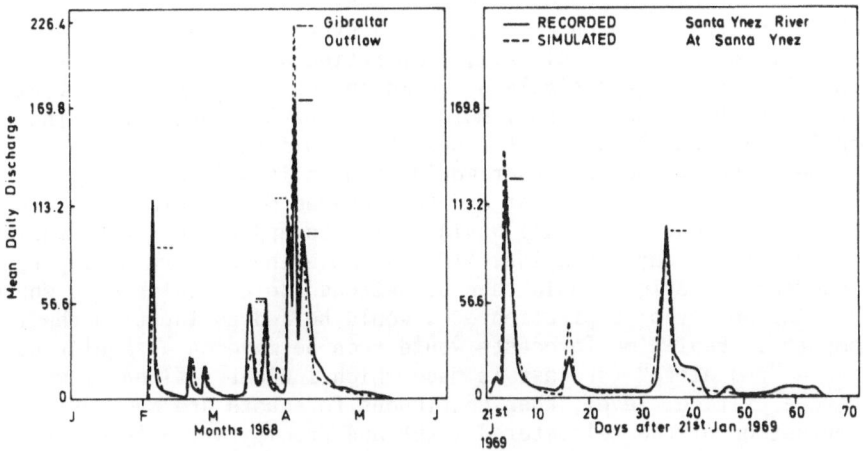

Fig. 13.2.6. Comparison between recorded and simulated daily streamflows

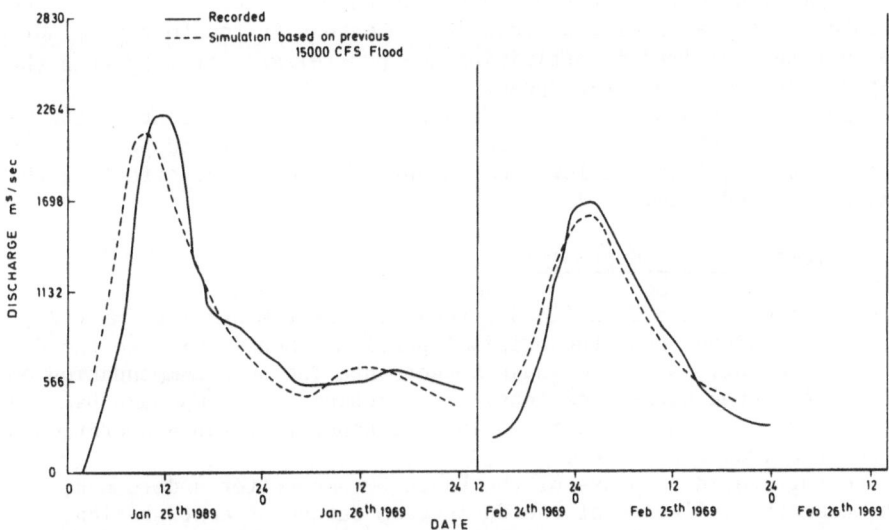

Fig. 13.2.7. Hourly Hydrograph Comparison January and February 1969 Floods

a computer system, suitable for both the model and the user. In the case of HSP this required an IBM computer, with a PL/1 compiler and 256 K bytes of storage, and virtual memory capacity, which was nearest to the user, namely the Santa Barbara Flood Control and Water Conservation District. A back-up computer system was also selected in case of failure of the first computer during emergency forecasts.

Computer access was by means of a Video Display Terminal. The user would follow a standard procedure as follows:

(a) Regular monthly sign on to forecasting system to update the data
    bank with the month's rainfall, evaporation and flow data. At this
    time the land surface simulator would be run to update the catchment
    state variables such as soil moisture, reservoir level etc. The
    model would then be keyed at a new start-up date.

(b) In the forecast mode the user would sign on in a "green alert" situ-
    ation where weather forecast and telemetered data indicated a possible
    flood situation. The model would be run to update the state variables
    to the current day situation. The telemetering network of input
    measurement stations would then be switched to a regular scan whereby
    incoming hourly precipitation etc. would be fed as input to the
    computer. Real-time forecasts would then be made on a regular basis.
    When a "red alert" forecast is made which indicates flood flows
    above a critical level, then continuous forecasts are made using a
    combination of the telemetered input and precipitation forecasts
    based on weather radar reports. Alternatives are tested to optimise
    pre-releases from Cachuma Reservoir, and to maximise storage levels
    in the reservoirs.

If the flood forecast of inflows and outflows from the reservoir are
greater than actually happen, then the pre-release strategy to minimise
downstream flows may allow too great a drawdown in the reservoir. Then
valuable stored water will be lost due to errors in the forecast. Great
care must be exercised in optimising the pre-release strategy when the
reservoir system is multi-purpose.

Figure 13.2.8 shows a typical example of a flow forecast on the Santa
Ynez with optimisation of pre-release and downstream flood peaks. Note
the contribution from the lower catchment has a critical effect on the
downstream flood peak.

13.2.8 Re-defining the problem

By the time the Santa Ynez flood forecasting system was implemented
insight was gained about the original problem. Sensitivity to the data
coverage limitations of the gauge network had led to recommendation on
new or rearranged gauge locations. The problem of the Cachuma Dam radial
gate operation had been clarified and the benefit of rapid operation and
control realised and proven.

The very rapid response of the Basin became better understood, in
particular the influence of the downstream catchment contribution.

The need was signalled to introduce automatic telemetered data
scanners and translators for direct update of incoming data at the
height of flood forecasting. This would save time and allow a step
towards automatic optimisation of the control of the water resource
system.

And finally the knowledge that it could be done, and that the forecasting ability was of reasonable accuracy, brings a new confidence that
the time and effort involved is well spent.

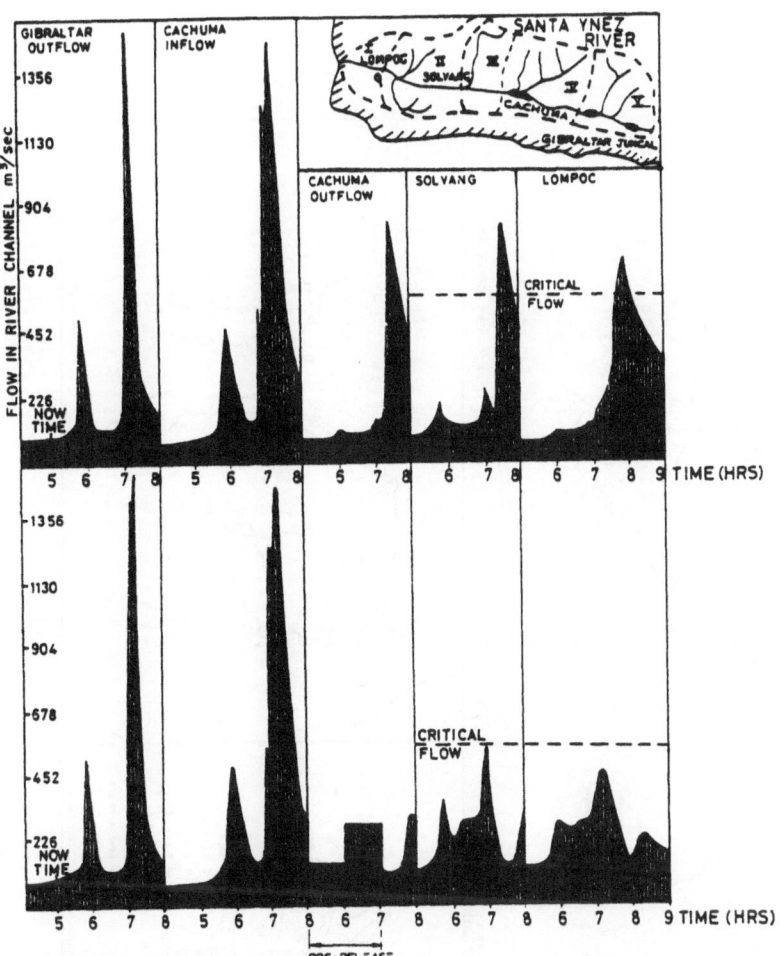

Fig. 13.2.8.   Real-time flow forecasting and optimisation of reservoir release (Santa Ynez River)

13.3   Derwent River system, England

The Derwent River system has its source in the English Lake District, an
area of relatively high rainfall. It drains a 663 km² catchment as
shown on Figure 13.3.1. The River Basin contains five large natural
lakes of which two, Thirlmere and Crummock, are used for water supply
diversions. Thirlmere Lake has been developed as a regulatory storage
with a maximum capacity of 49.6×10^6 m^3 and with release capability in
addition to its inter-basin transfer of water. The remaining lakes
including Derwent water, Bassenthwaite, and Crummock and Buttermere
combined have storage capacities of 23.66, 25.42 and 10.5 million cubic
metres respectively and in 1976 did not have discharge control.

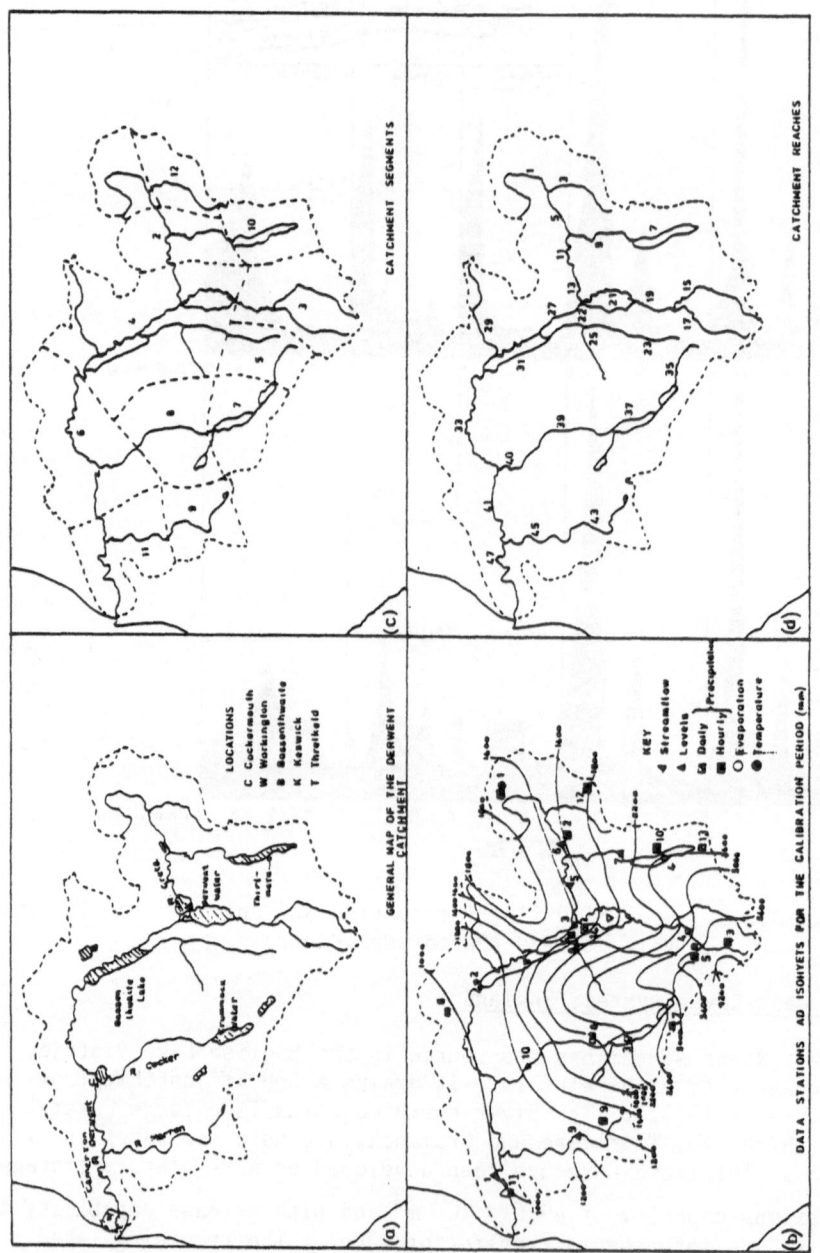

Fig. 13.3.1.   The Derwent River basin

The area was extensively glaciated during the Pleistocene resulting in steep tributary streams discharging onto a flat valley floor. The highly variable topography results in a considerable variation in precipitation (Figure 13.3.1). The vegetation is predominantly sparse moor grass and heather in the higher elevations of the upper catchment, wood and conifer in the lower slopes and marshy meadows giving way to arable farmland in the valley floor and lower catchment. Urbanisation is greatest at the outlet of the river at Workington, where water supply and water quality requirements are dominant.

13.3.1 The forecasting problem, River Derwent

A study of the River Derwent Basin was undertaken in 1976 (Hydrocomp 1976) with the prime objective of assessing the practicability of regulating discharges from Thirlmere Reservoir (Fig. 13.3.1) in order to sustain low flows at Workington. Water Quality considerations at that time required a sustained minimum flow at Workington of 2.75 m^3/sec. When natural flow recession was likely to pass below this rate, the releases, from the only regulatory storage at Thirlmere were required to compensate the natural recession. The problem of assessing the complex response of the river basin and of optimising the magnitude and timing of the required releases, led to the application of a conceptual river basin simulator.

In practice, full scale river basin tests had been conducted to study releases from Thirlmere, in order to assess the timing of flows and losses within the basin. However it was found that response from tributary streams complicated the response. For example isolated rainfall events centred over the River Cocker were of common occurrence and would cause rapid augmentation to the flow at Workington, often after valuable water had been released from Thirlmere unnecessarily. The ability to test various combinations of events on the Derwent by means of a real-time flow forecasting model was considered vital for optimising regulatory controls on the Basin.

In 1976 after the model for the Derwent had been set up, a drought sequence occurred which seriously affected the British Isles as well as other parts of Europe. The real-time flow forecasting model was then verified on the eighteen months of data between completion of the study and the "Now-Time" of the required drought forecast and a series of forecasts made.

13.3.2 Data sources, River Derwent

Main sources of data for the River Derwent came from the North West Water Authority, the Meteorological Office, the Institute of Geological Sciences and the Ordnance Survey.

Review of these sources indicated good coverage of physiographic and geological data. Adequate hydrometeorological data existed, although the period of record available, which included good coverage of hourly observations, was limited to the years 1973-1975.

13.3.3 Assessment priorities and model selection, River Derwent

As for the other river systems of this type, the assessment priorities
included: the rapid prediction of hourly and daily flow events including
peak rate and low flow recessions; the ability to represent one regulated
reservoir and three distributed uncontrolled natural lakes combined with
a complex channel network of highly varying longitudinal slope; and the
need to represent variable land surface response from twelve identifiable
sub-basins, of different physical characteristics.

Table II. Hydrometeorological data used for River Derwent

1. Precipitation

Map key	Station	Title	Segment
(a) Hourly			
1	Mungrisedale	PRECO1	1
2	Threlkeld	02	2
3	Seathwaite	03	3
4	How Farm	04	4
5	Honister	05	5
6	Linskeldfield	06	6
7	Bleaberry Tarn	07	7
8	Cornhow	08	8
9	Lamplugh	09	9
10	Dalehead Hall	10	10
11	Stainburn (Synthetic)	91	11
12	Groove Beck	92	12
(b) Daily			
11	Stainburn	PREC41	11
13	The Nook	42	10
14	Braithwaite	44	4
8	Cornhow	48	8

2.   Streamflow

	Station	Title	River	Segment
1	Camerton	FLOW56	Derwent	1-12
2	Ouse Bridge	54	Derwent	1-5,10,12
3	Portinscale	57	Derwent	1-4,10,12
4	Mountain View	53	Derwent	3
5	Low Briery	52	Greta	1,2,10,12
6	Threlkeld	51	Glendermackin	1-3,12
7	Thirlmere	50	St. John's Beck	10
8	Braithwaite	55	Newlands Beck	4,5
9	Ullock	59	Marron	9
10	Southwaite Bridge	58	Cocker	5,7,8

At the time of the study (1976) a good selection of river basin models existed (Fleming 1979) which are discussed in Chapters 2 and 3. For the Derwent catchment the choice of assessment method and the selection of a suitable model were made on commercial grounds. The proprietary Hydrocomp model (Hydrocomp 1969) was selected for the study on the basis that it was fully developed, had complete documentation and training facilities, was available on a contractual basis, but most important it was suitable for the assessment problem, as discussed above.

13.3.4 Data base assembly

The available data were assembled from the various sources for the gauge locations shown in Figure 13.3.1(b), and Table II. Isohyetal maps for the study period were prepared and consistency tests applied to the hydro-meteorological data. From analysis of the physiography, geology and hydrometry of the basin, twelve sub-basins were identified as having sufficient variability in response as to require separate analysis and representation. The resulting segments are shown on Figure 13.3.1(c).

Hourly precipitation data were available at twelve stations, but due to the recent nature of these records, and initial data tape translation problems, gaps existed in the time series. These gaps were completed by cross-correlation with surrounding gauges.

During initial data collection and analysis, the lack of hourly data in the North Western part of the catchment around Camerton, necessitated the generation of an hourly rainfall series using the nearest daily gauge and the time distribution of the nearest hourly gauge.

Estimates of potential evapo-transpiration were prepared for key stations based on meteorological data and maps for the area. In a low flow forecasting system, the importance of reliable estimates of potential evapo-transpiration (PE) is greater than for flood peak estimation. For the Derwent the seasonal variation of PE is pronounced, as shown in Figure 13.3.2.

The river channel network was inspected and sub-divided into 25 reaches, including 3 uncontrolled lakes and 1 controlled lake/reservoir, as shown on Figure 13.3.1(d). The HSP model uses a kinematic routing method for river channels and a level pool routing method for reservoirs. Each routing is undertaken for a variable time step to a minimum of 15 minutes. The physical data for input to the channel network are shown in Table III and were obtained from a full survey of the channel and lake cross-sections.

The average length of each channel reach for the Derwent was 6.72 km. Upstream and downstream elevations of the reaches from the Derwent survey if used directly to obtain the slope of the reach would lead to a biased estimate such as line AA on Figure 13.3.3. Since stream velocity is proportional to the square root of slope a better estimate for the reach slope could be obtained from a relationship such as given in Equation 13.3.1.

$$S_R = \left\{ \sum_{i=1}^{n} \frac{(S_i)^{1/2}}{n} \right\}^2$$

(13.3.1)

Table III.   Parameters for the CHANNEL Network, River Derwent

Parameter	Description
(a) For reach stream channel	
RCH	Reach number
LIKE	Reach number of identical reach (if any)
TYPE	Type of reach (RECT)
TRIB-TO	Reach number of downstream reach
SEGMT	Segment number
LENGTH	Length of main channel in km.
TRIB-AREA	Tributary area in km^2
EL-UP	Upstream elevation of channel bed in metres A.O.D.
EL-DOWN	Downstream elevation of channel bed in metres A.O.D.
W1	Channel bottom width (m)
W2	Channel bankfull width (m)
H	Incised channel depth (m)
S-FP	Slope of flood plain (dimensionless)
N-CH	Manning's n for the channel
N-FP	Manning's n for the flood plain
(b) For each lake or reservoir	
RCH	Reach number
DAM-	Lake or reservoir number
TYPE	Type of reach (DAM)
TRIB-TO	Reach number of downstream reach
SEGMT	Segment number
MAX-ELEV	Maximum pool elevation (m)
TRIB-AREA	Tributary area in km^2 – excludes water surface area
SPILLWAY-CREST	Elevation of Spillway crest (m)
MIN-POOL	Elevation of minimum pool (m)
NAME	Name of lake or reservoir
STORAGE-MAX	Maximum storage (m^3)
STORAGE-NOW	Current storage (m^3)
CONTROLLED	Maximum turbine discharge (m^3/sec)
SURFACE-AREA	At full pool (km^2)
RULES	Number of rule curves to be entered
PRINT	Min. flow for detailed output
USE-RULE (*,*)	Rule curve to be used for each day of the year (if RULES 1), 31 values for each month are specified
ELEVATION	Pool elevation, storage and corresponding dis-
STORAGE	charge for each rule curve (m^3)
DISCHARGE (rule 1),	(only one rule for lakes). Up to
DISCHARGE (rule 2),	75 elevations may
DISCHARGE (rule 3),	be specified.
TIME-RELEASE	Release time in hours
RELEASE	List of hourly discharges

where S_R = slope of the reach

n = number of sub-divisions of the reach

For many subdivisions of the reach the typical slope would be given by line AC on Figure 13.3.3. In the Derwent two subdivisions in each reach were made resulting in a typical line AB on Figure 13.3.3.

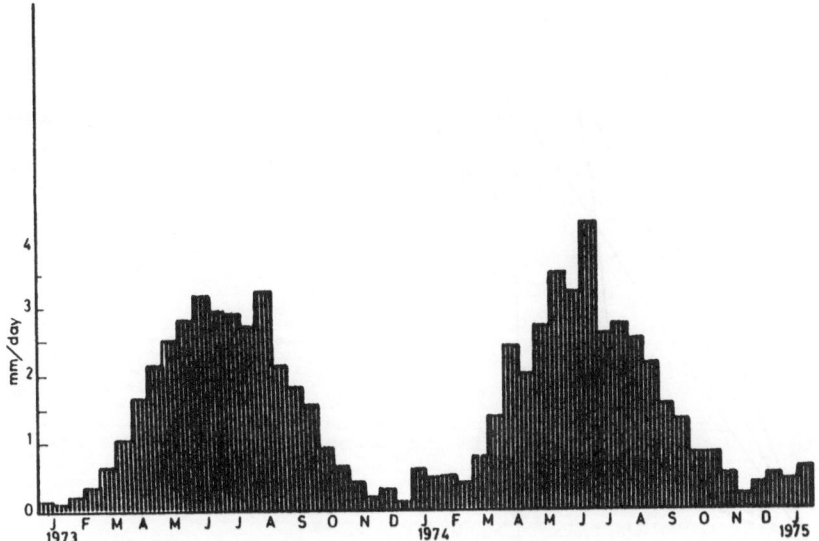

Fig. 13.3.2. Semi-monthly evaporation River Derwent

Reservoir and lake data were well defined for Thirlmere and Bassenthwaite. For Derwent Lake the elevation-storage curve was derived from planimetry of a 1:10,500 scale by hydrographic map. Crummock and Buttermere were combined and treated as one lake in the model and an appropriate discharge-elevation relationship estimated from a weir formula of the type shown in Equation 13.2.1.

13.3.5 Calibration of the model

Calibration of the model for a twelve segment system requires a logical, progressive approach whereby the upland segments are calibrated first, (segments 9, 7, 5, 3, 10, 12 and 1) then followed by the combination of upland and intermediate segments and finally the full segment combination. In each case the procedure outlined in subsection 13.2.5 is followed. In the case of the Derwent, only two years of data were available for calibration, hence it was not viable to subdivide the calibration period.

During the calibration stage, sensitivity tests are usually undertaken to examine the water balance and the main contributions to runoff. Figure 13.3.4 shows the sensitivity of the Derwent for high and low infiltration parameter values. This sensitivity analysis is essential

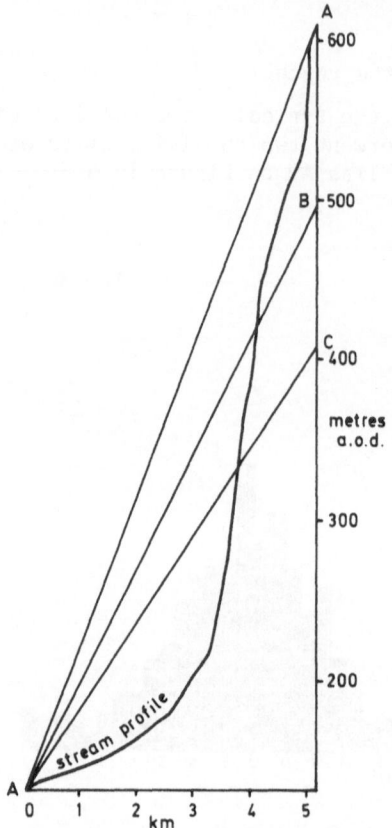

Fig. 13.3.3. Slope calculation for River Derwent

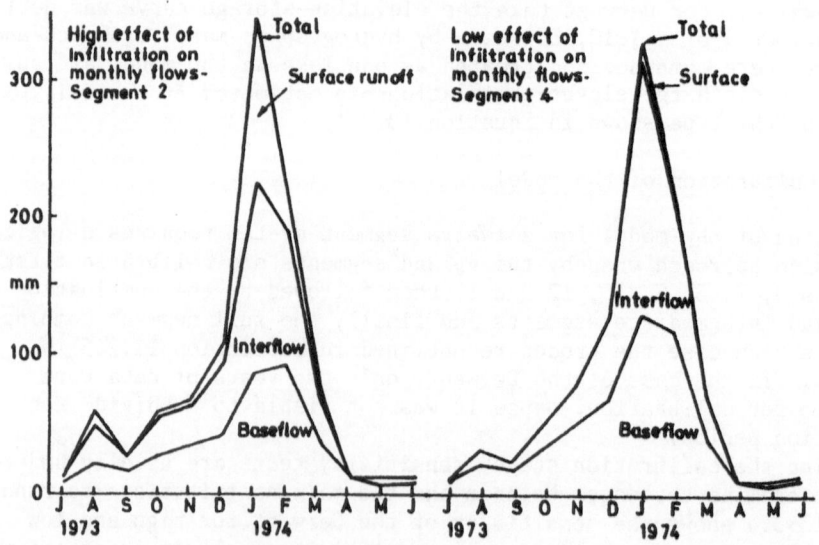

Fig. 13.3.4. River Derwent sensitivity of runoff components to
              infiltration changes.

in the calibration process and provides the data necessary for optimising the final set of parameter values.

For low flow forecasting the simulation of daily runoff volumes is as important as the correct simulation of hourly flow rates.

Calibration of the Derwent achieved a satisfactory standard for the available data. Examples of the monthly runoff volumes achieved are shown in Figure 13.3.5; of the daily runoff volumes in Figure 13.3.6; and of water levels at Thirlmere, in Figure 13.3.7.

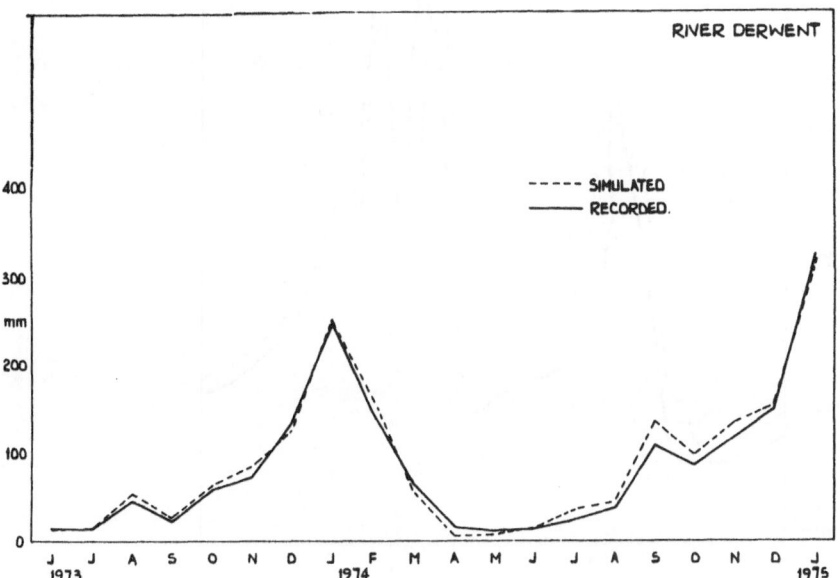

Fig. 13.3.5.   Simulated and recorded monthly flow volumes at Ullock (after Hydrocomp 1976).

13.3.6   Verification of the model

The calibration of the HSP model for the Derwent River was completed using data up to January 1975. The model was mothballed at the end of the study, but in July 1976 concern was developing as a result of a prolonged drought experienced at that time. The model was reactivated and using the parameters derived for the initial calibration and input data for the time series January 1975–July 1976, the model was run to update the state variables and verify calibration accuracy. No adjustment of parameters or recalibration was permitted and output data was not supplied to the model operators.

The results of this simultion are shown in Figure 13.3.8 for inflows to Thirlmere Reservoir. Good verification of the model permitted flow forecasts to be made of the likely effect of a continuation of the drought sequence experienced at that time.

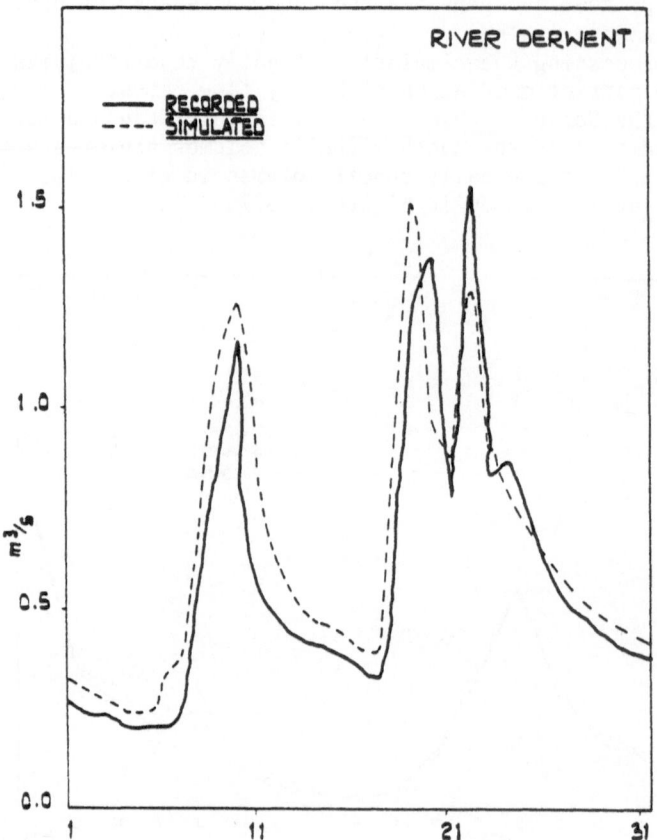

Fig. 13.3.6. Simulated and recorded daily flows at Ullock (after
              Hydrocomp 1976)

13.3.7 Application of the model

The application of the calibrated model was carried out in two different
modes. The first was the real-time application for the assessment of
the effect of continued drought sequences on water yields, and the
second was the application of the model to determine the effect of
releases from Thirlmere on the flow recession at Workington.

In the first mode the model was run with all input data series to
update the state variables to the 'now-time'. In the example the now
time was July 1976 and the update was for eighteen months of data. The
model was then run for a number of alternative conditions. The following
represent three scenarios.
(i)   Average rainfall occurring on "now-time" state variables of soil
       moisture etc.
(ii)   75% of average rainfall on present state variables
(iii) 75% of average rainfall on state variables usually expected at
       "now-time".

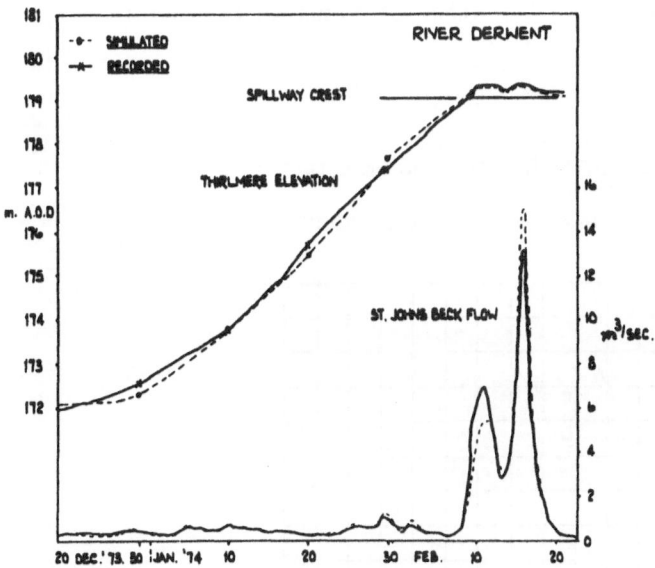

Fig. 13.3.7. First stage verification (after Hydrocomp 1976)

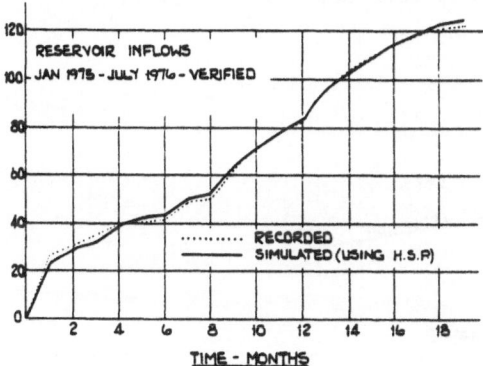

Fig. 13.3.8. Water yield forecasting for drought sequences, river
Derwent

The results of these three conditions are shown in Figure 13.3.9
and provide data on which to base decision making on whether regulation
of the water supply should be altered and rationing of water imposed at
specific times.

In the second mode the current forecast at "now-time" for flows at
Camerton is shown in Figure 13.3.10. The minimum flow for water quality
standards is set at 2.75 m^3/sec. The forecast uses series of pre-
releases to optimise the release required from Thirlmere to achieve the
desired minimum flow. The results of the releases from Thirlmere shown

in Figure 13.3.10 are shown for other stations on Figures 13.3.11, 13.3.12 and 13.3.13.

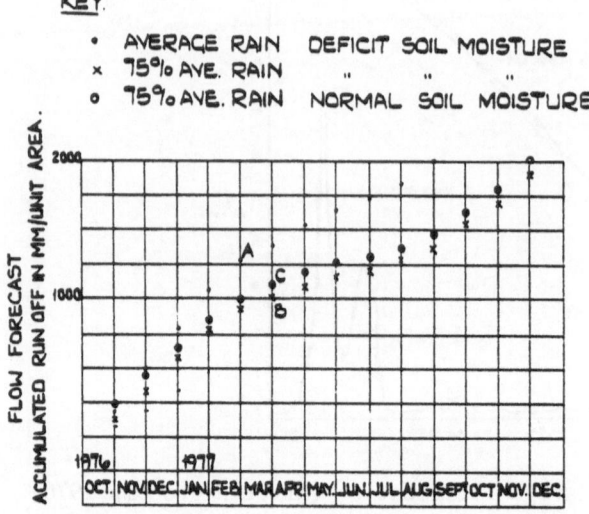

Fig. 13.3.9.   Drought sequence forecasts river Derwent

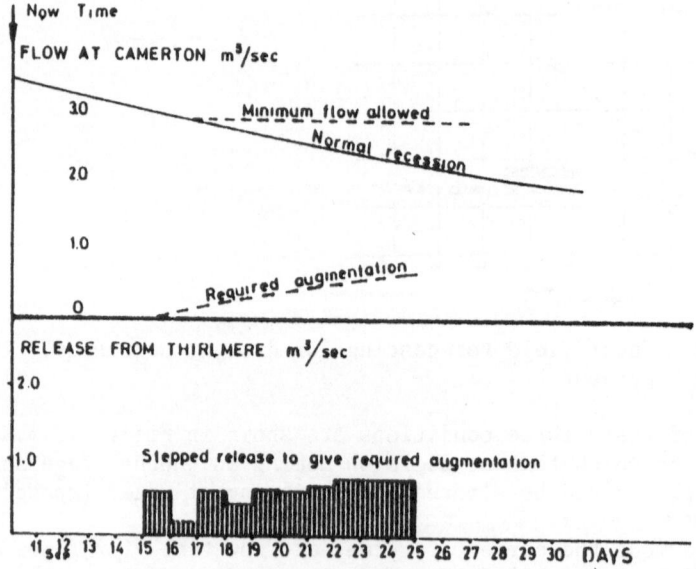

Fig. 13.3.10.   Required augmentation at Camerton and stepped release from Thirlmere

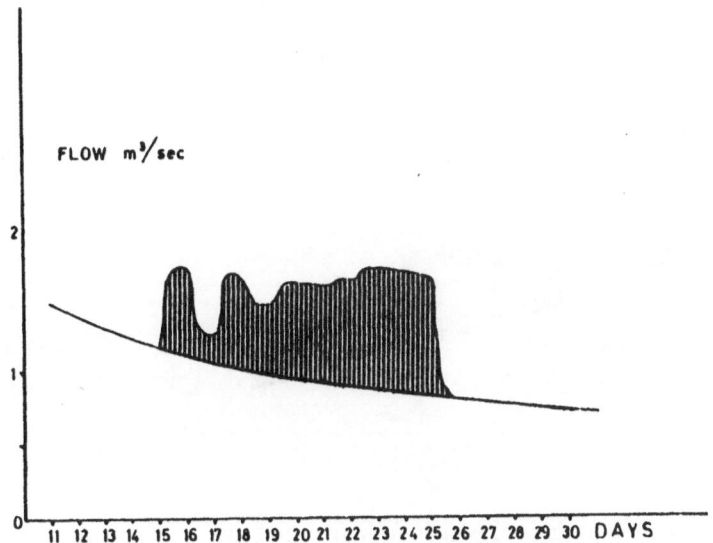

Fig. 13.3.11.   Stepped release at Portinscale

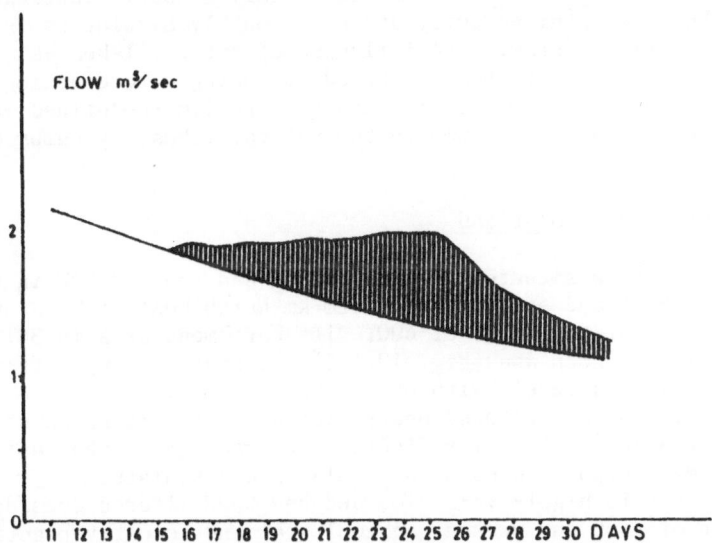

Fig. 13.3.12.   Stepped release at Ouse Bridge

13.3.8 Re-defining the problem

With the use of a flow forecasting system on the Derwent the limitations
in the data coverage are confirmed and specific limitations highlighted.

The importance of sub-basin contributions both in volume and timing
are clarified and the sensitivity of state variables on the longer term

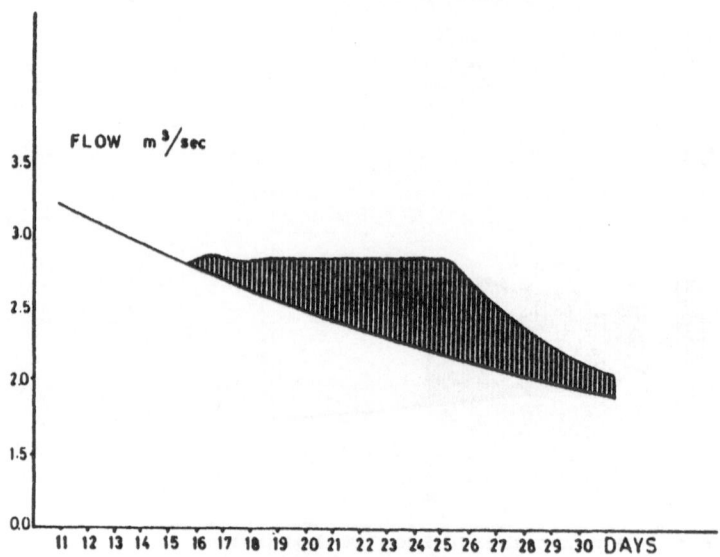

Fig. 13.3.13.   Stepped release at Camerton

forecast is better understood. The need for detailed hourly routing
combined with the less detailed daily and even monthly simulation is
appreciated in practical terms. The influence of natural lakes as
attenuators of flow has always been accepted but using a forecasting
model the attenuation can be quantified and the problem re-defined in
terms of introducing control systems on the natural lakes, by pumps or
regulating weirs.

13.4 Orchy River system, Scotland

The River Orchy Basin is situated at Lat 56°30'M and Long 4°50'W in the
West Highlands of Scotland approximately 120 km North West of Glasgow,
and has a population of the order of 400. Its catchment area is 341 km^2
at its inflow point to Loch Awe (Fig. 13.4.1). Elevation ranges from
1074 m to 40 m above sea level, with an average of 381 m.
    The river basin has experienced heavy glaciation resulting in steep
denuded hill slopes and flat valley floors. The geology of the basin is
predominated by metamorphic rocks such as slates and schists.
    Vegetation cover is highly variable, and has been altered consider-
ably as a result of past and present agricultural and forestry practices.
Bare mountain tops give way on the hill slopes to a mixture of heather
and rough grasses together with recently planted conifers. On the
valley floor deep alluvium exists with a mixture of vegetation types
ranging from mosses and reeds on the marshland to a variety of grassland
and arable land used in agriculture.
    Two natural lochs exist in the catchment, Loch Tulla in the upper
part and Loch Awe at the outlet. Loch Tulla is uncontrolled and undev-
eloped whereas Loch Awe is regulated for water level and storage as
part of the Cruachan Hydro Electric scheme.

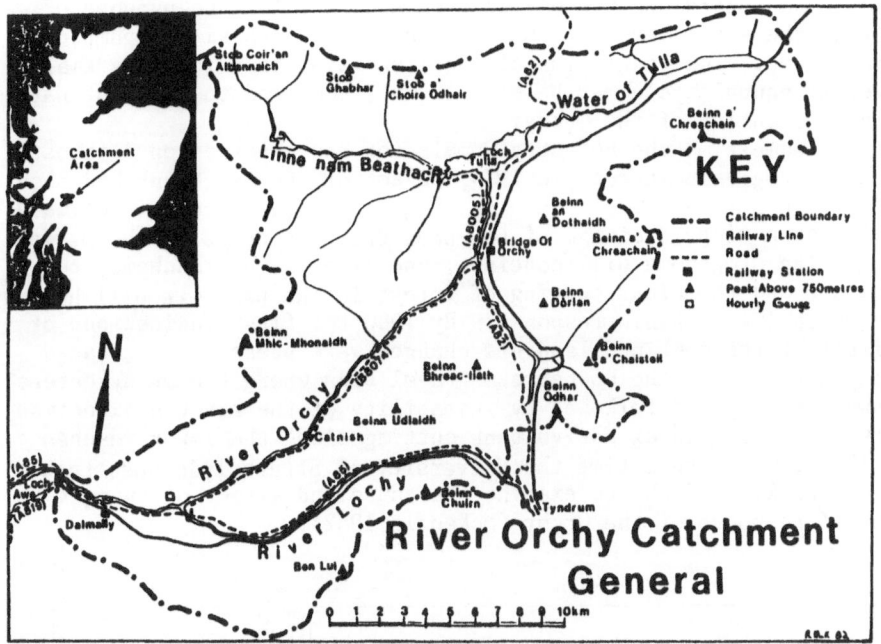

Fig. 13.4.1. The River Orchy Basin

13.4.1 The forecasting problem, River Orchy

Real-time flow forecasting is usually defined as the calculation at a
given time (the "now time") of the flow response of a river system based
on measurements or calculations of inputs to the river basin and state
variables of the system up to the given time and forecasts of inputs
beyond the given time. The technique has been most commonly used for
flood forecasting on large well gauged systems for 24 hour warning and
to a lesser extent for drought sequence or snow-melt forecasting for
periods of days, weeks and occasionally months in advance. The problem
experienced in the River Orchy is a completely different concept of real-
time flow forecasting. The problem is forecasting at now time what will
be the effect of physical changes to state variables of the river basin
on the future runoff response. The future in this case can be weeks,
months or years.

The River Orchy basin has undergone progressive land use changes
over a considerable number of years. The natural forest was originally
cleared and rafted down the river to Loch Etive where it was manufactured
into timber and charcoal. The land use was progressively altered to
agriculture where the slopes, soil fertility and drainage of the land
permitted. This change will have caused an alteration in the response
of the basin which was not measured or reported. In recent times in
addition to agriculture, two other major activities have been introduced
which also influence the response of the basin.

At the lower end the North of Scotland Hydro Electric Board have

established a control on the outlet from Loch Awe for the purpose of
regulating the water level in the loch in conjunction with a pumped
storage scheme. The regulation of the loch level is relatively small
but over an annual cycle may be significant, compared to the past water
level at the outlet of the Orchy.

At the same time the Forestry Commission has embarked on a forest
replanting programme whereby in 1974, 50 km^2 of the catchment had been
reseeded with spruce, larch and pine (Fig. 13.4.3(d)). This involves
extensive down slope drainage of the poor draining soils in the basin.
This planting programme will continue, and like the hydroscheme, the
agriculture and the clear cutting of forest in the past, it will have an
effect on the river basin response. By 1980 the first indications of
the effect of the combined land use changes were present. Increased
shoaling activity of the downstream gravel beds where the Orchy enters
Loch Awe was observed followed by instability of the river's direction
of flow, resulting in extensive bank cutting along the lower reaches
(Fig. 13.4.2). At this time the University of Strathclyde undertook a
study of the River Orchy to examine the cause and effect of changes to
the runoff response of the Orchy (McKenzie 1982).

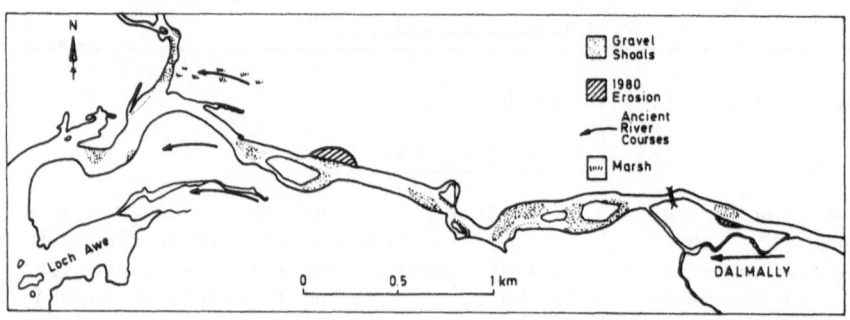

Fig. 13.4.2. Lower Orchy Channel at Dalmally 1870

The problem could be summarised as the forecast of changes to flows
that might be expected when converting the River Orchy from one group of
land uses to another. The problem is a common one and is usually complicated by a scarcity of physical and hydrometeorological data and the
lack of an adequate methodology for assessment.

13.4.2 Data sources, River Orchy

Data sources for the River Orchy included the Meteorological Office, the
Clyde River Purification Board, the Forestry Commission, The University
of Strathclyde, the Institute of Geological Sciences and the Macaulay
Institute for Soil Research.

Good data coverage of physiographic, soil, geology and vegetation
information were available on a range of map scales. Hydrometeorological
data for the basin, however, were scarce, as shown in Figure 13.4.3(a).
There was reasonable coverage of flow data from October 1978 but these

were collected for low flow response and calibration of the gauge for high flows was suspect.

   Precipitation data were limited to rainfall on a daily or monthly interval from as long ago as 1952. The general data coverage is shown in Table IV.

Table IV. Data available, River Orchy

Evaporation	Met' Office	Monthly values from Sloy 1966–1978*	
Flow	Clyde River	Daily values for River Orchy Oct '78 to date	
	Purification	Daily values for River Lochy Oct '78 to date	
	Board	Daily values for Lin Nam Oct '78 to date Bethach	
Precipitation	Met' Office	Dalmally	1952–1968 Daily
			1969–1970 Monthly
		Corryghoil	1972–1975 Monthly
			1977–date Daily
		Glen Lochy ⎤	1956–1961 Daily
		Crossing ⎦	1962–date
		Aindh Castulaich	1956–date Daily
		Bride of Orchy	1956–1965 Daily
			1966 Monthly
		Inveroran Hotel	1960–1964 Daily
			1965–1970 Monthly
			1971–date Monthly
		Tulla Cottage	1975–date Monthly
		Allt Coira A' ⎤ Ghabhalach ⎦	1958–1960 Monthly
Temperature	Met' Office	Kilchrenan	1979–date Daily*
	Met' Office	Corpach	1977–1980 Daily*
	Forestry Com'	Corryghoil	1977–date Daily (incomplete)
Wind	Met' Office	Rannoch	1977–date Daily*
		Kilchrenan	1979–date Monthly*
		Corpach	1977–1980 Monthly*
Soil	Field	To be obtained from catchment	
Size	Samples		
Vegetation	Forestry Com'	Approx. areas of afforestation given on request	
Reach	Field Meausrements	Cross-sections surveyed	

13.4.3 Assessement priorities and model selection

Forecasting changes of the runoff response is a difficult assessment
problem made more difficult in the case of the Orchy by a scarce data
situation. The first priority in this study was to examine the existing
data base in detail for sources of error and to secure an improvement
of the data coverage. The second priority was to examine the River
Basin sensitivity to the input data available. A proportion of the
precipitation on the catchment occurs as snow and the importance of this
to runoff generation had to be assessed.

Once understanding was gained of the dominant mechanisms in land
surface response, the flow routing processes required assessment,' since
the direct observations of changes in channel geometry were the first
indications of a problem in the Orchy. The full Orchy study would not
be complete without consideration of the sediment processes. However,
these have been excluded in this presentation although they did affect
the choice of model for the study.

The assessment priorities required: a full river basin model capable
of rain/snow simulation of land surface runoff for variable river basin
characteristics; the calculation of both water and sediment response;
flow routing in a network of channel reaches, including one uncontrolled
loch; and the ability to conceptually/physically represent land surface
conditions such as drainage paths and vegetation cover.

The model chosen for the study consisted of a combination of: a
watershed model based on a modified version of the Stanford Watershed
Model (Crawford and Linsley 1966) and a kinematic flow routing model
and a sediment erosion, transport deposition model (Walker and Fleming
1969; Fleming and McKenzie 1982) which were collectively called the
River Basin Model (RBM).

13.4.4 Date base assembly

The existing data for the River Orchy were collected, from the available
sources shown on Table IV. Much of these data were in the raw state,
comprising charts which still required abstraction. During the early
stage of data abstraction an hourly rainfall gauge was established in
the catchment at Corryghoil (Fig. 13.4.3(a)). This was operated for a
two year period to obtain representative data for use in the watershed
model. Synthetic hourly distributions of the recorded daily rainfall
data for 1978-81 were necessary for the initial calibration and recorded
hourly data used for verification for 1981-82.

All existing rain gauges were checked for consistent record and an
isohyetal map of the basin was prepared based on Meteorological Office
data (Fig. 13.4.3(b)). Physical data on slope and vegetation cover were
prepared for the assessment of model parameters and these are shown in
a generalised form in Figure 13.4.3(c) and (d).

The flow data were processed from their original charts using an x-y
digitiser and the results were processed to give daily flow data at the
respective stations. From the start of the study it was realised that
the flow data required detailed checking and this was undertaken at the
calibration state of the watershed model.

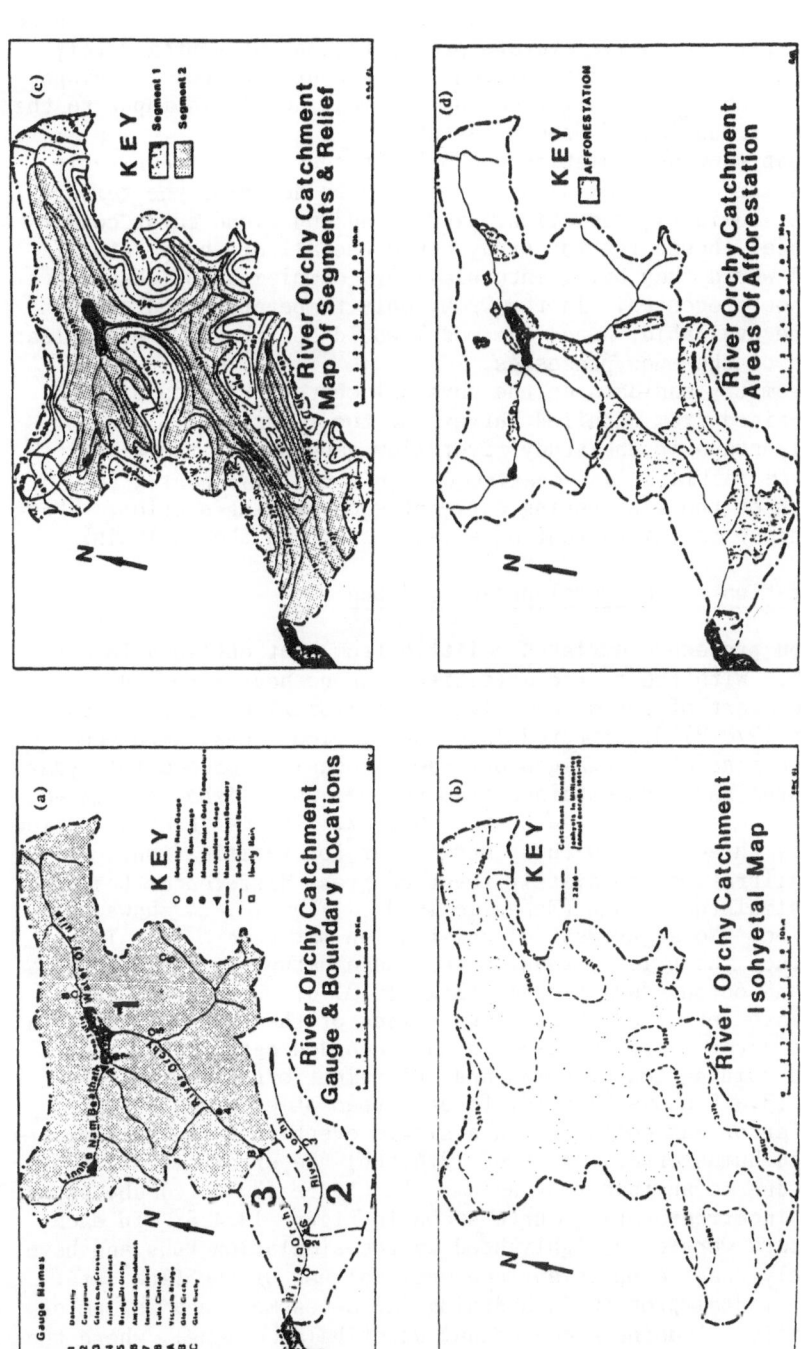

Fig. 13.4.3.   Gauge locations and physical data – River Orchy Basin

Evaporation data were calculated, based on meteorological data outside the Orchy Basin. Annual potential evapo-transpiration was found to
average 450 mm per year while average precipitation was approximately
2800 mm per year. Under most conditions throughout the year the evapotranspiration potential is satisfied and the most sensitive input to the
model is the distribution of precipitation.

To represent snow accumulation and melt processes it is desirable
to use as much meteorological data as possible to describe the temperature, dewpoint, humidity, radiation, wind speed and cloud cover conditions. The model chosen for the study could use all of these data if
available, and would then bring into play more complex algorithms representing the snow processes. In the Orchy only temperature and wind
speed data were available, hence the model would operate on the simplest
representation of the snow processes.

Channel geometry and data on the physical characteristics of Loch
Tulla were obtained from detailed surveys of the river.

At an early stage in the study river flow gaugings by the velocity-
area method were initiated and continued over a two year period to
provide a check on the flow rating curve at each gauging section. These
data were to prove very important at a later stage in the analysis.

13.4.5 Calibration of the model and data debug

The calibration procedure deviated a little from that outlined in subsection 13.2.5. With the scarce data situation no hourly records
existed at the start of the study. Initial calibration was based on
daily data for 1978-81 distributed into hourly time steps, with the
hourly distributions of raingauges outside the Basin. Once a full year
of hourly records had been obtained from the new established raingauge,
calibration was verified for the period 1981-82. The calibration period
was essentially three years with a fourth year for verification.

Primary calibration immediately revealed gross differences between
recorded and simulated output (McKenzie 1983). Fig. 13.4.4 shows an
early simulation. No adjustment of calibration parameters is valid at
this stage, since the primary task is the "data debug" to reduce errors
in the input and output data streams (Fig. 13.3.7). Detailed checks
based on simulated error differences were made of the rainfall input but
no significant errors were revealed in the recorded data. What was
revealed was a combination of human and instrument error in the flow
data. Figure 13.4.5 shows three different types of error. In Fig.
13.4.5(a) the error was contained in a single event when the data abstraction involved human error. In Fig. 13.4.5(b) the error affected most
flows and was due to an incorrect datum. In Figure 13.4.5(c) the error
was due to an incorrect rating curve shown in Figure 13.4.6. In each
case these errors were only highlighted by the simulation runs and have
remained largely undetected in any conventional use of the flow data.
This is a fundamental problem in hydrological assessment and is one of
the major benefits of using a conceptual water balance model, where the
continuity equations of mass are used in the calculations.

Calibration and verification of the response of the Orchy Basin were
completed for the four years of record and Figure 13.4.7 is typical of

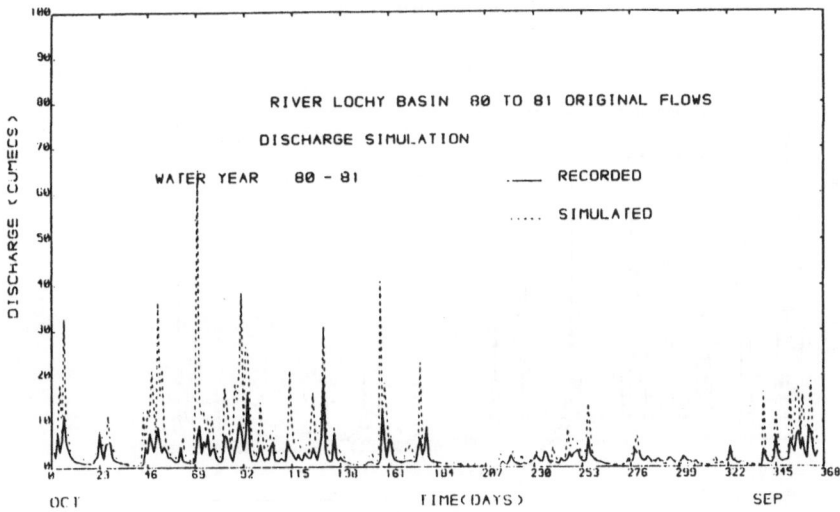

Fig. 13.4.4. Initial calibration - River Orchy

what was achieved. Errors remain as they will do in all hydrological
assessment, primarily because input and output information is not absolute and never will be.

13.4.6 Application of the model, Orchy River

As discussed above, the forecasting problem in the Orchy Basin was
aimed at understanding the sensitivity of the Basin to the basic hydrological processes as well as to changes in land surface processes.
    A series of tests were then conducted using the calibrated model
and changing the input or state variables.
    The results provided greater understanding of the processes and
their influence on hydrological response.
    Figure 13.4.8(a) shows a simulation with and without snow and demon-
strates the importance of this process yet indicating that it does not
dominate the overall response.
    Figure 13.4.8(b) shows the influence of infiltration rate on the
response and how high sensitivity exists on this catchment.
    Figure 13.4.8(c) shows the influence of soil moisture storage and
again the high sensitivity.
    Figure 13.4.8(d) examines the influence of interflow recession rates
and its significant influence on runoff timing.
    Considerable testing was undertaken to identify the dominant proces-
ses, affecting the runoff response both for the land surface and the
river channels and from this the components of the problem were better
understood.

13.4.7 Re-defining the problem

In the Orchy study, the first phase has revealed a number of important

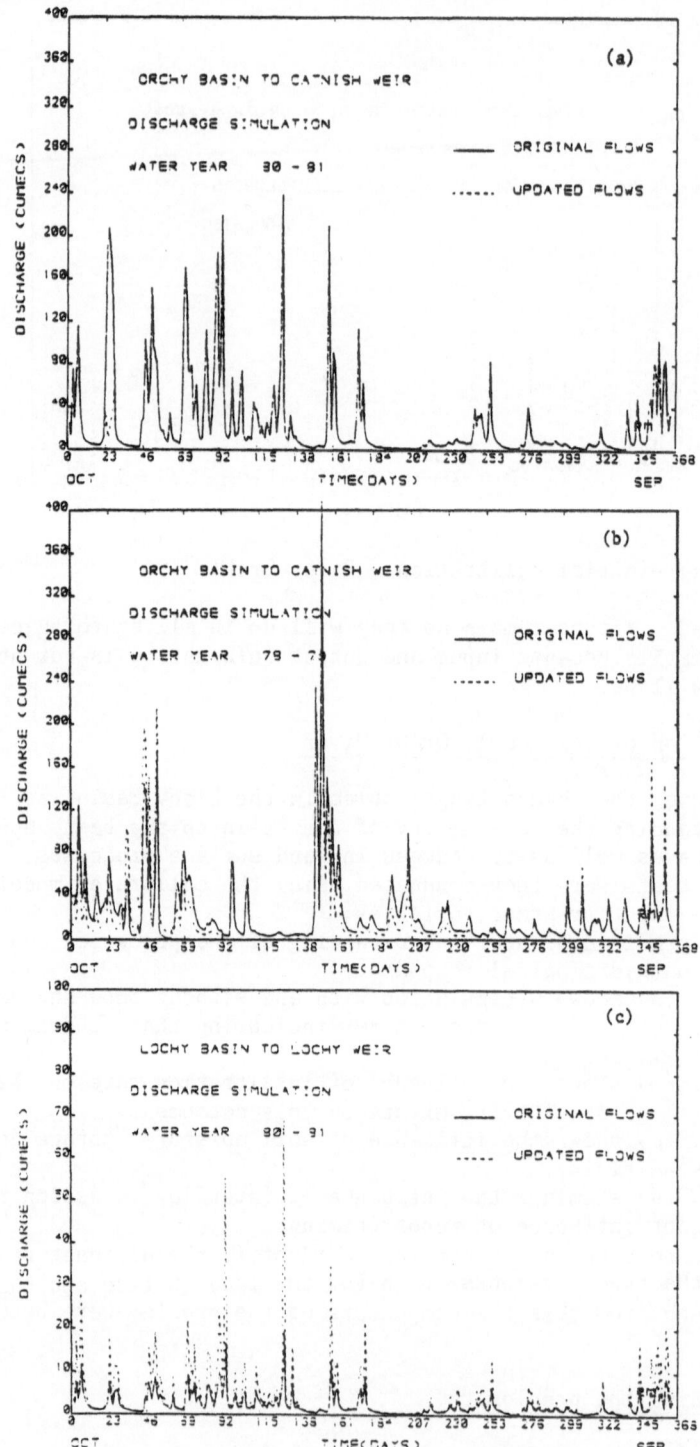

Fig. 13.4.5. Errors in flow data, River Orchy

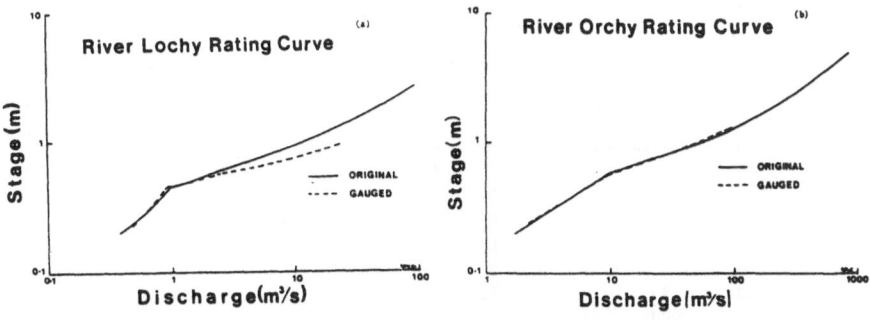

Fig. 13.4.6.   Discharge rating curves for the River Orchy

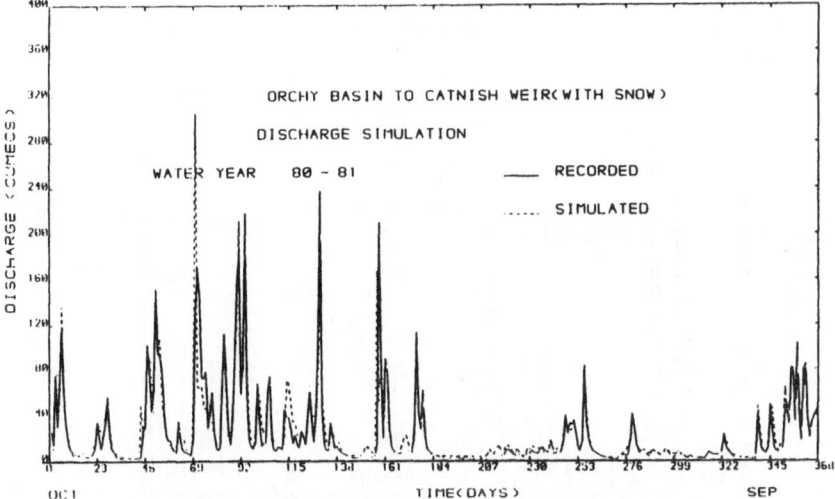

Fig. 13.4.7.   First stage calibration of River Orchy

factors. The data base although limited can be used, with care, to
identify sensitivity of response. It highlighted the danger of assuming
that recorded flow data are absolute, and demonstrated an important technique for testing the accuracy of both rainfall and runoff data.

At the start of the study the influence of the snow processes was
not understood. Testing indicated that snow, as assumed, played a significant part in the runoff response, for the years simulated, but was
not dominant. What was found to dominate the response was primarily the
rainfall distribution and the state variables of the system. The significant state variables included the infiltration, soil moisture and
recession rate parameters.

Tests also revealed that the river basin was sensitive to changes in
land surface conditions and indicated that rather than one specific
change in physical conditions causing significant shift in runoff response
- the instability recently experienced in the Orchy was the result of

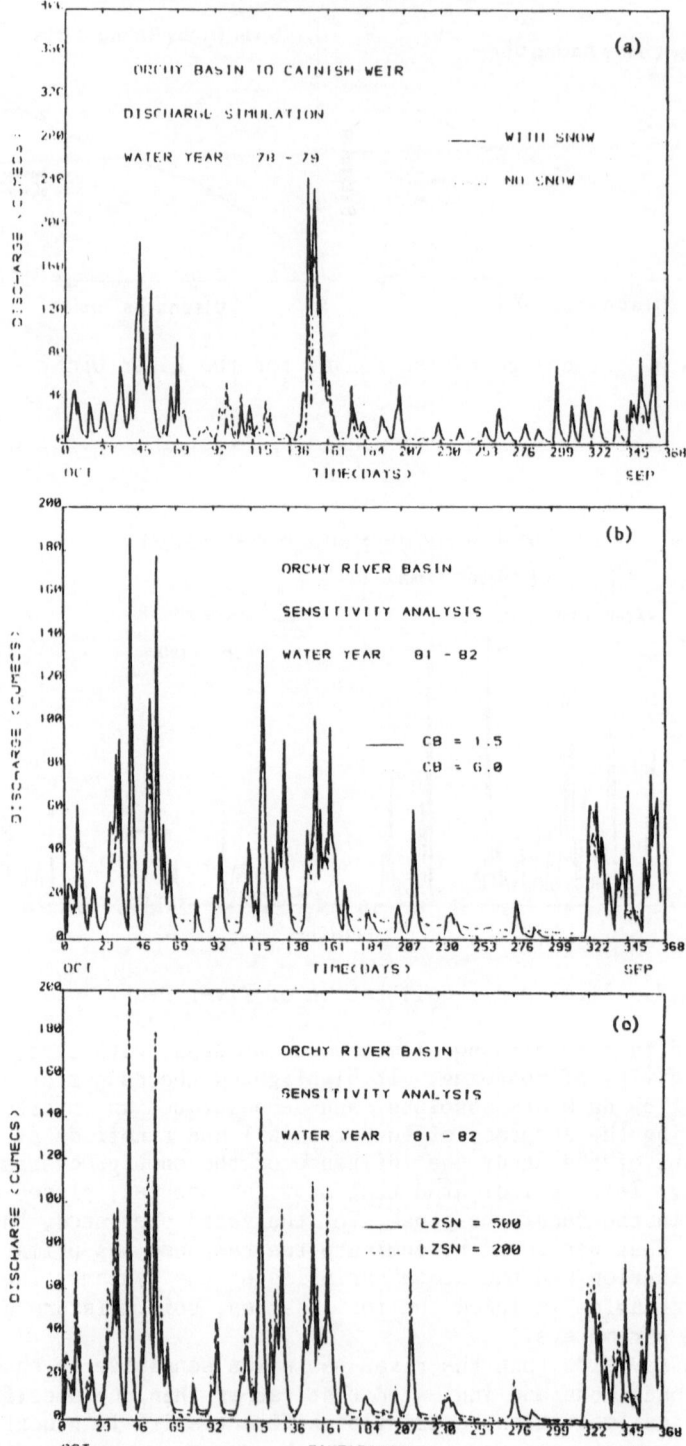

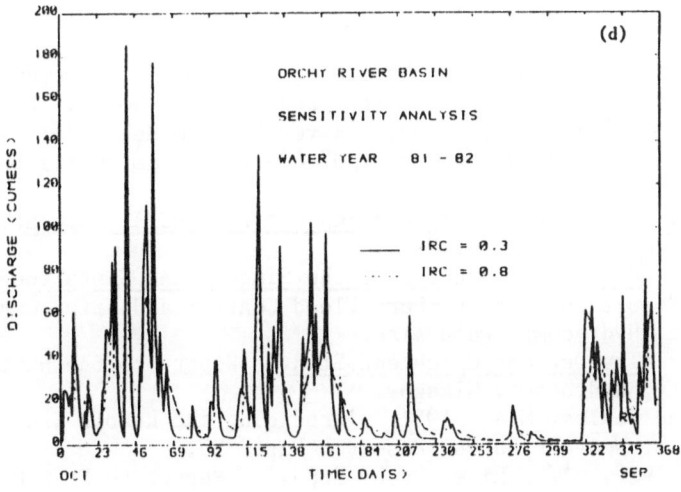

Fig. 13.4.8. Sensitivity of River Orchy response

a combination of changes. It was further demonstrated that the land
use and river regulation changes are time dependent and often one change
compensates another. However the instability in the lower channel
reach was the result of an exaggeration rather than a compensation of
the land use changes and once initiated is difficult to remedy without
considerable capital expenditure.

SYMBOLS

Q	spillway flow	L^3T^{-1}
C_d	coefficient of discharge	$L^{\frac{1}{2}}T^{-1}$
b	breadth of the spillway	L
H	Head of water over spillway crest	L
F^2_{min}	criterion of accuracy	1
q_o	observed flow	L^3T^{-1}
q_s	simulated flow	L^3T^{-1}
n	number of records used (13.2)	1
S_r	slope of reach	1
n	number of subdivisions used (13.3)	1

REFERENCES

Crawford, N.H. and Linsley, R.K. 1966. 'Digital Simulation in Hydrology:
    the Stanford Watershed Model IV'. T.R. 39, Stanford University,
    California.
Dawdy, D.R. and O'Donnell, T. 1965. Proc. ASCE Hyd. Div. 91; 123.
    'Mathematical Models of Catchment Behaviour'.

Fleming, G. 1975. Computer Simulation in Hydrology. Elsevier, New
  York or Den Haag.
Fleming, G. 1975. 'Deterministic Models in Hydrology'. United Nations
  Food and Agriculture Organisation, Irr. and Drainage No. 32, Rome.
Fleming, G. and McKenzie, R.E. 1982. 'River Basin Model: Vol. 1
  Watershed Model'. User guide, Report HY-82-7, Dept. of Civil Eng.,
  Strathclyde University, Glasgow.
Hydrocomp, 1969. The Hydrocomp Simulation Program (HSP): Operations
  Manual. Hydrocomp, Palo Alto, USA.
Hydrocomp, 1970. Hydrologic Studies of the Sisquoc and Santa Ynez
  Watersheds. Report to Santa Barbara Flood Control and Water Conserv-
  ation District, Hydrycomp, Palo Alto.
Hydrocomp, 1976. The Derwent Catchment Study. Report to the North West
  Water Authority, Hydrocomp, Glasgow.
Linsley, R.K. and Kohler, M.A. 1951. 'Predicting the Runoff from Storm
  Rainfall'. U.S. Weather Bureau Research Paper 35.
McKenzie, R.E. 1982. 'The River Orchy Study'. Report to NERC HHCD/82/
  3, University of Strathclyde, Glasgow.
McKenzie, R.E. 1983. 'The effects of snow and land use changes on
  River Basin Response'. Ph.D. Thesis, University of Strathclyde,
  Glasgow.
Rockwood, D.M. 1964. 'Streamflow Synthesis and Reservoir Regulation'.
  Engineering studies project 171, Tech. Bull 22, U.S. Army, Corps of
  Engineers, Portland, Oregon.
Fleming, G. and Walker, R.A. 1979. 'Sediment Model 1 user's guide'
  report HHCD-79-10, Civil Engineers, Strathclyde University, Glasgow,
  UK.

Aerial photogrammetry 104
Air temperature 99
Albedo 106, 107, 112, 116, 117
Alluvial rivers 261
Analysis 15, 20, 23, 299
– methods 12, 14, 19
Antecedent precipitation index 314
Aquifer 72, 78, 84, 96
ARIMA model 182, 291
Asymptotic bias 147, 148
Asymptotically efficient 165
Atmospheric counterradiation 108
Autoregressive moving average model 164

Backward shift operator 137
Base flow 16, 32, 282, 283, 284, 313, 314
Bed forms 261, 262, 263
– roughness 263
Benefit cost analysis of hydrological
    forecasting 318
Black box 59, 79
– approach 14
– model 266
Bottom roughness 260, 261
Boundary condition 241, 250, 252, 259, 290,
    291
–, downstream 44
–, external 251, 253
–, in flood routing 42
–, internal 251, 252
–, temporal 42
Break frequency 135, 140

Calibration 260, 269, 270, 308, 311, 337
– of linear model 60
– of the model 336, 347, 360
– period 309, 338
Capillary retention capacity 120
Celerity 254
Channel, main 257
Channel response 57
Channel routing 53, 54, 200
– model 203, 204, 332
–, non-linear equations for 44
Characteristic equations 41
– length 57
– methods 44
– network 46

– path 41, 42
– variable 41, 42
Chézy coefficient 259, 260, 261, 262, 289
Coefficient of determination 316
Compartment, well-mixed 158
Computer 241, 299
– hardware and software 198, 199, 211, 217,
    225, 227, 307
–, main-frame 178, 307
Confidence interval 166, 284
Confined ground water 85, 86
– systems 72
Conservative form 41, 248, 249, 250
– scheme 45
Continuity equation 39, 248
Convection-diffusion analogy 289
– model 290
Convergence 253
Convergent 45, 246
Convolution 26
– integral 4, 18, 63
Courant number 245, 247, 254
Courant-Friedrichs-Lewy criterion 47, 246
Cumulant 58
– matching 60

Damping, numerical 254
Darcy's law 84
Data acquisition 130
– base 333, 336, 345, 358
– collection 302, 306, 324
– collection and processing 301
– collection and transmission processing 316
– processing 299, 306, 324
– processing filing and retrieval 307
– processing storage 302
– storage 301
– transmission 183, 301, 302, 306, 324
– transmission and processing 182
Dawdy-O'Donnell model 33
Degree-day factors 114
– method 106
Delay time 60
Depletion curve 87, 96
Detection 15
Deterministic model 5, 7, 11, 29, 30, 165, 290,
    308
– variable 140

Difference equations 244
- molecules 244
Diffusion analogy 50, 52, 58, 59, 308
-, numerical 52, 245, 247
Dike-guard 273
Dikes 273, 274, 299
Direct runoff 16, 31–32, 282, 283, 313, 314
Disaster-prevention operations 300
Discharge rating curve 333
Discretization error 245
Diskin model 21
Drainage, urban storm-water 302
Drought 73, 74, 75, 77, 86, 87, 329
Dry channels 257, 268
Dunes 261, 262, 263.
Dynamic wave 254, 308, 322
- wave speed 50

Eddy diffusion 109
- method 108
Effluent conditions 69, 70, 71
- discharge 223, 224, 225
Energy balance 121
- budget method 106
Equation of continuity 79, 249, 252, 269
- of motion 249
- of motion (St. Venant) 289
- of motion and continuity 251
-, characteristic 41
-, difference 244
-, linearised difference 253
Equivalent sand roughness 261
Errors, numerical 254
Evaluation, criterion for 287
Evapotranspiration 75, 78, 87, 99, 313, 345, 360
Explicit finite difference schemes 46, 47
- formulation 290
- method 44, 269
- schemes 244, 246
Exponential memory 155
Extended Kalman Filter 174, 175, 176, 293

Fickian diffusion model 154
Finite difference equation 45, 250
- method 44, 241, 242, 250
Finite period unit hydrograph 18
Flood alarms 220
- control 209, 299, 331
- control reservoirs 209
- control schemes 266
- forecasting 87, 153, 184, 209, 213, 237, 287, 299, 329
- forecasting model 324
- mitigation 198, 208
- plain 257, 259, 263, 264, 266, 267, 268, 299, 301, 330, 331, 337

- routing 39, 153, 325
- routing in a prismatic channel 49
- routing model 5, 154, 182
- warning 187, 195, 212, 274, 321, 323
- warning system 210, 274, 321
- wave 288, 303
Flow forecasting, univariate approach to 173
- and quality model 229
- forecasting 164, 168, 170, 176, 220
- forecasting, self-tuning 168
- routing 237, 358
- routing model 229, 232, 332
-, overbank 44
-, unsteady in an open channel 39
Forecast in real-time 9
- lead-time 3
- real-time 4
-, criteria for 3
Forecasting glacier run-off 117
- methods, evaluation of 316
- model 155, 237
- model, deterministic 295
- model, empirical 274
- model, stochastic 294
- objectives 2
- of low flow, long-term 79
- of meltrates 115
-, flow 164, 168, 170, 176, 220
-, long-term 96
-, long-term and seasonal 101
-, medium-term 97
-, melt 114
-, meltwater 100
-, real-time 8, 206, 313
-, real-time flow 181, 186, 196, 287, 331, 338, 343, 355
-, seasonal 121, 122
-, self-adaptive 157, 168
-, short-term 77
-, short-time 101
-, snowmelt 355
-, snowmelt run-off 101
-, stochastic 308
-, unit hydrograph 313
-, water quality 195, 196, 221, 227
Forgetting factor 155
Fourier linkage equations 25
- series 24
- series transform method 25
- transforms 24
Frequency response characteristics 135

Gain term 144
Gamma sources, radioactive 102
Gamma-ray snow gauges 106
Gauss-Markov process 169
Glacier hydrological computations 111

Glaciers 99
Ground water 69
- extraction 86
- head 84, 85
- level 85
- model 78, 83, 94
- recharge 85
- storage 283, 284
- table 82, 87, 9
-, unconfined 85, 86

Hardware 217, 307
Harmonic analysis method 28
- series 24, 25
Heat budget 111, 113, 116
- transfer in the snow cover 111
Human factor 192, 193
Hydro-electric power generation 208
Hydrologic method 53
- model 59
- storage equation 54

Identification 14, 15, 19
Implicit finite difference method 250
- finite difference scheme 47, 48
- method 44
- scheme 244, 247
Implicity factor 254
Impulse response 17, 19, 20, 24, 150
Infiltration 313, 363
- index 314
- parameter 347
- rate 361
Influent conditions 69, 70, 71
Initial condition 259
Instabilities 269
Instrumental variable approach 147
- variables 148
Instrumentation 182, 198, 211, 225, 301
Inter-basin transfer 330, 341
Inversion 25
Isolated event model 34, 219

Jones' formula 260, 291

Kalinin-Milyukov 56, 57, 58, 60
- model 7, 137
Kalman filter 168, 169, 170, 172, 173, 174,
    182, 210, 288, 289, 293, 294
Kinematic approach 333
- routing method 345
- speed 50
- wave 182, 308
- wave equation 242
- wave solution 52, 56
Kronecker delta function 140

Lag and route method 131, 219
- and route model with gain 133
- and single reservoir 58
Latent heat 108, 112, 113, 115
- from freezing 111
- of melt 107, 110, 116
Lateral channel 257
- discharges 268
- fluxes 249
- inflow 40, 44, 57, 133, 238, 288, 289, 291
Lead time 8, 166, 167, 176, 184, 209, 274,
    287, 288, 295, 300
Leading negative characteristic 42
- positive characteristic 42
Least squares algorithms, recursive 146
- cost function 144
- estimate 142
- regression algorithm 146
- regression analysis 141
Level of aggregation 159
Linear catchment response 62
- channel 206
- channel response 55, 60
- conceptual model 55, 57
- diffusion analogy 51
- model 129
- reservoir 79, 206, 283
- stochastic system 168
- systems 4, 12
- time-series model 144
Linearised difference equations 253
Linkage equations between the transforms 24
Long wave 112, 113
- net radiation 107
Low flow 67, 75, 95, 97, 343
- forecasting 75, 87, 89, 94, 208, 282
- model 5, 279
- warnings 187
Lowest possible flow 284
Lysimeters, meltwater 106

Management, water quality 196
Man-made effects 77
Manning coefficient 269
Manning's n 261, 263
Matrix inversion method 28
- technique 26
Medium-term forecasting 97
- prediction 77
Melt forecasting 114
Meltrates 106, 111, 112, 113
Meltwater forecasting 100
- lysimeters 106
- run-off 99, 123
- run-off model 5, 121
Meteorite scattering paths 306

Method, matrix inversion 28
−, transform 23
−, unit hydrograph 16, 17
−, temperature index 113
Microcomputer 178, 195, 212, 213, 220, 307
Microprocessor 130, 195, 210, 216, 218, 221,
    235, 236, 237, 306
Microwave systems 105
Mini-computer 222, 226, 227, 228
Mini-microcomputer 237
Model choice 309
 − classification 308
 − order identification 154
 − parameters 20
 − structure (order) identification 148
−, aggregated dead-zone 154
−, convection-diffusion 290
−, Fickian diffusion 154
−, flow routing 229, 232, 332
−, hydrologic 59
−, multiple linear regression 277
−, over-parameterized 150
−, recursive 283
−, stochastic 129, 155, 165
−, variable parameter diffusion 204
−, water quality 222, 228, 232, 237
−, Stanford 30
Moisture accounting 313, 314
Moment matching 60
 − relationships 20
 − of area 19, 20, 22
 − of the linear channel response 56
 − (or cumulants) 57
Momentum equation 39, 40, 248
Monin-Obukov similarity 109
Monitors, water quality 221, 222, 224, 227,
    235
Multiple correlation 117, 119
 − linear regression model 277
 − regression analysis 115
Muskingum coefficients 61
 − method 62
 − model 55, 56, 57, 58, 59, 60, 135, 137, 138,
    150, 151, 152, 153, 155, 167
 − parameters 62
Muskingum − Cunge 322
 − model 52

Nash 56
 − model 7, 8, 20
 − reservoir cascade method 22
Natural gamma radiation 102
Network, characteristic 46
Network design 299, 301, 305
Noise 138, 139, 154
 − model 165, 293
−, coloured 165

−, white 140, 141, 146, 164, 165, 170
Non-divergent form 40
Non-linear difference equations 253
 − equations for channel routing 44
 − estimation 176
 − reservoir 62
 − stochastic systems 176
 − storage reservoirs 232
 − systems synthesis 28
 − time series 157
Non-steady flow in open channels 241
Non-uniform flow in a cross-section 256

Open channel, non-steady flow in 241
−, unsteady flow in 39
Operational forecasting, evaluation of 317
Over-parameterisation 149

Pade approximation 151, 153
Pandora's box approach 14
Parameter adjustment 28
 − estimation, recursive 174
 − fitting 35
 − fitting techniques 30
 − optimisation methods 32, 35
 − optimisation procedure 203
 − sensitivity 35
−, empirical 260
Parametric efficiency 150
Photographic methods 104
 − survey, terrestrial 104
Pollutant 234
 − loads 225
Pollution control 195
 − events 227, 235, 237
 − incident 228
 − monitor 236
Precipitation 78
 − forecasting 340
 − forecasting model, quantitative 288
 − forecasting, quantitative 301, 302, 332
 − measurements, systematic error in 101
Prediction 14, 15, 19
 − error 144
−, medium-term 77
−, short-term 75
Predictions of low flow, long-term 76
Pre-release 189, 351
 − strategy 340
Propagation, velocity of 254
−, wave 258, 259
Pumped storage scheme 356
Pure time delay 132, 137, 155, 166, 167, 206,
    207, 210, 219

Quality monitoring 237

Radar 183, 184, 209, 305, 306, 323
- raingauge 320
-, weather 195, 196, 203, 324, 340
Radiation 112, 113
-, net 99, 107, 111, 114, 115
Radio 213, 306, 324
- data transmission 305
- equipment 212
- links 220
- receiver 217
- telemetry 212
Rainfall discharge relations 74
- forecast, stochastic 184
- forecasting 200, 209
- forecasting, qualitative 301
- forecasting, quantitative 184
- histogram 18
- measurements 200
- radar 209
-, net 313
Rainfall-runoff model 5, 7, 182, 200, 288, 301, 313, 321, 324, 332
- modelling 211, 237
- relation 81
Random walk 171
- model 156, 169, 170, 174
Rating curve 260, 336, 360
Rational formula 11
Real-time data collection 315
- flow forecasting 181, 186, 196, 287, 331, 338, 343, 355
- forecasting 8, 206, 313
- forecasting model 199
- forecasting of meltrates 113
- forecasting of water quality 222
- forecasting system 210
Recharge 95, 96
Recursive algorithm 143, 155, 157, 168
- calculation 284
- estimate 8, 129, 138, 144, 150, 162, 168, 315
- estimation algorithms 148
- least squares algorithms 146
- methods of time-series analysis 154, 176
- model 283
- parameter estimation 174
- solution 141
Remote sensing 183, 299, 301, 305
Reservoir constant 79
- management 195, 196
- sedimentation 330
- systems, multipurpose 302
-, non-linear 62
-, non-linear storage 232
-, water supply 223
Response function 11, 14, 21
Ripening of snowcover 120

Ripples 261, 263
Risk and uncertainty 189, 190, 192
River flow forecasting system 269
- flow simulation 241
- forecasting, adaptive procedure 175
- regulation, multipurpose 208
- systems, operational management of 221
Riverbed forms 259
Roughness 337
-, bed 263
-, bottom 260, 261
Runoff, meltwater 99

Salinity 158, 159, 160
Satellite 183, 184, 209, 305, 306, 323
- data collection 322
- images 268
- snow survey 105
Saturated zone 70, 82
Sediment processes 358
Sensible heat 99, 108, 112, 113, 115
- latent heat 116
Sensitivity 35, 260, 358, 361
- analysis 258, 279, 315, 347
- of response 363
- tests 347
- to data coverage 340
Shape factor 58
- diagram 59
Short wave 112
- net radiation 107, 113
Simulation 12, 14, 15, 270
Snow 99
- courses 101, 102
- depth, stereophotogrammetric survey of 104
- pillows 102, 106
- survey 104
- water equivalent 103, 104
Snowcover 103, 120
- and its water equivalent 100
- assessment 123
-, spatial variation in 103
Snowgauges, radioisotope 102
Snowlines on glaciers 106
Software 198, 199, 217
Soil moisture 340, 361, 363
- accounting 322
- indices 314
Solar batteries 221
- panels 214, 221
- radiation 113
Solution, updated 141
Split-record testing 29
St. Venant equation 40, 46, 182, 204, 241, 266, 270
-, complete non-linear solution 62

−, linearisation 48
−, linearised 57, 58, 59
−, parabolic solution 50
−, simplification 50
Stability 246, 254
− criteria 247
Stable 45
Stage-discharge relation 263, 265, 291
Steady-state gain 133, 134, 135, 148, 149, 152, 153, 155
− rating curve 44, 259
Stefan − Boltzmann constant 107
Storage characteristic 22
− constant 206
− delay time 55, 130, 132
− equation, hydrologic 54
Structure identification 140, 149
Superposition 13
− of responses 17
Synthesis 12, 14, 15, 19, 299
− approach 19, 23
− methods 14
− techniques 30
System approach 4
− function 7
− gain 57
− identification 4
− modelling 12
− operation 4
− time-invariant 13
− time-variant 13, 14
−, non-linear 13, 14

Telecommunications 299
Telemetering 305
− network 237
Telemetry 198, 211, 218, 225, 226, 227
Telephone 212, 213, 214, 226, 227, 228, 305, 306
Temperature index methods 113
Terminal boundary conditions 42
Thornthwaite and Holzman 109
Time delay 134, 151, 152, 238
− of concentration 303
− of travel 21, 207, 225
− Series Methods 8
− series, non-linear 157
Time-area diagram 21, 210
Time-invariant linear 19
− analysis 22
− system 17
Time-series analysis 129, 130, 132, 140, 164
− analysis, recursive methods of 154, 176
− analysis, univariate 154
− modelling, univariate 174
− representations, univariate 168
−, stochastic 138

Time-variable parameter estimation 155, 156, 157, 158
Time-variant linear synthesis 22
Transform method 23
Translation delay time 22
Transmissivity 72, 84, 85
Travel time 206, 208, 230, 232, 234
Truncation error 244, 245

U.K. Flood Studies Report (NERC 1975) 28
Unconditionally stable 255
Unconfined ground water 85, 86
− systems 72
Uniform mixing 230
Unit hydrograph approach 54
− hydrograph forecasting 313
− hydrograph method 16, 17
− impulse response 134
− step response 133, 134
Univariate analysis 138
− approach to flow forecasting 173
− time-series analysis 154
− time-series modelling 174
− time-series representations 168
Unsaturated zone 70, 81
− outflow from the 87
Unsteady flow in an open channel 39
Urbanisation 343

Variable, characteristic 41, 42
−, stochastic 138
Verification 260, 261, 269, 311, 360
− criteria 311, 312, 317
− of the model 338, 349
− period 309
Volterra series 62, 63
von Karman constant 109

Warning system 273, 325
− of water quality deterioration 187
Water equivalent 104, 122
− equivalent of snow cover 99, 122
− quality deterioration, warnings of 187
− quality forecasting 195, 196, 221, 227
− quality management 196
− quality model 222, 228, 232, 237
− quality monitors 221, 222, 224, 227, 235
− supply 67, 208, 224, 274, 282, 330
− supply reservoir 223
Wave celerity 204
− propagation 258, 259
Weather radar 195, 196, 203, 324, 340